Y0-ARS-098

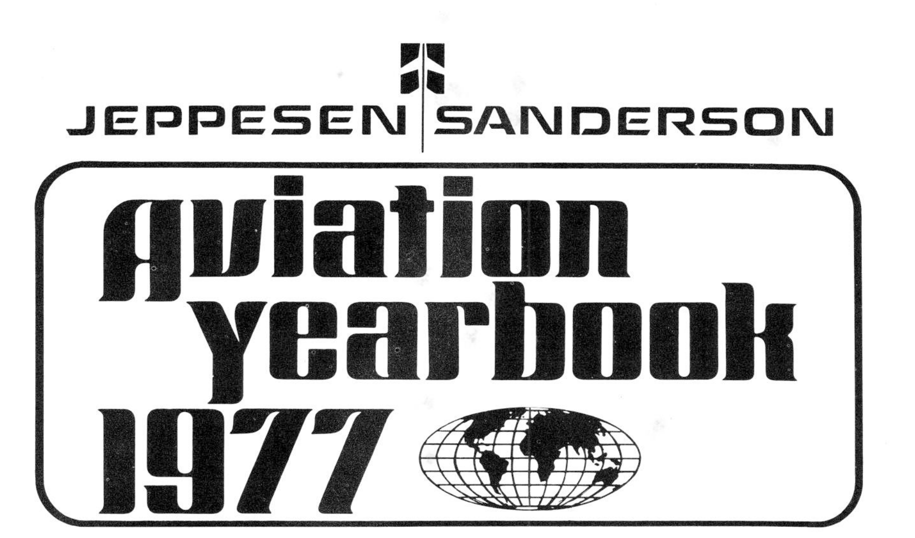

(Surveying the Period November 1975 through December 1976)

8025 E. 40th Avenue, Denver, Colorado 80207, U.S.A.

AV313205A INTERNATIONAL STANDARD BOOK NUMBER 0884870197

AVIATION YEARBOOK

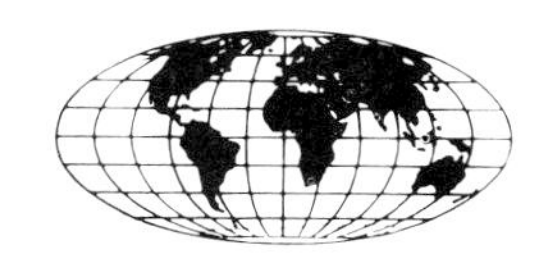

EDITOR IN CHIEF: Ed Mack Miller

SECTION EDITORS
GENERAL AVIATION: F. T. Stent
AIR CARRIER: Barry Schiff
MILITARY/AEROSPACE Allan R. Scholin
SPORT AVIATION Dan Manningham

MANAGING EDITOR: Gordon Girod

PRODUCTION
ART DIRECTOR: Ronald M. Jones
PRODUCTION SUPERVISOR: Richard L. Hahn
COVER DESIGN: Arlee J. Gaeddart
STAFF ARTISTS: Frank G. Davis
Leslie K. Johnson

PUBLISHER'S NOTE

Your JEPPESEN SANDERSON AVIATION YEARBOOK 1977 reviews the period from November 1975 to December 1976. Articles dealing with significant topics in the world of aviation that occurred during this period have been chosen from a number of English language aviation publications. Where published material was not available for specific topics, original material has been supplied by aviation writers or the YEARBOOK staff. Certain reprinted articles have been edited for space considerations. Similarly, comparable photographs or illustrations have been substituted in the few cases where the original material was not available.

ACKNOWLEDGEMENTS

Acknowledgement of source is made with each article. The cooperation of the many publications, organizations, manufacturers, and writers represented in this volume is greatly appreciated. Organizations cooperating included the Air Line Pilots Association (AIR LINE PILOT); The Aircraft Owners and Pilots Association (THE AOPA PILOT); the Experimental Aircraft Association (SPORT AVIATION); Soaring Society of America (SOARING); the United States Hang Gliding Association (GROUND SKIMMER); and the United States Parachute Association (PARACHUTIST).

TABLE OF CONTENTS

THE EDITORS

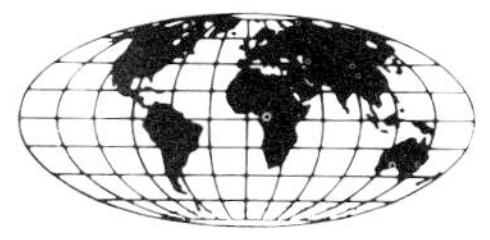

Capt. Ed Mack Miller, editor-in-chief

CAPTAIN ED MACK MILLER, recently retired 747 Training Captain for United Air Lines, has been flying and writing for more than half the history of aviation. He has been rated on everything from DC-3s to the current generation of wide-bodies, and has flown transports to practically every corner of the globe. As a widely read aviation author, Captain Miller has published five books, some 1800 articles and stories, and has won 14 state and national writing awards. He has also taught several journalism and creative writing courses at the college level.

F. T. Stent, general aviation editor

F. T. STENT is an ATP-rated pilot with an extensive flight background that includes a combat tour in Vietnam as a Navy fighter pilot, several thousand hours in General Aviation aircraft, and both domestic and international flying as an airline pilot for a major trunk carrier. A graduate of Yale University (degree in Literature), Mr. Stent also holds a Master's Degree from Harvard University and has been actively writing and consulting in aviation for the past several years.

Barry Schiff, air carrier editor

BARRY SCHIFF, a Boeing 727/707 captain for TWA, began flying at the age of 14 and, at 21, became the youngest holder in the U.S. of the Airline Transport Pilot Certificate. One of the few pilots in the U.S. with all seven flight instructor ratings, he also has an aviation teaching credential from the State of California. The author of several books and numerous training manuals, he has had more than 300 articles published in aviation magazines and journals around the world. Presently, he is editor-at-large for The AOPA PILOT magazine. As the holder of numerous world class speed records, Captain Schiff was selected by the Federation Aeronautique Internationale in 1969 to receive the Louis Bleriot Air Medal. He has also received a congressional commendation for his contributions to aviation, and a special writing citation from the Aviation/Space Writers Association.

Allan R. Scholin, military/aerospace editor

ALLAN R. SCHOLIN is considered to be one of the country's leading military aviation and space experts. A professional journalist for 30 years, Mr. Scholin served as an information officer in the Air Guard and Air Force Reserve during World War II and Korea, and later worked as a writer in the Washington bureau of the National Guard. For the past 12 years he has also been a contributing editor to AIR PROGRESS magazine. He is a member of the Aviation/Space Writers Association and the Society of Professional Journalists.

Dan Manningham, sport aviation editor

DAN MANNINGHAM's expertise in aviation runs from sport parachuting to flying Boeing 727s as an airline pilot. A widely read free-lance writer and contributing editor to BUSINESS & COMMERCIAL AVIATION, Mr. Manningham first learned to fly helicopters in the military and later transitioned to heavy transports and a variety of general aviation aircraft. A graduate of Tufts University, Manningham has thoroughly covered the broad spectrum of aviation's most unique activities and is singularly qualified to serve as sport aviation editor.

Gordon Girod, managing editor

GORDON GIROD is a former United States Air Force Academy instructor who has also served on the faculties of several major civilian universities. An Air Force officer for several years, Mr. Girod was a frequent lecturer at the Squadron Officers School and the Air Command and Staff College at Maxwell Air Force Base in Montgomery, Alabama. He holds Bachelor and Master's degrees in English and has done work on a doctorate degree. Mr. Girod has edited several major book projects in the past, and currently serves as the managing editor of the JEPPESEN SANDERSON AVIATION YEARBOOK 1977.

INTRODUCTION

The editor of the Farmer's Almanac noted a sign in a gardening center that said: "Plant our seeds . . . and jump back!"

Well, that's the way it is with aerospace these days. A few years ago we planted some ideas, and now the developments are coming along so fast it's hard to keep up with them.

This second edition of the JEPPESEN SANDERSON AVIATION YEARBOOK is our attempt to keep up. General Aviation starts things out. One thing you will notice in this section, as well as in the others, is the process of refinement. Aviation doesn't give us too many absolutely, blood-stoppingly new things each year. Rather, it's a process of honing, stretching, polishing and perfecting the old. For instance, this marvelous Learjet instrument gets a new wing and longer legs, and sets a new record, squeezing the old world into an ever smaller ball. At each of the annual shows, from NBAA to Reading, from Hanover to Paris/Farnborough, you see the new wrinkles, true, but what you mainly see is the constant process of stretching and refining. It would be so steady, this process of perfection, that it might become dull if it weren't for a few lumps in the batter—for instance, last year the so-called "ELT Mess," which you'll read about in Delta pilot Terry Stent's General Aviation section.

Accidents, of course, we always have with us, and we try to include something on those where the reader needs to be warned of a potential trap or can learn something helpful (in a highly sophisticated form of hangar-flying). But we try not to dwell on the gloomy and seamy sides because, of course, they're only a small part of the whole.

Still a provocative nuance this year is the continuing question of wind-shear accidents that suddenly snapped into the forefront of our collective consciousness in 1975. Something really new in the old witches brew of weather? The final decision isn't in yet—especially in light of a recent Texas International DC-9 takeoff crash at Denver that happened almost at the same time as last year's Continental B-727 takeoff crash, and possibly for similar reasons. A new factor in this type of problem has an oriental slant—"the Fujita postulate," which you'll read about in the Air Carrier section, edited this year by that "ubiquitous homologator," veteran TWA pilot Capt. Barry Schiff.

Barry also has some interesting material on the SST, which, as the agonized and irresponsible howls of the critics fade in the distance, seems to be doing a pretty darned dependable and safe job of carrying people quickly around the globe. The hard facts are beginning to emerge. As Bill Magruder, President of Piedmont Airlines and former boss of the U.S. supersonic effort, noted, "A Volkswagen Beetle—just one—crossing a city does more to hurt the ozone layer than an SST crossing overhead." The anti-SST arguments are beginning, indeed, to sound as thin, in the words of Abe Lincoln, "as the boiled shadow of a homeopathic pigeon."

The gals are getting bigger jobs in the biz, too. The Air Carrier section tells you about our good friend, personable Emily Howell, now a Frontier captain (and recently married), and shows you a nifty picture of the lissome and lovely four-striper.

As we noted, it is a by-product of progress that each year sees more of the stretch-and-refine technique. But 1975 had a nifty switch on this when Boeing shrank the 747 into the "SP" (for Special) model, thereby giving this big doll Seven League Boots (and now allowing for the wives of Emirs and Caliphs and Shahs to come nonstop to Uncle Sugar land for shopping sprees—returning home with JumBoeings full of Cadillacs and candy, and all the other nice his-and-hers things that Nieman-Marcus proffers.)

Oh yes, and there is an imaginative new approach to airfoils, the scissor wing, so maybe there are some new things under the sun.

Along the military hardware line, one of the nations's most qualified writers, Al Scholin, has assembled a potpourri of articles that gives a nice crosscut of the aerospace spectrum.

The gals? Sure, they're in the military picture too, and by the numbers. At the Air Force Academy, since September, the first contingent has had a lower attrition rate and higher grade averages (not surprising) than the male Doolies . . . and now the first bids are out for gals to ride the hot tin floors in the astronaut program. (So now we'll have to worry about possible hanky-panky on the first manned . . . woops . . . "personned" Mars landing?)

Three months after Lt. Viktor I. Belenko defected to Japan from Russia in his MiG 25, he was brought to Langley Field, Va., where, according to Associated Press, "he won high marks as a pilot from U.S. officials." In the military bailiwick, you'll find what the U.S. experts determined when they probed the Foxbat's construction and its combat potential.

In this section you'll also get to fly the B-1, somewhat vicariously—from your easy chair; be able to take a look at the Space Shuttle, and find out what really happened that night at Entebbe, among other interesting features.

I've been working on "the definitive" book on Sport Aerobatics for the Jeppesen Aviation Book Club this year, and was terribly saddened by the death of two good friends, who were nonpareil in this corner of aviation. Manx Kelly, the handsome British champion who put the Rothmanns and Carling precision teams together, was killed when his bird came to pieces in a California airshow; and Dave Rahm, the equally handsome Seattle "Flying Professor," was killed when he pulled out of a downward snap too low in an airshow for King Hussein in Jordan.

Both had contributed much to sport aviation, which is growing "like to bust its britches." Last year we covered this burgeoning area of aviation in a section called Special Operations. This year, just to show you we can stretch and polish, too, we've changed the section name to . . . you guessed it, Sport Aviation. In it United pilot Dan Manningham gives you a pretty good cross-cut of everything from hang-gliding to the bigger air races, from the man-powered aircraft to the Angel Derby, and from the Smirnoff glider classic (sorry about that: for "glider" read "sailplane") to the "hometowning" the U.S. Acro team got at the hands of the Russ in Kiev. (Another cute gal, gutsy Betty Everest, "came back with the bronze" from that meet—and with plenty to say. "If you want to see what the Soviet pilots thought of their own victories, look at photos of the award presentations," she said, showing us photos of the medals being pinned on some pretty sheepish-looking "winners.")

In building such a book as this (and where else can you find an instant encyclopedia of what happened yesterday in aerospace?), some subjects defy classification.

For instance, I made four trips in 1975 to the National Air and Space Museum in Washington just because it's such a super treat for anyone who has wings in his heart. It's so complete, colorful and imaginative that, as one wag aviation buff put it, "If you lead a good life, when you die, you go to NASM." You get a good report on it in our NOTAMs section, where we pick up these hard-to-classify subjects. Here, also, you'll get a look at a guy whom a lot of people considered the ultimate kook, Howard Hughes, but who, in his younger days, contributed greatly to aviation. You'll also find a salute to that gallant pioneer, Grover Loening, who also died last year.

Humor? History? Plenty of both in these pages, too. My favorite in the former category is something called, in French, "The Airplane Crashed Normally." *Naturellement!* It's funny.

But why should I keep on selling you on how great our new book is? Sell yourselves, amigos. Read on! Read on! and next year we'll have an even better book—Lord willin' and the crik don't rise!

Ed Mack Miller
December 8, 1976

GENERAL AVIATION

AVIATION YEARBOOK

INTRODUCTION

1976—the Bicentennial year of stars and stripes; stickers and pins; patches and peanuts . . .

In General Aviation, golfing great Arnold Palmer and golfing-not-so-great K.O. Eckland deserted the links to commemorate the nation's 200th birthday in planes. Palmer streaked to a new round-the-world record in a Learjet 36 and noted author Robert Serling wrote eloquently about it in "The Saga of 200 Yankee." Pilot K.O. Eckland, on the other hand, staggered to no known records in a "garden-variety Cub" known as the *Spirit of N76* and author K.O. Eckland writes hilariously about it in "The Cub As a Traveling Companion."

But the year went on and stories poured forth. There was much to select from this year. Many fine writers and a multitude of fascinating reports. They ran the gamut: from pilots and planes, to careers and controversies. Only the best are here:

—Stories chosen for their extraordinary interest. A group of enterprising young teenagers making like the "Blue Angels" by flying their Cherokees through low-level stunts while "only a handshake apart."

—Stories selected because they inform. "How to Build a Career in Aviation," "The Challenge of Ag Flying," and "How to Get Into Aviation by Flying a Desk." For those readers who look skyward today and see a future tomorrow these articles will have particular importance and value!

—And stories that teach. For most pilots, conditioned by training to expect relatively safe landing areas, crashing is a subject quickly ignored. But should it be? NTSB Investigator, Gerard Bruggink has written a fascinating and long-needed paper, "How to Crash an Airplane," on such tabooed subjects as impact loads, energy absorption, deceleration forces, and egress. And as a follow-up article, author Steve Knaebel provides a look at the real thing in his story, "L'Avion C'Est Crashe' Normalement" (The Airplane Crashed Normally). Knaebel begins on a chilling note: "Roger, we've lost our engine now, and we . . . looks like we've got a fire in there . . ."

Then there are stories on the major issues of the year. Fuel problems were exacerbated by the gradual disappearance of an old and trusted friend, 80/87 octane. The new kid on the block was called Low Lead 100 and it sparked "The Continuing Avgas Controversy."

And other familiar problems resurfaced in 1976. BFRs (Biennial Flight Reviews) and ELTs (Emergency Locator Transmitters)—two of the FAA's imposed programs currently suffering from a marked degree of unpopularity—became significant issues during the year, and stories covering their status and shortcomings are included.

In the continuing dialogue (shouting contest?) between the Regulator and the regulated, things heated up considerably when acting Associate Administrator for the FAA, Mr. Fred Meister, suggested in late 1975 the possibility of future constraints on the growth of General Aviation. AOPA Senior Vice President Victor Kayne took vehement exception to Meister's comments and his reply is presented in its entirety. With the government's role constantly expanding outward in every direction, aviation—and in particular General or private Aviation—faces an uncertain future at best. Meister's statements and Kayne's response sharpen the issues and draw the lines. They deserve a close reading.

In another effective counterattack, Robert Serling, in a speech before the Flight Safety Foundation, went after such stalwart opponents as the *Washington Post*, William Proxmire, and the Federal Energy Administration. Entitled "Whipping Boy," it hit dead center so many times that it has been singled out as one of the year's most important and substantive pro-aviation statements! Serling does not mince his words; no punches are pulled. He pins down some of the most serious threats posed to General Aviation today and his thoughts deserve the widest distribution!

Many other stories broke for General Aviation '76 and deserve mention, if not full articles. The Short Notes section, new in this year's Yearbook, provides such mention. Aircraft introductions like Mooney's 201 and Cessna's Hawk XP are here. New bizjets still on the drawing boards (LearStar 600); those already rolled out (Lockheed Jetstar II, AJI's "Hustler") or certificated (Israel Aircraft Industries 1124 Westwind) are covered. The new generation of lightplane avionics, new members of the NTSB, and similar topics deserving of comment will be found here.

In selecting the articles for this year's JEPPESEN SANDERSON AVIATION YEARBOOK, the intent, in part, was to record for future years the events that counted in 1976. But it was also to provide the reader with something of value, a selection that could amuse and educate, inform and stimulate.

. . . I think General Aviation – '76 does just that . . . and I trust you do also!

F. T. Stent
Fall 1976

THE SAGA OF 200 YANKEE

Reprinted by Permission
FLIGHT OPERATIONS MAGAZINE (September 1976)

by Robert J. Serling

A bunch of normal business jet trips laid end to end, Arnold Palmer's record-breaking 'round-the-world flight posed no special problems for the pilot or his crew. But how it was put together provides this commentary by Bob Serling, who went along as official observer.

Arnold Palmer, one of the most respected pilots in business aviation, has nearly 5,000 hours of logged flight time.

Demonstrating business jets on all continents, James E. Bir rarely gets excited over an international trip. But a message from his office at the Learjet factory in Wichita, which caught him in Africa back in December, 1975, interested him so he couldn't wait to get home: It informed him that he would be co-pilot to Arnold Palmer, the golfer, on a 'round-the-world flight to break the speed record for bizjets set by Arthur Godfrey, Dick Merrill and two others in 1966 in a Jet Commander. Sponsors would be the Aviation/Space Writers Association and the flight would be a part of the U.S. bicentennial celebration.

Bir made it back in time for the first preliminary planning session on January 10, 1976, with James Greenwood, Chuck Dyas, Richard Ross and Al Higdon of the Gates Learjet staff. A blueprint laid out by Greenwood would dictate most of the operational decisions, ranging from the route to be chosen to the time of final landing. As Greenwood explained it, the plan was as follows:

1. On May 17 the flight would leave from, and eventually return to, Denver, site of the AWA 38th annual news conference being held those three days; arrival would be in mid-afternoon of the 19th, preferably no later than 1530. That, Greenwood explained, would allow time for a welcoming ceremony at the airport, a short news conference and sufficient rest for the crew before the AWA awards banquet at which Palmer would be the main speaker.

2. The routing and the various refueling stops had to take into consideration the flight's status as an official bicentennial event. This would require assurances that the flight would be welcome in each country along the route, that there would be maximum news media coverage, and that there would be no unexpected delays generated by sheer politics—such as the one encountered by the Godfrey crew, which had wanted to overfly India but was forced to land there on orders of the Indian government.

3. To receive the official sanction of the National Aeronautic Association, U.S. representative of the Fédération Aéronautique Internationale, the flight had to cover at least 22,859 statute miles. The aircraft itself was to be a standard Model 36 devoid of any special equipment, extra fuel tanks or unusual electronics gear.

4. The crew would consist of four men: Palmer as the command pilot, Bir as copilot, L.L. (Bill) Purkey of Learjet as the third pilot and general troubleshooter, and a pool reporter who would serve as NSS's official timer/observer and also as a pool reporter for AWA (this assignment was to be mine).

To Bir, a veteran of more than 30 Atlantic flights and two across the Pacific, went the job of coordinating operational planning. His philosophy of preparation for any long-distance flight could be summed up in one sentence: "What's gonna rear up and bite me?" Starting with the assumption that the aircraft had to fly west to east to avoid headwinds and, hopefully, to take advantage of tailwinds, the task was to finalize a route and choose an alternate.

Engineering began pumping out various computerized flight plans. From Boeing's highly regarded wind and temperature historical records (which offer a high degree of accuracy) came a mildly optimistic long-range forecast: modest tailwinds over most of the route that would finally be chosen. The exceptions were Colombo-Jakarta and Jakarta-Manila.

The nav and approach charts Bir began stuffing into a couple of large boxes made the Manhattan telephone directory seem

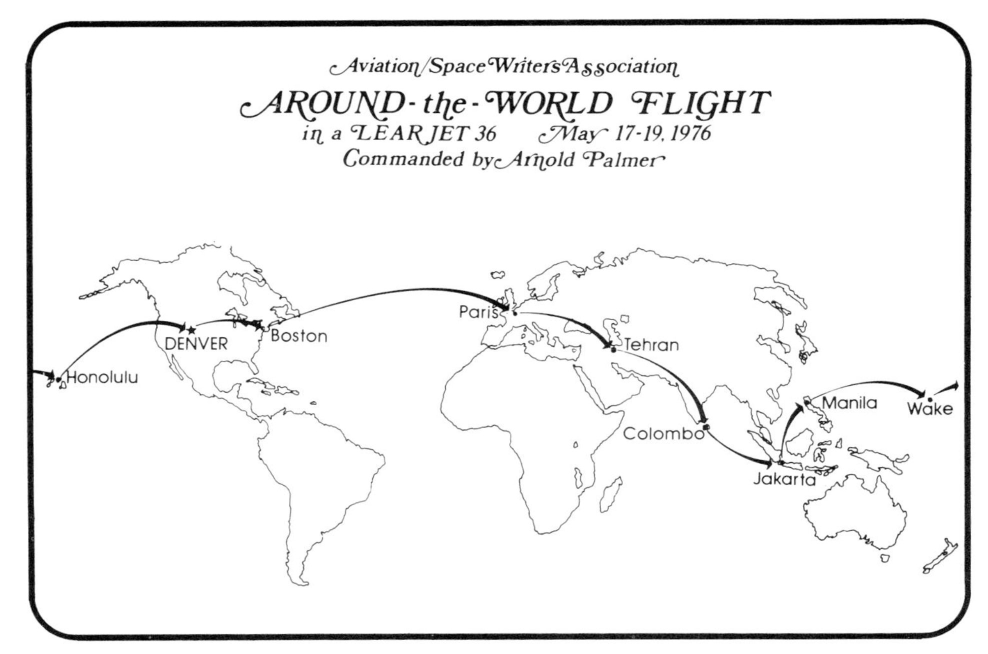

as thin as a volume of sonnets. He himself described the flight as "a bunch of normal business jet trips laid end to end," but he knew that this particular adventure, with its patriotic and publicity overtones, could not really be considered routine.

He had no worries about the crew. Bir, in 25 years of flying—some of them in the Marines—had logged some 7,500 hours. He knew Palmer was no aviation neophyte, either—Arnie owned a Learjet and the golfer's own logbook showed more than 4,500 hours. Purkey was not only an experienced pilot (some 5,000 hours) but was an expert on powerplants and airframes—if anything went wrong, "Mr. Fix-it" would come in handy. As it turned out, Palmer and Bir did most of the flying, Arnie himself making all the takeoffs and all but one landing; Purkey's main task was to supervise refueling operations. I went along as the flight's official observer and timer.

In the early planning stages, some consideration was given to stopping in Moscow and/or Peking but this was discarded, along with several other suggested stops, in favor of the final route chosen: Denver-Paris, Paris-Tehran, Tehran-Colombo, Colombo-Jakarta, Jakarta-Manila, Manila-Wake Island, Wake-Honolulu, and Honolulu-Denver. Jakarta, where the political climate was uncertain, remained a question mark even after the flight left Denver. Singapore was picked as an alternate refueling stop, although the Learjet would have to overfly Jakarta on the way to meet mileage requirements. The "just-in-case" alternate route would take the flight from Paris over Greece, Lebanon, Jordan, Syria and Saudi Arabia to Bahrain.

The route finally selected involved only three critical fuel legs: Boston-Paris, Manila-Wake and Honolulu-Denver. But they posed no real problems. Operationally, it would have made more sense to fly from Denver to Gander, and then refuel for the next leg to Paris. But Gander obviously offered no media coverage advantages and Boston-Paris nonstop appeared routine—particularly in view of the tailwind expectations.

Originally, it was planned to take a brand-new Model 36 off the Wichita assembly line and give it a special paint job for the flight. But the aircraft earmarked for the flight never got into Bir's hands—a customer grabbed it as soon as it was finished. So Greenwood began looking for a Model 36 owner willing to lease his aircraft for the historic trip. He got plenty of offers but each had strings attached—either the owner would have to go along on the flight, or his own pilot or pilots would have to be in the crew. Such conditions were unacceptable, but Greenwood finally found a 36 owner willing to lease his airplane with no strings attached. Herbert Hamilton, who in 1964 had become the first man to put a Learjet into business service, offered a Model 36 owned by Yankee Leasing of Cincinnati, one of his subsidiary companies. All Hamilton asked was that the aircraft be returned to him in its original paint scheme—a brown and white fuselage with an accenting gold stripe. What made his Model 36 even more attractive was the fact that he owned aircraft registration number 200Y.

Thus was born 200 Yankee, an identification so appropriate to the bicentennial theme. Learjet actually owned number N76GL which would have been used if the Yankee Leasing plane hadn't become available. Gary Sale of Learjet's design department came up with the final paint scheme after considering a dozen different ones: an all-white fuselage with a band of red stripes and blue stars. "Spirit of '76" was supposed to go on the tail—until Greenwood happened to read

James E. Bir

Lewis L. "Bill" Purkey

Robert J. Serling

A traditional Aloha at Honolulu International Airport.

about a Miss Ellen Harness of Litchfield, Conn., winning a nationwide contest for a bicentennial slogan. He wrote her, asking permission to use her slogan on 200 Yankee. It was quickly granted and on the tail went the red lettering: "Freedom's Way–USA."

Bir, meanwhile, made up waypoint coordinate logs for every stop . . . graphs which showed in a split second whether fuel consumption at every checkpoint was too high or too low . . . pre-drawn ICAO flight plans for each stop to enable faster filing . . . even a personal rest schedule (which he didn't follow).

Help for Clearance and Service

From all the computerized flight plans and Bir's own calculations, the projected goal was a 53-hour flight including eight refueling stops of no longer than one hour each. Publicly, the goal was "under 60 hours" which was more than sufficient not only to break but demolish the Godfrey-Merrill record of 86 hours and nine minutes. If 200 Yankee left Denver May 17 no later than 1030, a 53-hour flight would get it back to Denver about 1530 on May 19. The planned cruise altitude was 37,000 feet Denver-Boston and 41,000 feet the rest of the way. Bir, after studying all the data, privately hoped to circumnavigate the globe in under 50 hours, assuming a general absence of headwinds.

Pan American World Airways agreed to handle all overseas arrangements but this plan changed, partially because of a golf game. Some weeks before the flight, Tom Evans of Universal Weather and Aviation, Inc., Houston, Texas, was on the links with Jimmy Demaret, one of Palmer's closest friends. Demaret told Evans about Arnie's 'round-the-world jaunt and Evans immediately made contact with Greenwood and Bir.

"We have the facilities to arrange your clearances, help with flight plans and set up refueling anywhere you want to fly," he told Bir.

"We've already contracted with Pan Am," Bir apologized, "but if the situation changes, I'll be in touch."

The situation did change; Pan Am became involved in its own world flight, that of a new 747-SP, and Learjet decided to shift at least partially over to Universal. Bir informed Evans 200 Yankee would utilize Universal's services at Jakarta and Colombo, and possibly Wake—later the Houston firm was to help at all stops except one that wasn't planned.

Bill Purkey was brought into the planning picture when the subject of logistics acquired some priority. For example, what kind of spare parts should be carried? "If it takes more than one hour to change, let's leave it off the plane," Purkey urged. Others felt a two-hour replacement time limit was more reasonable, so the spare parts list included: A generator, voltage regulator, cabin flow control valve, main spare tire and wheel, fuel control computer, box of engine seals, and an ignition exciter (transformer.)

The engine manufacturer (Garrett) added welcome logistics support by assigning a representative to meet the globe-girdling Learjet at each stop. Major aid also came from Collins Avionics, which agreed to supply the services of its high-frequency, single-sideband (HF SSB) radio station in Cedar Rapids, Iowa, to facilitate communications between the airplane and the AWA convention headquarters.

Collins analyzed the proposed flight plan and determined the best frequency for each time and place along the route. Collins provided a chart dividing the earth into six sectors based on longitude, and a primary and secondary frequency was assigned to each sector for every two-hour time period. Aboard 200 Yankee itself was a 718U HF SSB transceiver.

The communications plan was relatively simple in concept—once contact was established between 200 Yankee and Cedar Rapids, radio operators at the latter would set up a phone patch between their station and a standard speaker phone at the convention, so the crew could broadcast messages, progress reports and even conduct interviews with radio and television people.

The same system had been provided for other, including Godfrey's flight and the 1965 polar circumnavigation of the earth by a Flying Tiger 707. But in all previous cases, Collins was given permission to use Air Force frequencies in areas of extremely poor propagation signals. Collins and Learjet requested the same permission from the Air Force for 200 Yankee, and the Air Force agreed, but with one important difference—final authority would have to come from the Department of Defense. Right up to takeoff time, DOD never replied to the request one way or the other, and the result was an almost total communications blackout between Paris and Manila.

The navigation equipment aboard 200 Yankee was a VLF Omega with a Loran unit as a backstop. Bir later summed up the navigation aspects in these words: "As far as route planning was concerned, we didn't do anything different than what we'd do for any demonstration tour, with one possible exception. Since we had a VLF in the airplane, I did make a log of geographic coordinates of all fixes we'd be flying over, so we wouldn't have to try to determine enroute what the coordinates were for some of them if they didn't show up on a particular chart. Otherwise, the enroute nav planning portion was pretty straight-forward."

It was not a true demonstration flight except by inference. Gates Learjet's president, Harry Combs, had put the Model 36's reputation on the line by authorizing the project, but right from the start Learjet kept the commercial aspects not only in low key but virtually out of sight. Palmer was the star, not the airplane, and the script was written in accordance with the interests of the two organizations most directly concerned: AWA and the American Revolution Bicentennial Administration.

Headwinds But No Problems

The global trip itself was remarkable for the absence of "things that reared up and bit," to paraphrase Bir. In nearly 50 hours of actual flight time over a distance of 22,958 statute miles, there was exactly one mechanical "malfunction"–the selector knob on the fuel quantity indicator worked loose; Purkey tightened it with a screwdriver. So much for mechanical requirements.

The same was not true of the enroute weather. Once 200 Yankee left Boston, it ran into constant headwinds until the flight was well past Wake. The worst encountered were over the North Atlantic, between 40 and 50 knots, and both Bir and Palmer decided to divert to Rhoose Airport at Glamorgan, Wales, instead of attempting to fly nonstop to Paris. The diversion set the flight back almost two hours and the time was never made up. But Bir still believes it was not only the right decision but the only one.

"When we did cross the Channel enroute to Paris," he recalled, "the French controllers–as usual–gave us an early descent clearance and this would have made the fuel situation even more critical if we had tried to come in without refueling."

The only other departure from routine occurred at Manila. It, too, was a weather situation. Typhoon Pamela was hitting the outskirts of Manila International Airport as 200 Yankee took off for Wake. Palmer and Bir were warned to expect icing during climbout, and this could have been a problem. The Learjet's deicing system draws its power from the engines and this could have meant higher-than-desired fuel consumption. If the airframe accumulated ice in abnormal quantity, resultant drag also would boost fuel consumption. But ice accumulation was less than anticipated and 200 Yankee reached Wake without difficulty. Flying through Pamela, however, was quite an experience. Purkey later compared it to "flying in a submarine with wings" and Bir commented that climbing through such heavy rain was "like swimming underwater."

Fatigue was an ever-present factor, even with three pilots aboard. It became particularly bad on the Wake-Honolulu leg, when Palmer actually became dizzy and Bir once dozed off reading a chart. But once Honolulu was reached, the adrenalin began pumping again, and it was really gushing when 200 Yankee reached the West Coast and headed home. Nowhere was national pride in the mission exemplified more than by the U.S. controllers handling the flight. As the Learjet raced eastward, the various enroute centers sandwiched messages of congratulations in between official transmissions–plus frequent invitations to Arnie "to come down and play some golf."

At 1950, 200 Yankee streaked by the Stapleton tower as pre-arranged and NAA official John Slattery in the tower recorded a total elapsed time of 57 hours, 25 minutes and 42 seconds, breaking the Jet Commander's record by nearly 29 hours.

It had been decided to stage the flyby instead of an actual landing at Stapleton, so as not to disrupt the airport's normal traffic with a welcoming ceremony. The latter was held at nearby Arapaho County Airport where Harry Combs and others greeted the triumphant, if tired, crew. It was Combs who furnished the best commentary on a feat that was truly a tribute to private enterprise and business flying.

Ex-astronaut Wally Schirra had been kidding 200 Yankee's crew, wondering why they were "dragging your feet when it took us only an hour and a half to fly around the world." "Wally did his trip," Combs growled, "at government expense."

Editor's Note: Another remarkable event connected with the flight of 200 Yankee occurred after landing. That night, going almost directly from Arapahoe County Airport to the AWA banquet, Palmer gave a noteworthy speech—delivered without advance preparation, and without notes. A quotation from that speech (found in the National Aeronautic

Despite initial uncertainty whether 200 Yankee would be cleared into Jakarta, the welcome turned out to be a warm one.

After boarding this elephant at Sri Lanka (Colombo, Ceylon) at 1:00 a.m. Arnie has a wee bit forced grin.

Association commemorative book "This is 200 Yankee..."*) says much about the man and about flying:*

"Something that happened between Iran and Colombo as we were flying, and for those of you who are interested in flying and have flown a great deal internationally, will understand what I'm talking about. For those of you who haven't, you'll not understand, probably, and there is no way I can transfer from my lips to your mind what I'm trying to say. But in the flight, it was one of those things I guess makes flying very interesting.

"We were flying over the Indian Ocean and it's the middle of the night. There's not any light at all. The stars aren't out, the moon's not out, the clouds are coming up under the airplane, sort of scratching the bottom of it, and there are four people in the airplane and three of them are asleep and you're flying that airplane and you think, gee, isn't this great? And then you think, what in the

hell am I doing here flying over the Indian Ocean, not sure whether this navigational equipment is working, not sure whether this airplane's going to work, not sure what's going to happen next, who's going to greet me at the next station, what's really going to happen?

"Then you find out a little bit about yourself. You think and you think about the people who are with you and you think about your family, you think about a lot of things. I thought, and right now think, that in my life I've had a great many experiences. Playing golf, I've been scared of a shot I've had to make, or something that was going to happen to me on a golf course. And I've flown a lot. I've flown airplanes around the United States and I've been scared in airplanes. But I think possibly this flight around the world gave me an insight into life, as well as flying, that I've never seen before..."

Learjet estimated direct flying costs for 200 Yankee's global voyage at 20 cents a mile. The aircraft averaged 1,104 pounds an hour in fuel consumption and averaged 486 miles an hour in speed despite the almost constant headwinds.

The Reading Air Show a few weeks after the world flight was 200 Yankee's last appearance in its bicentennial colors. Late in July, it went back to Wichita for restoration of its original paint scheme and is now flying regular charter trips for Yankee Leasing.

The commander of 200 Yankee with His Royal Highness, Crown Prince Reza Pahlavi of Iran.

THE CUB AS A TRAVELING COMPANION

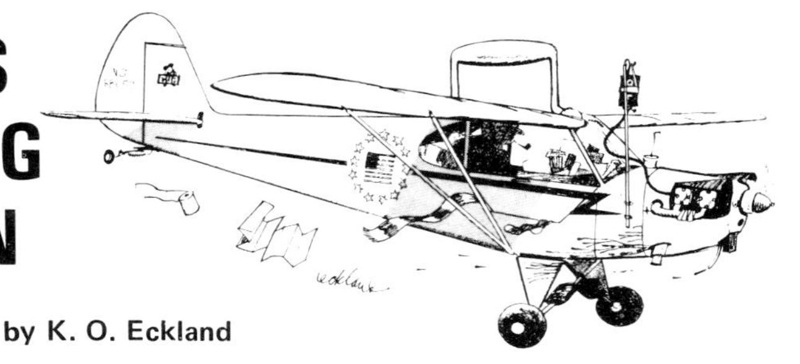

Reprinted by Permission
PRIVATE PILOT (May 1976)

by K. O. Eckland

A common, garden-variety Cub can provide an interesting partnership over an extended period of time. One hour in the less-than-capacious cockpit of a Cub is the rough equivalent of two hours in a dentist chair or four hours watching the Howard Cosell show.

This point became painfully apparent to me as I was approaching Yuma, some six hours into my nationwide flight. The Cub seemed to want to drift off to the right, a symptom which I attributed to a back yard rigging job. I slowly became aware that no given amount of left rudder would keep her going straight—then I realized that all the pressure I was applying to my left foot was only going as far as the kneecap. My foot had gone to sleep!

"Come on foot," I implored, "this is no time to sluff off! We have a nation to conquer and I can hardly be expected to do it on one foot."

By pulling my lower leg with both hands, I managed to get the great slab of stone into a vertical position between the cabin wall and the box occupying the front seat; the blood began to flow about as fast as liquor at a company party. In my mental checklist, I added a cross before "Check Foot Every So Often." Later on, I was to add such items as "Check Spine Every Two Hours" and "Run Around a Lot During Refueling." This last ploy made for a lot of interesting comments around the gas pump and, even, an occasional direction to the men's room.

"No, it's not that," I would explain. "I tend to cramp up a bit and have to work the old joints out every now and again."

Not much of a knight-image, you say? You could be right.

In defense, I would like to challenge anyone, short of a professional contortionist, to sit in a Cub for an hour and then leap out and start playing tennis. And, on top of that, not just any old Cub but the *Spirit of N76*, herself. To begin with, the plane had to be mickey-moused in order to carry the creature comforts I would need to fly for six months as well as to have room for those gewgaws I would doubtless pick up along the way. Of prime importance were my two 35mm cameras and 100 rolls of film (more would be mailed to me at predetermined stops), emergency food and water, clothing with life-sustaining cold-weather gear, portable VHF-FM-AM radio, tape recorder, tapes, road maps and sectionals, sleeping bag, blanket, personal effects, an ELT that would emit only a pitiful little squeak when shot with a .45, and miscellaneous other items.

For three days prior to departure, I experimented with various stowage positions, assigning stations on a note pad to each item. One day before departure, I was looking for a DC-6 as a replacement.

The Cub is an uncompromising mistress. She has just so much cubic space to offer and that's it. The rear seats went out and I folded the sleeping bag, doubled many times, to serve as a seat cushion. The blanket and tarp and pillow fit nicely on the front rudder pedals and I double-checked for pedal clearance. The front stick was removed and over the stub went my camera bag, foodstuffs and some clothing. The box that occupied the front seat was stuffed to overflowing. Charts went on the dash where I found I couldn't reach them and they eventually were consigned to under the front seat. Behind me, in the "50-lb. baggage" well, went that and then some—clothing, flight bag, things. Under the rear seat I wedged in shoes, towels, unmentionables and miscellaneous, some of which jogged their ways back to the tail post. I stood back after cramming what I could on board and surveyed the assortment of valuables that still were on the ground. These went back home.

Knowing full well that I would never get this great mass airborne, I decided, mainly for the benefit of the wellwishers, to go through the procedures of calling the tower for clearance, taxiing out, aborting the takeoff, taxiing back and going home amidst the tumultous boos of my supporters. Needless to say, I was astonished to find us airborne after a modest distance and establishing a 10,000-foot-a-minute climb rate. The maps slid off the dash and my foot began to go to sleep. We were on our way.

A word here about seating. Remember the sleeping bag I was sitting on? Soft, downy, sitting on a cloud, right? Wrong. Like sitting on a clod. Lumps, bumps and a zipper whose location I could pinpoint through four layers of kapok. I did more wiggling than a nine-year-old at a PTA meeting.

Then there was the CG. The fuel tank played with the G more than the C could keep up with. I learned to stuff a lot of loose equipment behind me in the baggage area until the gas supply dwindled and then would start metering things forward to keep the Cub on a trim status. The trim handle was merely a placebo. Once in a while, I would forget or become entranced with some local grandeur and find I was flying full stick forward. Usually, I remembered.

Turbulence did interesting things with my load. Once, I caught a Pentax drifting by my head on its way out through the windshield. Or my milk bottle would open and begin redecorating the interior. Turbulence at Fredericksburg, Texas, put my thermos between the cabin wall the the front control stick stub, making for an interesting no-left-stick landing.

Rain is nice. Flying a Cub in the rain is akin to flying at the small end of a funnel. Bill Piper must have had a chuckle or two when he approved the drawings for the doors and wing roots. There was the same effect with cold winds which would play upon my knees. They were a good test for a sleeping foot. If I could feel the intense pain that foreshadows freezing, my foot was not asleep. Providence works in strange ways. Hint: a Sectional wedged in the crack between the right cabin door and the frame will sort out a good fifty percent of the drafts. Also, Sectionals, wrapped around the legs, will delay freezing by 30 minutes to an hour.

After an hour and a half in the Cub, no amount of cockpit calisthenics will ready you for the Grand Exit. There is some TV footage I would like to see burned, at San Angelo if memory serves me, where I made one of my typical exits while newsmen stood around filming and chuckling. I think I came out head first that time (as well as a few other times), bringing everything that was loose with me. For an encore, I walked around for several minutes like I had on an athletic supporter three sizes too small until I could once again stand up on my hind legs. At the more uncouth airports, I could usually count on a round of applause.

The Cub itself (or herself, if you prefer) had her own set of ground rules that I familiarized as we wended our way. Even with a fully-alert foot, there was that built-in tendency to sneak over to the right. Oil consumption was an Arabian's delight—a steady one-quart-per-hour for the entire trip, even despite the quantities of additives that were the consistency of tapioca pudding. This was the least of my problems.

There were the mags and the number one cylinder. Original logs were gone on the Cub and it was anybody's guess as to the total time. One mech told me: "Probably somewhere between 300 and 3,000 hours. Best thing is, if it ever quits—overhaul it!" I thought about that comment many times over swamp lands and mountains. I learned to become an expert left mag man from Florida to Pennsylvania, where I finally found someone who spoke Cubese. Some day, when you have a lot of time, I'll tell you about the "experts" that lurk at every airfield. The number one cylinder finally threw up on takeoff at Easton, Pennsylvania, and by virtue of superior pilotsmanship and a lot of coarse language and dumb luck, we limped into Van Sant airport—where I spent a week waiting out weather and watching Bill Smela work miracles on the Continental.

There it was—wrapped up in two paper log books—over 17,000 miles in a crampy Cub and I loved every moment of it. That's easy to say now, sitting here before a roaring fire (I almost wish I had a fireplace) and doing my contemplation bit. Given the bad mags and the chills and the ridiculous crosswind components and that draft that refused to let me light my pipe and all those other things that were the Cub, that were the *Spirit of N76*, given them all, I'd do it again. Maybe the Cub and I understand each other.

Traveling companions have a way of learning to put up with each other. We were two souls stuck on an aerial raft for 264 hours and we learned not only to live together but to gain a mutual respect and trust in each other. Only, I can't help wondering if the Cub had as much fun as I did. Maybe, if she could hammer out an article, she'd have a few words to say about Eckland as Flight Director. Maybe it's just as well she can't write.

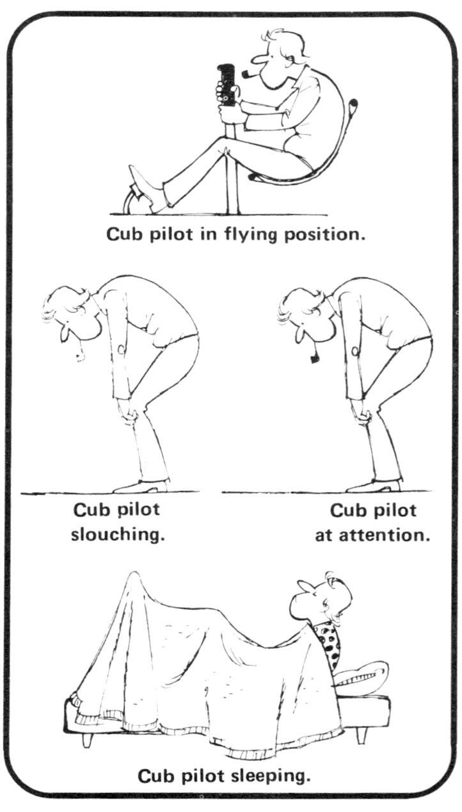

Cub pilot in flying position.

Cub pilot slouching.

Cub pilot at attention.

Cub pilot sleeping.

Illustrations: K. O. Eckland

EXPLORER CHAIRMAN FLIES IN CLOSE FORMATION SQUADRON

Reprinted by Permission
BSA NEWS
(February 1976)

Three small planes, their wing tips a handshake apart, brush through the blue southern sky.

A thumbs-up signal and broad smiles weld a triumvirate of teenaged pilots just a little closer together. The planes, including one piloted by 19-year-old Tom Morgan, of Pascagoula, chairman of the National Association of Aviation Explorers, cut into a descending spiral, emitting red, white, and blue smoke; a parachutist floats to earth cradling the Stars and Stripes; and another formation flying demonstration of Explorer Squadron 514 draws to an end.

The Blue Angels they are not. But the Blue Angels they have met and the routine they go through can be every bit as challenging.

While most of Morgan's contemporaries are stepping on an accelerator cruising Main Street somewhere, chances are good Morgan can be found around a Cherokee 140. He flies right wing for Squadron 514's demonstration team out of Pascagoula, Miss. A serious, quiet sophomore at the University of Mississippi here, he won't bother telling you that he is also the elected head of more than 12,000 young people, male and female, ages 15 through 20, who participate in Scouting's Aviation Exploring program.

Morgan, as chairman of the National Association of Aviation Explorers, is marked as one of the leading spokesmen of the new generation of fliers.

"I've been interested in flying ever since I knew there were airplanes," reflects Morgan. He earned his private pilot's license at 17 and, later this year, will complete his commercial and instrument ratings. He's more typical than atypical of the young flying fraternity.

The congenial Ole Miss student joined the Pascagoula Air Explorer squadron back in 1973 for only one reason: it offered him the chance to fly. He has learned to fly. Someday Morgan may combine a career interest in medicine with his love of planes. "I want to help people and I think I might like to do it through aerospace medicine." There may be a detour, though, he admits, while he checks out the field of corporate aviation. And an interest in law is still competing with medicine.

His exuberance is contagious.

He'll tell you that earning his license and ratings has cost him—not his parents—perhaps $7,000 so far. To earn that money, Tom, a husky 190 pounds with a red belt in karate and a sports background in football, has worked stacking bricks for a company that builds smokestacks. He's seen more than his share of flicks as a projectionist in theaters. And along with other members of Post 514 he has done so many service and fundraising chores either at Pascagoula Airport or for Gulf Coast Aircraft Sales, which is the post's "partner organization," that he's quit counting.

He'll tell you, too, that the "crazy

idea" of kids flying tight formation in a bunch of Piper products (the other planes are another Cherokee and a 1973 Piper Arrow) wasn't only his, but also had its genesis with fellow post member Greg Johnston of Pascagoula, now 18 and also an Ole Miss student. Johnston, who holds both commercial and instrument ratings, flies lead for the group and also doubles in brass as chairman of Aviation Exploring in BSA's Southeast Region.

At left wing in the formation is Donald Theisen, 18, post president and a licensed commercial pilot.

To the best of their knowledge they are the only teenaged, close formation aviation demonstration team in the country. They helped open Stennis International Airport in Bay St. Louis. They've drawn cheers at July 4 celebrations in Pascagoula. The air shows they've hosted now number five and attract thousands.

Morgan and his colleagues practice at least 10 hours a month. "We watch the (professionals) and then arrange our own maneuvers," explains Morgan, who gives credit to post advisor Obie Young for working hard with them on their basics and to both Young and Gulf Coast Aircraft owner Ronnie Hogue for moral and physical support.

"Everything we do," continues Morgan modestly, "everyone who trains for a private license does; we just do it in close formation." Like two feet from wing tip to wing tip. Their most difficult maneuvers are cross-overs or formation landings, which can be complicated by turbulence.

To date, no mishaps! And if Morgan and Johnston, Young and Hogue have anything to say about it there won't be any.

The maneuvers are not really that difficult, but they do require training, practice, and teamwork," points out Young. The young men add that they have cooperated with, and have the cooperation of, the FAA. "This program fulfills a need of young adults to develop and to conduct a quality program as a proving ground where they can test their capabilities 'on their own' in the adult world," Young also notes.

In Room 224 at West Twin Towers on the Ole Miss campus, which Morgan shares with a non-flying colleague, there's not even a picture of a plane. The late karate hero, Bruce Lee, glares down in a menacing portrait pose; the Stars and Bars of the Old Confederacy can't be missed; and amid the general neatness of the room (beds even have spreads!) empty milk shake containers vie for shelf space with footballs, just like they do in college dorms everywhere.

But mention three small planes, their wing tips a handshake apart, and you'll soon learn there's nothing like getting out there in the blue southern sky—and flying.

Air Explorer Squadron 514 Flight Demonstration Team (left to right) Right Wing - Tommy Morgan; Lead - Greg Johnston; Left Wing-Donald Theisen; and Slot - Ronnie Wriborg.

HOW TO BUILD A CAREER IN AVIATION

by Keith Connes

Reprinted by Permission
AIR PROGRESS (July 1976)

At first, I was a little leery about talking to flight school directors about career potentials in aviation. After all, they have axes to grind, right? It's in their best interest to paint a rosy picture in order to attract new students, right? Well, I needn't have worried, because I picked two of the really reputable aeronautical academies, and their directors were candid and conservative in their viewpoints. They spoke of the past, the present and the possible future, and even though each school was on an opposite coast of the country, the views expressed were remarkably similar.

My first visit was to Oakland, California, where Skip Everett heads up the Sierra Academy of Aeronautics.

Oakland International is a large facility with separate tower and runways for general aviation. Sierra is housed in what was once, a long time ago, a hotel building. The facilities are a trifle worn, but the atmosphere is homey. Skip Everett is friendly, energetic yet relaxed. He and Dave Esler have collaborated on a number of "careers in aviation" articles for Air Progress in the past.

Here's what Skip had to say.

AP: I imagine that the amount of hiring activity that the airlines are doing affects the hiring practice in general aviation.

EVERETT: That's right. When the airlines get active, some of the better jobs with the FBOs go begging. At the same time, the FBOs tend to look for older people who will stay with them and not go running off to the airlines.

Interestingly enough, though, we've found that in the period following the fuel crisis—which, of course, was devastating to the airline industry—the general aviation positions, the CFI and charter pilot jobs, have been in abundance. In fact, during this period, we've had more job leads than we've had graduates. We haven't been able to put our finger on exactly why that's happened, but apparently there's been some job growth or turnover.

I think perhaps—and let me emphasize that word "perhaps"—we're entering into a new era. We can look at certain periods in the past when the airlines would talk to pilots with private licenses and say, "as soon as you get your commercial instrument, we'll take you." And we can also look at periods when people with 5,000 hours couldn't get on.

AP: Okay, but what about this new era you're talking about?

EVERETT: First of all, in airline hiring, there's a shift in emphasis from military to general aviation pilots. I'd say they're hiring about 50% general aviation pilots and 50% military pilots, with the empha-

sis going more towards general aviation.
AP: What's the reason for this?
EVERETT: First of all, the military services aren't producing that many pilots anymore. For example, the ROTC graduates are now finding that there aren't any flight jobs available for them, even though this commitment was made many years ago. These people are now being offered two alternatives, one, they can enter the service as a commissioned officer in a non-flying spot, or, if they don't want this, they don't have to fulfill their obligation to the military.

On the other hand, the military is trying to keep from losing its pilots to the airlines by offering very attractive benefits. And indeed, it's very attractive if you're oriented to the military lifestyle. You've got great pay, great equipment, great retirement benefits—things that just aren't available in the civilian world even with the airlines. So I think the military is hoping that more people will view it as a career. This, of course, will affect the number of military pilots that are available for the civilian market.

Another big factor affecting civilians today is the absence of the draft. In the past, a young man had to spend two non-productive years, whereas now he can take those two years and build up two thousand hours with an FBO. Conversely, a man who comes out of the military at, say, age twenty-seven will have maybe only three good years' time in which to find a spot with the airlines.
AP: When do you think the airlines will start hiring again?
EVERETT: I think over the next five years we're going to see considerable activity. In the year and a half prior to the fuel crisis, the airlines were having a real boom. This was reflected in hiring, and it was really starting to build. In the 12 months prior to the fuel crisis, we had 386 students that got airline jobs through our school. The projections for future hiring were fantastic. Then the fuel crisis came and it was just like turning a faucet off.

Now we're back in a rebuilding period. Unfortunately, there is such a direct connection, between the airline and the general economy that it's hard to predict the future but we can also look at the fact that most of the people that joined the airlines following World War II will be reaching mandatory retirement age in 1978-81. The attrition that will occur during that period is estimated to be in excess of fifteen percent. So even if the airlines aren't expanding, they'll have to replace that many people, unless they're on a backward slide.

One thing that young people today frequently don't take into account is the time lapse factor. I've talked to a lot of people 18 years of age who may be down on the airlines if they're not hiring right now. They fail to realize that it takes some time to prepare for a career with the airlines, and what's happening today has nothing to do with what may be going on by the time they're ready for a job with the airlines. You just can't predict one way or another, but what you can say to yourself is, "If I start at an early enough age, and if I don't have to go into the military, I'm going to be potentially employable at such an early age that if there is no hiring taking place when I'm ready, I still have six or seven years and something is bound to happen in that time if I really go about this seriously and remain aware of everything that goes on."
AP: What are the airlines looking for these days in terms of experience?
EVERETT: Okay, let's go back to that period a year and a half before the fuel crisis, when the airlines were starting to

hire again after a long period of nothing going on. Most people figured that the people who got hired at that time would have five or six thousand hours, and indeed, there were plenty of people available with that experience. We found, however, that the average experience level of those getting hired was about 2,000 hours.

AP: How do you account for that?

EVERETT: Well, for one thing, we were having success at placing people who had flight engineer's tickets. This represented quite a change from the days of DC-7, when the personnel departments didn't even know what a flight engineer was. Today, an airline pilot must start as a flight engineer unless his carrier does not have that position on the airplane.

AP: Let's take a look at the job situation with FBOs.

EVERETT: Okay, 90% of the jobs open to civilians in aviation are with the FBOs. Most FBOs now look for a commercial certificate plus a flight instructor's certificate with an instrument instructor's rating. Now, a person flying as an instructor and charter pilot can log 800 to 1,000 hours a year. This means that in two years, he can accumulate enough flight time to be of interest to the airline. At that point, he might want to get a flight engineer's ticket on a 707 or 727. This would increase his competitive position and make him aware of what happens on a large aircraft.

AP: Working conditions with an FBO aren't the greatest.

EVERETT: It's generally a matter of low pay, long hours and weekend work. Your customers are frequently members of the business establishment who aren't available for instruction until five o'clock on weekdays. Mostly it's a no-guarantee, pay-by-the-hour arrangement and most FBOs will get a take on the flight instructor's pay. They may charge the student $8.00 an hour for dual and then they'll rake off $2.00 or $3.00 from that.

AP: The instructor may also be teaching ground school, right?

EVERETT: That's right, and to increase your marketability, it's not a bad idea to get your ground instructor's rating and indicate your willingness to do this. Now, it's a sad commentary on our educational system, but most flight instructors don't like ground school. You need to take an interest in it, maybe working with a slide projector and following a programed course–which is not really teaching as it should be.

AP: I get the feeling you're not too happy with the audio/visual programs?

EVERETT: Audio/visual is great if used properly, but unfortunately, it's generally used as a crutch. We've got 3,000 flight schools in the country and most of them are in the business not because they want to be but because they have to be, in order to sell planes. So most FBOs regard it as a necessary evil. They don't make any money on it; they're dealing with a small amount of people, so how can they have a good ground school? They can't; there is no way they can do it economically, so they get one of these programed courses. Now, there are some very good materials available, and if a person were to use those materials incorporated into some well-developed personalized instruction, then that's fantastic. But what does happen is that a person sits down with a projector and pushes a button and just follows along and it's really less than desirable. The people marketing these courses present it as a great innovation in education, but it's really just an improvement over a bad situation.

AP: You mentioned the low pay of the flight instructor. Recently, I interviewed an aircraft dealer whose flight instructors doubled as airplane salesmen, which gave them a greater earning potential. What do you think of this concept?

EVERETT: Well, I'm not sure that it's a completely ethical way to go and I don't think too many flight instructors are cut out for the role of salesman. On our application forms we ask our students about their objectives, including aircraft sales, administration, business flying and I'd say maybe a quarter of one percent check any of these administrative functions. These people are strictly flight oriented.

We do try to educate our students on various facets of the aircraft industry. For example, we'll put on little evening programs on aircraft sales, but they're not well attended.

AP: Well, even if an instructor doesn't want to sell airplanes, he certainly ought to know how to sell himself.

EVERETT: Yes, and this is where a lot of people fall down. A person has to be able to understand the business and how it operates. He's got to be able to speak intelligently to a prospective employer. Most people don't. Time and time again, I've asked FBOs what sort of an impression the typical job item seeker makes and he'll invariably say, "terrible." The guy just doesn't seem to understand what an FBO does. Now, the same holds true if someone wants to move from an FBO into the airlines or corporate flying; he has to understand those operations.

AP: What's the best way to go about doing this?

EVERETT: Basically, aviation is a word-of-mouth industry. Suppose you work for an FBO for a number of years and you're tired of it and you want to break into corporate flying. You should make it your business to shoot the breeze with every corporate pilot who flies into your airport. Part of a corporate pilot's job is to stick around and wait at airports, and if you approach him in the right manner, chances are he will be glad to give you lots of useful information. And you would be making a valuable contact. Let's say the guy flies in every Wednesday in a Sabreliner and you know the guy for two years. You're an old buddy! It takes that kind of drive, and I think that's something a lot of people fail to realize.

You should also do such basic things as making Xerox copies of your applications and keeping notes of your interviews for future reference.

It may surprise many people, but the failure rate of military pilots to transition to the airlines is quite high. You can take a pilot with high time in good equipment and he gets out and fires off his applications to the airlines and maybe to the corporations and nothing happens and he doesn't know what to do. Well, what he *can* do—and this applies also to the airline pilot who's been furloughed—he can go to a flight school, small or large, get his flight instructor's license and instrument instructor's rating, get to know what FBOs are all about. It's a hard pill to swallow, because what I'm suggesting is that a person who has a good background has to play it humble for a year or two, but this lets him continue his job search while he's an active member of the aviation community. He can keep on making contacts at the airports he flies in and out of. This is so much better than taking a job on the outside, selling insurance or shoes or something while the wife sends out resumes; you never get a job that way, operating outside of the industry, not making any contacts or getting any feedback.

Mental attitude is very important. In this industry, people are going to have their setbacks and I've seen people with all kinds of credentials roll over and play dead; all they tell you is how terrible things are. This guy has eliminated *himself* from the job market. I've also seen guys with marginal backgrounds who've said, "Hey, I'm going to stay at this thing," and those guys usually end up getting on.

AP: Are the airlines doing any hiring at all these days?

EVERETT: It's starting to pick up. Several classes—groups of thirty—have been hired lately.

AP: What about corporate flying jobs?

EVERETT: This is one of the least understood segments of the industry. Those jobs are not easy to get; they come almost certainly by word of mouth.

AP: What about the personnel agencies that specialize in aviation placement?

EVERETT: There may be some good ones, but I don't know of any. Actually, I think the employment agencies provide a disservice. You have to learn to rely on yourself and if you think you can pay an agency $150 or so and then sit back and wait for things to happen, you're deluding yourself.

AP: How about the people who have a pilot's license and would be willing to take a job involving little or no flying just in order to stay in the aviation industry?

EVERETT: There are many opportunities for someone who has a good sales or administrative background. Take this school for example. We have many administrative positions here, and we like our people to have a pilot's license, but it's hard for us to find pilots with good administrative experience. Then there are administrative and sales jobs with FBOs, aircraft and parts distributors and so forth. One mistake a lot of people make when applying for these jobs is that they stress their flying background, which is exactly the wrong thing the interviewer wants to hear. The applicant should emphasize his business background, so it doesn't just look like he's trying to find some way to keep flying.

AP: Tell us something about Sierra.

EVERETT: Well, 95% of our students are professionally oriented. Our emphasis is on the academics with full time programs. I think it's perhaps the most

diversified school in the country. We have programs on the 707, 727, Sabreliner, seaplane and helicopter training. We're the largest school on the West Coast, with 64 staff members, the majority of whom are instructors. We try to screen our people very carefully in terms of background and enthusiasm for their work; we pay them substantially more than they could earn with an FBO. Our instructors do nothing but instructing; they don't do any charter or selling, so they can concentrate full time on the students.

AP: How many students are enrolled with you now?

EVERETT: About 250. We do a lot of contract training for foreign airlines, as well as some domestic military training. Well over half of our training is in jet equipment, so our emphasis is in the advanced programs.

AP: Do you have a student placement department?

EVERETT: No, I think "placement" would be a misleading term. We give our students all the assistance we can; sometimes we get inquiries from airlines, corporations and FBOs when they're hiring and we pass this on to our students, but realistically, the student will have to get employment largely through his own efforts.

Burnside-Ott Aviation Training Center is something of a contrast to Sierra. It is located in the soft climate of Southern Florida, at Miami's Opa Locka Airport. The building is more modern, and the ear is greeted by a combination of piped-in music and excited chattering in unfamiliar tongues. (Burnside-Ott has a high percentage of foreign students.) Bob Ott, one of the partners, is a dynamic, busy type. He and his twin brother are also in the restaurant business.

AP: Tell us about Burnside-Ott.

OTT: We started this school in 1964. Up to that time, Don Burnside had been director of flight training for Embry-Riddle. When Embry-Riddle moved to Daytona, Don wanted to stay in this area, so my brother Bill and I, who were students of his, got together with him and we started this school.

AP: Are you planning to develop in the same direction as Embry-Riddle, with a college curriculum?

OTT: No, we're in the continuing process of increasing our curriculum to cover advanced training programs, especially type ratings and ATP ratings in big equipment for the airlines like L-1011s, 707s, 727s and the like. Also intermediate programs like the Twin Otter.

We were very heavy in training for domestic airlines when they were doing a lot of hiring in the late '60s, and then that tapered off around 1970. By that time, we had gotten into training for foreign airlines and foreign companies, and now we represent well over 125 countries. About 60-70% of our students are from other countries.

AP: How large is your student enrollment at present?

OTT: Including our facility at Tamiami, and including part-time students, I'd say we have about 600.

AP: And how many aircraft do you operate?

OTT: We have approximately 40 150s, 20 172s, 8 Twin Comanches and a couple of DC-3s. We also have a Lear Jet program.

AP: What do you think of the job prospects in the airline industry?

OTT: Well, it's going to take some time. They're going to have to run out their furlough lists and bring those people back first, but it's just like the condominium market here in Florida; they've overbuilt and it's going to take some time before they sell those and then they're going to start building new ones. It's the same with airline pilots; a very high supply and very little demand, but those factors should change in time.

AP: What about corporate flying?

OTT: I think the corporate market is very bright. Of course, a lot will depend on the recession, which I think we're still in. A lot of companies will be very careful about coming back with their full company airlines, like they used to have. I do feel, though, that business eventually has to get better. Another factor that's going to require corporations to have their own aircraft is the fact that a lot of airlines are dropping a lot of their shorter runs which have been money losers. So the corporations will have to pick up those short-haul routes. These short routes make more sense to the corporations, just as the long routes make more sense to the airlines.

AP: What about the FBO market?

OTT: Well, again, this all relates to the airline industry. When the airlines aren't hiring, the instructors tend to stay put in their jobs with the flight schools and FBOs. When the airlines increase their hiring, I'm sure there'll be more opportunities for CFIs and charter pilots with the FBOs. I'd say right now, you have an oversupply of people who'll do anything to get in an airplane and build up more time for an airline job. There are guys flying air taxi or doing flight instructor work who'd give it up tomorrow to get an airline job.

Understand, I think that general aviation has a great future. Cessna, Piper and Beech are having record sales years and it's a very healthy market. I'm just saying that the employment picture would be brighter if the airlines were doing more hiring.

AP: When the airlines do hire, do you think they'll have different criteria than they had before?

OTT: Yes, it used to be they were looking for hours, hours, hours. Now

they're looking for higher ratings. The more you can do to cut down on the training they'll have to give you, the more desirable you'll be to the airlines.

* * *

If you're interested in getting further information on career opportunities, and the investment in time and money required to get the necessary ratings, you should write to a number of flight schools.

THE CHALLENGE OF AG FLYING

by Mike Dillon

Reprinted by Permission
AIR PROGRESS (July 1976)

Taxiing onto the runway, I lower the goggles over my eyes and adjust my huge orange crash helmet. This is the big moment; my first flight in a 450-hp Stearman duster. But it should be light stuff. After all, am I not already an established ag pilot with seven hours of Super Cub spray time logged?

Now with the nose pointed down the runway, ease the throttle forward and hang on. Ride that right rudder and watch out for torque. The noise is horrendous. With the airspeed passing through 75-mph, ease the stick back and let the old girl fly herself off. With no windshield to stop it, the prop wash and the increasing airflow flay at my head and face, puff my cheeks with air, and threaten to suck me backward out of the cockpit. It hurts!

At 200 feet, pull the power back and start a left turn out of the pattern. To clear for the turn, I glance left. Instantly, the windstream rips the goggles from my face and slings them down around my neck, inside out! Good Lord! Now what do I do! Just stay calm and there's no reason you can't live for another 30 seconds.

Eyes watering, I trim for a shallow climb. Handling the stick with my knees, I duck down behind the back of the hopper, untangle the goggles and wrestle them into place. Cautiously, I raise back up into the windstream, being careful not to turn my head more than a few degrees to either side. It's as if I were balancing the goggles on my nose.

If this is what dusting is like, maybe

I've chosen the wrong profession. Fortunately, however, this is only a ferry flight for maintenance, so for the next hour and a half, I hide behind the hopper. From time to time, I peek out each side of the low-cut cockpit to keep the wings level and to watch the compass and altimeter (about two inches from my nose), in order to maintain heading and altitude. Every time I raise up to look over the nose, the windstream tugs at the crash helmet and fills my leather jacket like a balloon (everyone knows ag pilots wear leather jackets). The strain is just too much! I don't see how anyone could work a plane that causes the pilot this much neck and back pain.

It's an exhausted, scared pilot who finally lands his Stearman in Phoenix, Arizona. That was the first time I saw Phoenix and boy did it ever look good. But then, any airport that offered relief from the torture of that wideopen, wind-torn cockpit would have looked good.

That was my first flight in a Stearman duster; even though I had flown 20 or 25 hours in a stock 220-hp Stearman, this experience was a complete surprise. As it turned out, I only flew the big biplane about 20 hours that year, probably because Walt Davis, my wise and kindly employer, notices my lack of proficiency in the craft. Davis confined my flying mainly to his Super Cubs—a shrewd move on his part because the 450 Stearman is not an amateur's machine. That was my introduction to cropdusting.

During the next six years, I logged 2,000 hours, tore down five sets of powerlines, ran out of gas twice, sprayed the wrong field once, taxied into (and destroyed) a pickup truck, and, in general, did all those things that duster pilots do. I made some money and had a hell of a lot of fun. Ag flying has been good to the Dillon family—providing time and money for college plus the flying experience that led to employment with the airlines. Several pilots have inquired about the possibility of their following the same route. What follows is my perception of the truths and myths of cropdusting.

Actually, the name cropdusting is something of a misnomer as very little dust is used today. Most crop work now involves the application of spray, seed, pelletized fertilizer, or ground-up corncob impregnated with poison. Nevertheless, the anachronistic terms of dusting and duster pilot have stuck with us. Several years back, the industry began to refer to themselves as "aerial applicators," but that term hasn't received much public acceptance. In fact, if you want to find the telephone number of an "Aerial Applicator" you'll have to look in the yellow pages under "Cropdusting."

The first year in Safford, Arizona I logged about 120 hours spraying and dusting. For this I was paid $1,400, which, though not bad for a summer job, was hardly as much as I had expected. The following year, flying in Mesa, Arizona, my earnings were about $2,300—still not the big money one associates with ag flying. But I was going to college full-time, so it was still more money than any other summer job would have paid at that time. For the next three years, I not only flew cotton in Arizona, but rice in Eagle Lake, Texas. By my fifth year, I was earning $12,000 plus and still attending college two full semesters per year.

The job served my purposes perfectly by allowing me to earn during the summer and to learn during the winter. But as a full-time job, the first few years would have been pretty lean. I think my earning experiences were typical or better.

This explodes one of the myths of

cropdusting, fostered in recent years by certain ag schools that stand to make a quick buck "training" pilots for the ag business. The new duster pilot won't make a great deal of money his first few years simply because he won't get to work in the prime areas or for a first-class operator. These operators don't need to look for, or hire brand-new pilots. There are more good, experienced pilots than really good jobs. The job that falls to the beginner, with rare exception, is the job that no experienced man wants. One has to work his way up. Also, with the big operators, there exists the seniority system. That is, the senior pilots get the most and best work: the bigger fields, the bigger planes, and the higher-paying jobs.

There can be a great deal of difference in a pilot's pay for a day's flying. The pilot is generally paid 25 percent of the plane's gross earnings. So if the plane earns more doing a certain type of work, the pilot makes more. For instance, in rice work, a pilot spraying to kill grass can easily turn $200 or $250 in a day while another pilot planting or fertilizing the same fields would do well to earn $70 or $80 a day. You can bet that the old hands do the spraying.

The new pilot is also the last to be put to work at the beginning of the season and the first to be laid off at the end of that season. So, even after you land your first job, it's going to be a couple of years before you'll get your hand very deep into the till.

What about the danger, you ask. Is it worth it to risk your neck when it's going to take two or three years to work up into the money? This question is difficult to deal with, but I can relate my personal experiences and impressions. I've had many close calls while dusting, but no serious accidents. Since I began flying professionally in 1960, some 43 friends and acquaintances have been killed in airplane accidents. Only two of these were dusting. Another was an airline crash and the rest were sport flying—mostly in World War II military machines. From this, I can only conclude that ag flying is not nearly as dangerous as sport flying. On the other hand, FAA records for 1966 show that for every 100,000 hours of ag flying, there are 4.24 fatal accidents. This is compared to 1.45 for business and corporation flying.

What danger there is comes almost exclusively from flying into something you did not see; powerlines, standpipes, and so forth. They are the hazards that are always with you and that can never be completely avoided. The other trouble area is obvious: a heavily loaded plane. This in time, can be dealt with as you learn your machine's limitations and learn to plan ahead in such a way that you never need to crowd a turn or a takeoff. When in doubt, try to get your partner to carry the first load off any strip you're not sure of. If he makes it, then you can try. If he runs through the fence, you can try a smaller load.

How about chemical poisoning? With DDT legislated out of use, the deadly organic phosphates such as parathyon and phosdrin are being used more widely than ever. But there is no reason why a pilot should ever get poisoned if he takes one simple precaution: don't breathe the poison's fumes. This can be prevented with the use of either a good air filter system or an oxygen system charged with compressed breathing air. The latter is by far the most effective.

All right, that's the dim side of the business. Still interested? Let's look at the good features. Ag flying is a job you can get with fairly low time—300-400 hours usually being enough. It is a type of flying that gives the pilot a tremendous sense of satisfaction once he has learned to do it well. If you'll pardon the expression, it's *real* flying. To climb into a big Stearman, fly out to a rolling green pasture and land in the mists of early morning is like setting the clock back 35 years. You get a chance to see what your plane will do and what you're capable of making it do. Max weight takeoffs often require the utmost in technique to keep from running through the fence. In rice country, where wind is considered by the farmers to be no factor, the landings and takeoffs on short, narrow, often slippery, improvised strips keep a pilot on his toes. Many of these strips are only 15 to 20 feet wide and crowded by fences, canal banks or ditches. Add to this a 20 to 30 mph crosswind and things can get exciting. One large company in Beaumont, Texas, wrecked seven of its 30 Stearmans in one month of particularly rainy weather. Every morning their planes would leave the hangar yellow and come back (if they got back at all) black and covered with mud.

When a pilot has flown under these conditions, there is little he need fear when approaching a new plane or a different type of flying. Richard Bach, in an earlier issue of Air Progress, suggested that we'd do better to measure a pilot's time by his number of flights rather than hours. Well, an ag pilot working rice will land and take off an average of once every eight minutes. During the season, he will fly about eight hours a day, so that in only one month, he'll log almost 2,000 landings under the most trying conditions. That's a lot of experience.

When I say that a rice pilot flies eight hours a day, that's just an average. Twelve hours of flying in one day is common and 15 isn't considered unreasonable. After 60 days of this routine (there are no days off), I've seen even the nicest, calmest pilots turn into miserable bastards and that includes the author. It's a tough grind. Fatigue can become a major factor, especially in the modified Stearmans with their heavy control pressures and slow rate of roll.

I recall one rice season several years ago, sitting down for a quick "between fields" break with my flying partner, Bruce Loston. Bruce was one of the best airplane handlers I'd ever flown with—a damned hard man to keep up with. As we talked, I noticed Bruce's right hand was heavily calloused from the palm to the fingertips. I asked how come? He said it was from gripping the Stearman's stick so tight, and he noted with disbelief that my hand looked soft and white by comparison. The explanation was simple: not being as well-muscled as Bruce, I had to use both hands on the stick to last out the day. Also, by bracing my arms across my legs, I could use my legs to help power the ailerons. All this, plus gloves, kept me from getting tired or calloused.

Another effect of fatigue in the Stearman: it causes the pilot to crouch down and hide from the wind behind the windshield thus cutting down his ability to see. I much prefer flying the planes with canopies. A canopy allows the pilot to sit tall and really see what's outside.

Being too comfortable can create another sort of problem, however. Flying a Pawnee 235 one winter morning, I was spraying lettuce in the Harquahala Valley of Arizona. The fields were short, which means lots of turning for little accomplished. Having been unexpectedly routed out of bed at 4 a.m. after only two hours' sleep, I was generally in no mood for this nonsense. Back and forth we go across the short fields. Suddenly a violent, noisy, bumping shudder shakes the little plane. Instinctively, I come back on the stick and everything smooths out. I had fallen asleep at an altitude of two feet and had touched down in the heads of lettuce at 110 mph!

You say you're still interested? Okay. Let's look at the different areas and types of work you might want to try for. Rice work—seeding, fertilizing, and spraying—has a constant factor that gives it a degree of predictability, i.e., it has to be done every year at approximately the same time. Since the season is too short for a pilot to make his living on rice alone, most operators have to hire additional, temporary pilots to man their planes. This might give the beginner a chance. Also good for the beginner is the fact that rice planting and fertilizing are flown quite high: about 30 feet in Stearmans, Ag Cats, and Snows, and about 70 feet in Pawnees. This gives the material time to spread out more before reaching the ground, and also greatly lessens the danger of the pilot's hitting some object in or near the field being worked. Once the rice season has started, however, the pace is so fast and furious that a beginner might rapidly find himself being pushed beyond his limitations.

The best rice areas are Southern Louisiana, Central California and the Gulf Coast of Texas.

Spraying cotton would probably be a better way to start out. Here, the initial pace would be slow enough for you to handle, and would give you a chance to grow into it as the season picked up speed. It should be remembered, however, that this work requires the pilot to fly very low—about 3 feet above the crop. Most ag pilots tear down a set of powerlines every one to three years.

Cotton is grown in many areas, and one might look for a seat in Texas, New Mexico, Arizona, California, Arkansas, Louisiana, or Mississippi.

What kind of equipment you'll fly is something else to consider. With the new planes now on the market, the small operators have a better chance to own and provide even their rookie pilots with a good safe seat. The Stearman, properly modified and maintained, is still a fine working plane and certainly crashworthy.

The problem with the Stearman is that many were not properly modified and some small one-or two-plane operators lack the knowledge and ability to properly inspect and maintain the machine. This can lead to some weird incidents. A young ag pilot named Ray Rotge gained instant fame along the Gulf Coast of Texas the first year he dusted. He's still known as the only pilot to ever lose an engine and still land his Stearman without damage. Now when I say he lost his engine, I don't mean it failed—it fell off. The upper mount bolts broke at the firewall and the engine fell forward off the plane. As it went, the prop sliced into and blew both tires. Ray said, "I just pushed the stick forward and set up a glide, then flared out and landed." The FAA said the engine was still running when it hit the ground.

Heavy aircraft such as B-17s, TBMs, PV-2s, B-26s, DC-4s, DC-6s and P4Ys, are used in still another type of ag work. These big, heavy-load-carrying planes are used for fire-fighting, timber-spraying, and poisoning the pesky imported fire ant. The captains' seats on these big planes are some of the most sought-after jobs in the industry and simply are not open to a low time pilot. However, the job of copilot is one you might well look into. To be a copilot, you must, by law, have a commercial license, and must, by necessity, be a pretty good mechanic. Copiloting on these big old birds is a good way to build time, and a way to get your foot in the ag flying door.

I'm in a quandary as to how to tell

you to actually apply for a job. There is no clear-cut way. As best I can tell, the ag schools do little to equip a pilot for work, and the schools themselves are not held in high regard by the ag operators. They do however, have a good listing of ag businesses around the country, and surprisingly, seem to be able to place many of their students. Once placed, it's up to the pilot to learn his job almost as if he had never been to school.

The other method requires personal initiative and imagination. The pilot must ask questions, write letters, make phone calls. In general, he must haunt the industry until he finds an opening. This is the traditional method and it still works.

As to where to get the names and locations of the people for whom you want to work, I would write to the aeronautical and agricultural department of the various states in which you are interested in working.

You can pretty well count on being told "no" when you first ask an operator for a job. If you don't persist he'll know you don't really want or deserve the job.

If some summer morning you should find yourself strapped into the cockpit of an ag plane, remember one thing: that old guy with the long-hooded cape and scythe isn't a farm worker. He's looking for a chance to harvest you. Know your limitations and stay inside of them. Don't try any fancy stuff. Just do your job at an easy pace and you'll have a damned good chance of looking back on some of the most enjoyable flying on earth.

HOW TO GET INTO AVIATION BY FLYING A DESK

by Peter Lert

Reprinted by Permission
AIR PROGRESS (July 1976)

When people think of a flying career, they invariably think of piloting—whether as the gold-striped captain of an airliner, the natty pilot of a corporate jet, a weathered ag or bush pilot, a flight instructor, or any of the many other ways a person can earn a living driving an aircraft of some sort.

There are other jobs available, though—sometimes, in fact, more available than piloting positions—which can offer just as rewarding a career both personally and financially, and can also include almost as much flying, without actually being piloting jobs. These jobs—in which piloting ability and experience plays an important role even though it's not the primary job objective—exist both within and outside the aviation industry itself, and might be called "indirect" flying jobs.

Within the aviation industry itself—by which I mean, in this context, the general-aviation airframe, avionics, and accessory manufacturers—job openings are available in sales and marketing, distribution, public relations, engineering, and general management. Similar jobs are available, of course, outside the aviation world, but since our industry is a rather unique one, there are some special conditions which don't obtain in other areas.

Compared to a giant like, say, the automobile or home appliance industry, general aviation is relatively small, as well as being a comparatively new segment of industry. This results in a fairly small number of key people, most of whom know each other at least by reputation if not personally.

Moreover, since general aviation hasn't been around all that long, some of the top-echelon management people are men who've been in the industry since its first vague beginnings. It's only recently that the industry has begun to recognize the fact that it takes more than just being able to put out a good product to be successful in today's highly competitive markets; while younger people with appropriate expertise are beginning to infiltrate the upper levels of the industry, there's still a certain feeling of "I learned to fly in a Jenny and I'm not about to let some young feller from Harvard Business School tell me how to run my airplane company."

Going hand in hand with this, of course, is the tendency for personnel directors to hire people they know or know of, and who've been in the industry for some time and have learned the "right way" to do things. All of this makes it a bit difficult for the newcomer.

The situation is changing, though, and Air Progress staffers interviewed various industry representatives and spokespersons to get their views on general-aviation management careers and the qualifications required to enter them.

One almost universal comment was that there's almost automatic distrust of a young man or woman who's heavy on pilot credentials but light on management skills. Most personnel officers feel that these people are looking for jobs in which they can gain flying experience and mark time until a full-time flying-only position—e.g. airlines—comes along. The mere fact that you can fly a plane isn't all that impressive in the aviation industry, where a whole lot of other people fly as well; what counts are such things as a business degree, sales experience and training, and so forth. The general consensus was, "It's a lot easier for us to teach a good businessman to fly than to teach a pilot to be a good businessman." Unfortunately, the distrust mentioned above has been borne out by experience in all too many cases.

Why hire pilots at all, then? Because the general-aviation industry is such a specialized one that "inside knowledge" of the product is a necessity. This is perfectly obvious in the case of lower-echelon salespersons—the ones who actually take the prospect up in an airplane—but is just as important higher up the corporate ladder, and it can mean that the type of piloting experience can be important, too. A newly-retired military pilot, for example, may find that his experience has little or nothing in common with the job of marketing, say, Cessna 150s.

Other technical training can be a great asset when combined with piloting skills. Dean Humphrey of King Radio noted, ". . . one of our mottoes is 'Designed by Pilots for Pilots,' and we can really get excited about someone with piloting experience as well as an electronic engineering degree."

Dave Franson of Cessna noted that ". . . a lot of our regional sales managers are young guys now, guys in their early thirties, and it's essential that they know the product line from the pilot's point of view." Again, it should be stressed that these "young guys" have very strong backgrounds in marketing and management . . . flying, while important, is secondary.

This, in fact, may be the key to the whole matter: that in these careers flying—no matter how important it may be—is still secondary to the primary task. Ask yourself what you really want to do: if you're sure that your future lies in the cockpit and nowhere else, you should be resigned to battling for a "pure flying" post. If, on the other hand, you feel that your talents, experience, training, or expertise qualify you for a management or marketing position—and if you're sure that that's what you want to do, so that you can make a real commitment to the job—the management levels of the general-aviation business merit more than a cursory look. While the size of the industry may make it look like there aren't many jobs available, there are still likely to be more than there are in the "flying only" marketplace.

Of course, the aviation industry itself is not the only segment of commerce in which piloting skills can be a very definite asset, and in which one can find–or create–a job which will include considerable flying. Let's look at just a few of the possibilities:

Remember those airplane ads that show a group of civil engineers, for example, hopping into their shiny new airplane to buzz off to some distant construction site? You can believe it; there *are* engineering firms in which such travel is a frequent requirement. If you are, for example, an engineer out job-hunting, it may be worthwhile to let prospective employers know that you are capable of getting yourself–and, by implication, other employees–safely and rapidly to some location difficult or impossible to reach by the airlines. In fact, on some occasions a flying employee has not only been hired, but has caused the firm to purchase an airplane for him to fly, on the basis of a common-sense demonstration of his–and the airplane's–usefulness to the firm. Big construction firms–the ones that build freeways and such–are particularly good prospects, since a freeway construction site is often its own airport for company planes. Of course, the usefulness of an airplane–and an appropriately rated employee to fly it–isn't limited to engineering and construction firms. Almost any company that requires its personnel to travel extensively to any but the largest cities can probably save time and money if those employees can fly themselves, even in rented aircraft.

A suggestion here: again, don't play up your piloting skills too much. You're trying to get hired as an engineer or manager or whatever, not as an airplane driver; too much emphasis on piloting may make the prospective employer think that (a) you're just out to fly, and (b) that you're some sort of hot daredevil pilot who'll crash with a planeload of his top executives. There's still a lot of resistance at corporate levels to what top management considers "little planes," which usually means anything with less than two engines. It may be up to you to demonstrate the economics and safety of flying to a prospective employer.

Personal travel, too, can be a way of getting a "flying job." For example, any traveling salesman who doesn't need to carry large or heavy sample assortments could benefit from a light aircraft; even if many sales calls must be made by automobile, some can be made by rented airplane. If you're self-employed, this is a particularly attractive way of gaining flying experience, since the travel costs can be written off at tax time.

It should be noted that a commercial license is not required as long as any flying is, in the words of the Feds, "incidental" to the primary job, and as long as you receive no compensation for flying other than the operating expenses of the airplane. For example, if you're a real-estate salesman, your company can pay to rent or operate an airplane in which you fly prospects to inspect land you wish to sell, but cannot legally pay you to fly them there–although you'd still get your normal salary or commission, of course–unless you hold a commercial certificate.

To sum up, then, there are numerous positions available which can offer a greater or lesser amount of flying–and possibly some paid flight training, as well–as "fringe benefits" of employment in some non-pilot capacity. True, many of these jobs may not offer as much flying as a true piloting position; but there are a lot more pilots around than airplanes, and full-time airplane-driver jobs won't be getting any easier to obtain. Moreover, there are those who swear that full-time flying, particularly at the corporate or airline levels, consists largely of "hours of boredom punctuated with moments of terror." While this is an exaggeration, it's true that a "non-flying flying job" may offer considerably more in the way of variety. Getting such a job requires a real interest in, and commitment to, that job—not flying—but with that in mind, you can parlay your specialized skills and training, along with your piloting abilities, into a rewarding career that will also satisfy your craving for flight.

FLIGHT STUDENTS SHOULD ALSO BE TAUGHT HOW TO CRASH AN AIRPLANE

Reprinted by Permission
AIRPORT SERVICES MANAGEMENT
(May 1976)

**Based on a paper by
Gerard M. Bruggink, Air Safety Investigator, Bureau of Aviation Safety, National Transportation Safety Board**

It is not unusual for a general aviation pilot to find himself in a situation where his experience level provides no alternative but an emergency landing. Unfortunately, so much stress is placed during flight training on "a suitable landing area" that some pilots will not even entertain the thought of a precautionary landing unless they can save the aircraft. Too many fatal weather accidents, classified as "maintained VFR in IFR conditions," undoubtedly resulted from desperate attempts to get through because the underlying terrain did not fit the pilot's mental picture of an emergency landing area—a picture that was undoubtedly painted for him in his student days.

A pilot who has been conditioned during his training to expect to find a relatively safe landing area, whenever his instructor closes the throttle for a simulated forced landing, may ignore all basic rules of airmanship to avoid touchdown in terrain where aircraft damage is unavoidable. The desire to save the aircraft, regardless of the risks involved, may be influenced by two other factors: the pilot's financial stake in the aircraft, and the certainty that an undamaged aircraft implies no bodily harm. However, there are times when a pilot should be more interested in sacrificing the aircraft so that he and his passengers can safely walk away from it.

Avoidance of crash injuries is largely a matter of:

Keeping vital structure (cockpit/cabin area) relatively intact by using dispensable structure (wings, landing gear, fuselage bottom, etc.) to absorb the violence of the stopping process before it affects the occupants.

Avoiding forceful bodily contact with interior structure.

Energy Absorption

Accident experience shows that the extent of crushable structure between the occupants and the principal point of impact on the aircraft has a direct bearing on the severity of the transmitted crash forces and, therefore, on survivability.

Dispensable aircraft structure is not the only available energy absorbing medium in an emergency situation. Vegetation, trees, and even manmade structures may be used for this purpose. Cultivated fields with dense crops, such as mature corn and grain, are almost as effective in bringing an aircraft to a stop with repairable damage as an emergency arresting device on a runway. Brush and small trees provide considerable cushioning and braking effect without destroying the aircraft.

The second requirement—avoiding forcible contact with interior structure—is a matter of seat and body security (seatbelt and shoulder harness). Unless the occupant decelerates at the same rate as the structure surrounding him, he will not benefit from its relative intactness, but will be brought to a stop violently in the form of a so-called "secondary collision." In case of a partial restraint, such as the use of a seatbelt only, the same reasoning applies to the unrestrained body portions.

Since not all light aircraft are equipped with shoulder harnesses, the pilot should try to minimize this hazard by avoiding a nose-first impact against solid obstacles. (He should also make it a habit to insist on the routine use of seatbelts in his airplane.)

Speed vs. Stopping Distance

The overall severity of a deceleration process is governed by speed (ground speed) and stopping distance. The most critical of these is speed: doubling the groundspeed means quadrupling the total destructive energy and vice versa. Even a small change in groundspeed at touchdown—as a result of wind or pilot technique—will affect the outcome of a controlled crash. For example: an impact at 85 mph is twice as hazardous as one at 60 mph. It is three times safer to crash at 60 knots than at 104 knots (104-squared is about three times 60-squared). It is obvious that the actual touchdown during an emergency landing should be made at the lowest possible—but *controllable*—airspeed using all available aerodynamic devices (flaps, etc.)

Attitude and Sink Rate

The most critical error that can be made in planning and executing an emergency landing is the loss of initiative over attitude and sink rate at touchdown. Because the aircraft's vertical component of velocity will immediately be reduced to zero upon ground contact, it should be kept well under control. A flat touchdown at a high sink rate (well in excess of 500 feet per minute) on a hard enough surface can be injurious to occupants even though the cockpit/cabin structure is not destroyed—especially during gear-up landings in low-wing airplanes. A rigid bottom construction in these airplanes may preclude adequate cushioning by structural deformation. This characteristic, in combination with the rather limited human tolerance to vertical g's, has led to spinal injuries in hard "pancake" landings. Similar impact conditions in high-wing airplanes may cause structural collapse of the overhead structure. On soft terrain, an excessive sink rate may cause digging-in of the lower nose structure and a severe forward deceleration.

Controlling the Crash

The "school solution" to an emergency forced landing says:

Maintain aircraft control—establish a glide at the proper speed.

Select a field and plan an approach.

These actions may be combined with attempts to correct the emergency—however, attempts to troubleshoot the cause should be made only on a time-available basis. Under certain conditions the pilot may have a full-time job just controlling the aircraft. When losing one engine of a light twin during the critical take-off phase, a pilot may not have more than a split second to decide what is best: relying on the performance charts, or his impulse to reduce power on the good engine to maintain controllability.

Concerning the controversial subject of turning back to the runway following an engine failure on takeoff, each pilot should be taught to determine the minimum altitude at which he would attempt such a maneuver in his particular aircraft. Experimentation at a safe altitude should give the pilot an approximation of height lost in a descending 180° turn at idle power. By adding a safety factor of about 25 percent, he should arrive at a practical "decision height." The ability to make a "180" does not necessarily mean that the departure runway can be reached in a power-off glide. This depends on the wind, the distance traveled during the climb, the height reached, and the glide distance without power.

Terrain Selection

A pilot's choice of emergency landing sites is governed by:

The route he selects during preflight planning.

His height above the ground when the emergency occurs.

His airspeed. (Excess airspeed can be converted into distance and/or altitude. The student should be taught how.)

The only time that he has a very limited choice is during the low-and-slow portion of the takeoff. He should realize, however, that even under those conditions the ability to change the impact heading even a few degrees may insure a survivable crash.

If he is beyond gliding distance of a suitable open area when the emergency occurs, the pilot should evaluate the available terrain for its energy-absorbing capability. If he has plenty of altitude, he should be more concerned about first selecting the desired general area rather than a specific spot.

Terrain appearances from altitude can be very misleading, and considerable altitude may have to be lost before the best spot can be pinpointed. For this reason, the pilot should not hesitate to discard his original plan for one that is obviously better. As a general rule, however, he should not change his mind more than once. A well-executed crash landing in bad terrain can be less hazardous than an uncontrolled touchdown on an established field.

Aircraft Configuration

Flaps improve maneuverability at slow speed and lower the stalling speed. Therefore their use during final approach is recommended, time and circumstances permitting. However, the associated increase in drag, and decrease in gliding distance, call for caution in the timing and extent of flap application. Premature use of flap, and dissipation of altitude, may jeopardize an otherwise sound plan.

No hard-and-fast rule concerning the desired position of a retractable landing gear at touchdown can be given. In rugged terrain and trees, or during impacts at a high sink rate, an extended gear would definitely have a protective effect on the cockpit/cabin area. However, this advantage has to be weighed against the possible side effects of a collapsing gear such as ruptured fuel tank. Manufacturer's instruction—if given—should be followed.

When a normal touchdown is assured, and ample stopping distance is available, a gear-up landing on level, but soft terrain, or across a plowed field, may result in less aircraft damage than landing gear-down.

But positive aircraft control during the final part of the approach has priority over all other considerations, including aircraft configuration and cockpit checks. The pilot should try to exploit the power available from an irregularly running engine. However, to avoid unpleasant surprises during the touchdown phase, it might be best to switch the engine and fuel off just before touchdown. This not only insures the pilot's initiative over the situation, but also a cooled-down engine reduces the fire hazard.

Approach

When the pilot has time to maneuver, planning the approach should be governed by three factors: wind direction and velocity; dimensions and slope of the chosen field; obstacles in the final approach path.

These three factors are seldom compatible. When compromises have to be made, the pilot should aim for a wind/obstacle/terrain combination that permits a final approach with some margin for error in judgement or technique. A pilot who overestimates his gliding range may be tempted to stretch the glide across obstacles in the approach path (trees, powerlines, etc.). For this reason, it is sometimes better to plan the approach over an unobstructed area regardless of wind direction.

Experience shows that a collision with obstacles at the end of a ground roll, or slide, is much less hazardous than striking an obstacle at flying speed before the touchdown point is reached.

There may not be time to set up a pattern. The most important consideration is to get into such a position that the selected spot can be reached by using normal techniques such as playing the final turn—turning in early or late, depending on altitude—slipping, and moderate S-turns. If considerable altitude has to be lost while over the chosen field, it should be done in such a manner that the field remains within gliding distance. Speed control during all maneuvers is vital.

Crash-site Election

Confined Areas: The natural preference to set the aircraft down on the ground should not lead to the selection of an open spot between trees or obstacles which cannot be reached.

Once the intended touchdown point is reached, if the remaining open and unobstructed space is very limited, it may be better to force the aircraft down on the ground than to delay touchdown until it stalls (settles). An aircraft decelerates faster after it is on the ground than while airborne. Thought may also be given to the desirability of groundlooping or retracting the landing gear in certain conditions.

A river or a creek can be an inviting alternative in otherwise rugged terrain. But the pilot should insure that he can reach the water or creek-bed level without snagging his wings.

The same concept applies to road landing with one additional reason for caution: man-made obstacles on either side of a road may not be visible until the final portion of the approach. Road traffic must be given priority.

When planning the approach to cross a road, it should be remembered that most highways, and even rural dirt roads, are paralleled by power or telephone lines. Only a sharp lookout for the supporting structures or poles may provide any warning.

Trees (Forest): Although a tree landing is not an attractive prospect, it can be survivable.

Use the normal landing configuration (full flaps, gear down).

Keep the groundspeed low by heading into the wind.

Make contact at minimum indicated airspeed, but not below stall speed, and "hang" the aircraft in the tree branches in a nose-high landing attitude. Involving the underside of the fuselage and both wings in the initial tree contact provides a more even and positive cushioning effect, while preventing penetration of the windshield.

Avoid direct contact of fuselage with heavy tree trunks.

Low, closely spaced trees with wide, dense crowns (branches) close to the ground are much better than tall trees with thin tops; the latter allow too much free-fall height. (A free-fall from 75 feet results in an impact speed of 40 knots, or 4,000 feet per minute.)

Ideally, initial tree contact should be symmetrical–both wings should meet equal resistance in the tree branches. This distribution of the load helps to maintain proper aircraft attitude. It may also preclude the loss of one wing, which invariably leads to a more rapid and less predictable descent to the ground.

Always aim for the softest and, when possible, the lowest part of a tree or tree line. Judge trees by their ability to slow the aircraft's forward speed in the same manner as a firefighter's net catches falling people.

Once the aircraft is on the ground, if heavy tree trunk contact is unavoidable it is best to involve both wings simultaneously by directing the aircraft between two properly spaced trees. Do not attempt this "maneuver" while still airborne as recommended in some textbooks.

Mountainous Terrain: The variety and irregularity of mountainous terrain makes it impossible to list general rules. The pilot should be trained to instinctively avoid situations where an emergency would leave him without any choice. Flying needlessly low and slow over cragged terrain is an example of such a situation.

In mountainous terrain only a short glide may be sufficient to bring the aircraft over lower lying terrain, thereby increasing effective altitude and terrain choice. Maintaining a comfortable cruise speed helps assure the pilot of this advantage. He should also be able to visualize his glide angle at different altitudes to help see the possibilities in reaching the lower terrain.

Slope landings should be made upslope whenever possible with due consideration for the terrain conditions at the end of the slope. Avoid a situation where an excessive roll or slide would bring the aircraft to a sharp dropoff.

When landing on a pronounced upslope, enough speed should be maintained to change the aircraft's descending flightpath just before touchdown into a climbing one that approximately parallels the slope. (A descent at 50 knots and 500 feet per minute results in a 6° flightpath. In combination with an approach to a 24° upslope, an uncorrected 6° flightpath would lead to a ground "impact" angle of 6° + 24° = 30°.)

Water (Ditching): A well-executed water landing can involve less deceleration violence than a poor tree landing or a touchdown on extremely rough terrain. A fixed wing aircraft that is ditched at minimum speed and in a normal landing attitude will not sink like a rock upon touchdown. Intact wings and fuel tanks (especially when empty) provide flotation for at least several minutes even if the cockpit may be just below the waterline in a high-wing aircraft.

Loss of depth perception may occur when landing on a wide expanse of smooth water, with the risk of flying into the water or stalling in from excessive altitude. To avoid this, the aircraft should be "dragged in" when possible. Use no more than intermediate flaps on low-wing aircraft–the water resistance of fully extended flaps may result in asymmetrical flap failure and slewing of the aircraft.

Keep a retractable gear up.

Ditching downstream in a swift running river has the same effect as a headwind in that it reduces the relative groundspeed.

Snow: A landing in snow should be executed like a ditching, in the same configuration, and with the same regard for loss of depth perception (white-out) in reduced visibility and on wide open terrain.

An even snow layer, several feet thick,

may blanket smaller obstructions and make otherwise rough terrain more suitable. Pronounced "humps" that may hide larger obstructions should be avoided.

In summary...

A pilot who knows his aircraft and understands the what and why of the techniques that will insure a survivable emergency landing under adverse conditions has no reason for morbid preoccupation with the possibility of being forced down. The peace of mind associated with this knowledge should improve the pilot's overall performance which, in turn, may prevent an emergency or benefit its outcome.

What other responsibility is of greater importance to a flight instructor?

"Precautionary" landings...

A precautionary landing, generally, is less hazardous than a forced landing because the pilot has more time for terrain selection and planning his approach. In addition, he can use power to compensate for errors in judgement or technique.

Unfortunately, too many situations calling for a precautionary landing are allowed to develop into immediate forced landings when the pilot uses wishful thinking instead of reason—especially when dealing with a self-inflicted predicament.

A low-flying pilot who is trapped in weather and does not give any thought to the feasibility of a precautionary landing accepts an extremely hazardous alternative: inadvertent flight into an obstacle.

L'AVION C'EST CRASHE' NORMALEMENT

by Steve Knaebel

Reprinted by Permission
AIR PROGRESS (June 1976)

...En provenance de Porto-Rico l'appareil survolait le bourg de Deshaies a une altitude de 7500 pieds, s'appretant a venir effectuer une escale technique de ravitaillement au Raizet avant de reprendre sa route vers le Bresil.

Soudain, un debut d'incendie que l'on suppose du a une fuite d'huile se declara dans le moteur que le pilote s'empressa de couper pour eviter la catastrophe.

Avec beaucoup de sang-froid, jouant sur ses reserves d'altitude et de vitesse, le pilote rechercha un endroit ou il lui serait possible de posser son appareil...
(From a news item appearing in France-Antilles—Le Journal d'information de la Guadeloupe, Friday, February 21, 1975.)

Two hours and twenty minutes after takeoff from San Juan that February morning we were level at 11,500 feet overhead Saba Island, which juts mountainously out of the blue sea about 130 miles southeast of the St. Thomas airport. At some point during this part of the flight I mentioned to my companion that the latest edition of the *Airman's Information Manual (Part I)* had a revised and more thorough section on ditching techniques. I suggested he read it, which he did while I practiced instrument flying "under the hood." San Juan Center had been following us on radar until we were about 25 miles from Saba, and although the engine was running perfectly, I remember thinking how comforting it was to know those FAA men were constantly aware of where we were over all that water. ("Cessna Zero One Alpha, Center, advise for radar confirmation crossing the zero seven zero radial of the St. Croix VOR and say your altitude now, over." "Center, Zero One Alpha, Roger, and we're leaving niner point seven.")

Near Saba I switched to Coolidge (Antigua) Approach Control and gave them a pilot report of cloud tops at 11,000 to 12,000 ft. Coolidge asked us to give them winds aloft upon reaching Monserrat Island, which I was estimating at 1130. After passing the tropical beauty of St. Eustatius, St. Kitts and Nevis, we overheaded the Monserrat airport at 1126, and I reported to Coolidge that we were estimating winds at 11,500 ft. from 045 degrees at 8 knots. Shortly thereafter I switched to Le Raizet (Guadeloupe) approach control, advised them that we had been overhead Monserrat "at 26" and that we were estimating Guadeloupe at 1600 Zulu (Greenwich Mean Time), or noon local time.

I then began our descent at 1,500 fpm, by this time having Guadeloupe in sight far off, but close enough to see a cloud deck at about 2,500 ft., under which we would have to pass to reach the airport. As I eased the nose down, the increased airspeed pushed the rpm up to 2,650, so I throttled back. I must have been looking at a chart when Larry said, "Hey, keep your rpm up." The rpm had dropped to about 2,100. I pushed the throttle all the way in, but nothing happened. Thinking that with the high relative humidity and cool air temperature at our altitude (now about 8,000 ft.) the engine might be picking up some carburetor ice, I applied carburetor heat. The rpm dropped fur-

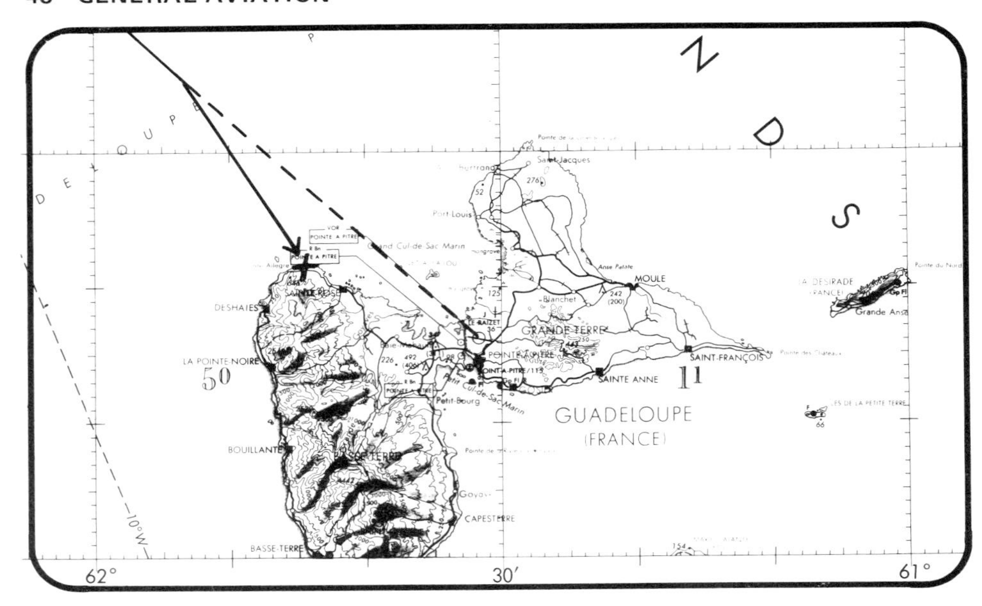

ther, as is normal, but did not pick up as it would following the removal of carburetor ice.

Larry started to fiddle with the mixture control; the engine began to run a little rough. Suddenly he said, "We've got no oil pressure!" I looked at the gauges: no oil pressure; the oil temperature, to my surprise, was normal (210°).

I didn't really believe it was anything serious, anything that a little "messing around" with the carb heat or mixture controls wouldn't fix in short order. After all, it was hardly the first time in my nearly 600 hours of flying that an engine on whose faithful behavior I depend so heavily has shown signs of what in humans would be called indigestion or temporary malaise. Just four days earlier this same engine had kicked up a fuss, only to calm down and complain no more almost as swiftly as it had begun disturbing our peaceful travels. Real engine failures seldom or never happen to careful pilots who follow the rules. And anyway, one always feels that accidents and tragic disasters happen only to others, or in novels and movies.

Six days before I had flown from Brazil to Miami to meet Cessna N3101A and its owner, Larry Holland. N3101A was a spiffy red and white 1954 Cessna 170B which had twice flown to inhospitable Tierra del Fuego on the southern tip of South America, had crossed the rugged Andes from Lima, Peru, to La Paz, Bolivia, and had made the long trip between the United States and Brazil ten times. By virtue of the many hours it had flown in all kinds of weather over every imaginable type of terrain, N3101A was an airplane for which one soon developed respect and in which confidence was almost automatic.

Larry Holland had owned N3101A for ten years and had logged most of his 4,000 hours of flight time in it. Larry, age 44, is a professor of physics at the Instituto Tecnico Aeroespacial, a prestigious aeronautical engineering school located in Sao Jose dos Campos, some 60 miles from Sao Paulo.

Larry has an FAA Flight Instructor's Certificate, and agreed to prepare me for the Commercial Pilot's flight test, which I took from the FAA attache in Rio de Janeiro in December 1972, flying Cessna N3101A. During the time I was receiving instruction from him, we often talked of flying together from the States to Brazil some day (I had dreamed of flying from the U.S. to South America ever since I was a college freshman 16 years before).

My opportunity came when Larry flew N3101A home for Christmas. The return was to be their eleventh one-way flight together between the U.S. and Sao Jose dos Campos. The total distance involved was some 5,100 statute miles, about forty-five hours flying time in the Cessna 170; most of these hours would be spent flying over either ocean, swamp or jungle.

Before leaving Brazil I re-read Ernest Gann's fascinating account of his experiences ferrying Lockheed Lodestars from the States to Rio just prior to World War II, over practically the same route we were to take. In *Fate Is the Hunter*, he tells how he spotted oil gushing from the filler hole of the right engine because he went back into the cabin to get some charts and happened to be looking out the window when the oil abruptly decided to leave the aircraft:

> "There is not time to think of a cause. That black stuff is our lubricating oil, our blood, our life. Without it the engine can seize solidly in a few seconds. The propeller will tear itself away most certainly, and the chances of structural damage and a fire are excellent."

(They made it back over the dense jungle to Belem on the remaining engine and found no technical explanation for the incident, concluding later that the cause must have been sabotage.)

Meanwhile Larry was in Denver investing over $2,200 to put N3101A in tip-top shape. Talking to him over ham radio one day about two weeks before our meeting in Miami, I responded to his proud comments about N3101A's new paint job by saying, "I'm glad we'll look pretty, but with all that water and jungle down there, how's the engine?" Later, at Miami's Opa Locka International and at San Juan, we were to receive numerous compliments from pilots and other airport denizens on the fine appearance and meticulous maintenance of this 21 year old airplane; the tower operator at Mayaguez offered over the radio to buy the plane from Larry when he landed there.

We lifted off Opa Locka early Friday afternoon, February 14th, ponderously if not sluggishly; in addition to our own weight, N3101A was carrying a lot of electronic equipment for Larry's physics labs, toys for his wife's nephews and nieces, a two-man raft and jungle survival gear, food, water and fruit juices, 22 lbs. of maps and charts, and about 400 lbs. of fuel (22 gallons were stored in each wing tank, 12½ more in an auxiliary tank in the back where two passengers usually ride, and the remainder in a ten gallon jerry can, the gas from which was hand-pumped into the aux. tank).

The flight down the Bahama chain at 11,500 feet was beautiful and uneventful. Turquoise, emerald and every other imaginable shade of blue and green contrasted serenely with the bright white of the beaches and the cotton white of the puffy cumulus line that accompanied the string of islands into the distance. We landed in the dark on the small island of South Caicos at 1835, five hours and 17 minutes after takeoff and 613 miles to the ESE of Opa Locka. Soon afterward we were eating lobster in the informal, friendly yachtsman's atmosphere of a cozy little hotel called the Admiral Arms.

On Monday afternoon, February 17th, we changed the oil, putting in seven quarts of Aeroshell 100W, since we were not able to obtain the lighter and more recommendable 80 weight. After the oil change the dip stick showed 6½ quarts; we decided to wait to Guadeloupe for gas (351 miles), until morning to see if all of the oil would have settled by then.

Tuesday morning we were up before six o'clock; we planned to fly then on to Martinique for lunch (120 miles). From Martinique we'd hop the 101 miles to the beautiful little island of St. Vincent, where we planned to spend the night before heading on to Trinidad. We checked the weather: a weak tropical convergence covering most of the Caribbean was causing local cumulus buildups and thunderstorm activity, with cloud bases estimated to be 1,500 ft. and tops over 12,000 ft. along our route of flight—strictly VFR with some weaving around the CBs.

At the airport we carefully checked the oil level and found that our assumption was correct: during the night the thick unheated oil had settled; the dipstick showed 7½ quarts, which we didn't top off to eight because this Continental engine tends to "throw off" the top quart in the first hour or two of flying time. After completing a thorough preflight check of the aircraft and carefully stowing our baggage, we took off at 0809.

It is exactly three and one-half hours later, as we are descending for our approach to Guadeloupe, when Larry suddenly exclaims that we have no oil pressure. My eyes instantly corroborate this disconcerting information: the oil pressure gauge reads "0"; the rpm continues to fall off. At the same time my ears confirm the bad news: the engine sound is increasingly rough and loud. Even my hands carry the same message to an unwilling brain, as the engine vibrations begin to be transmitted through the control wheel. I reach for the mike.

(11:39:05)

N3101A: Point-a-Pitre, Point-a-Pitre, mayday, mayday, mayday: Cessna November Three One Zero One Alpha, over.

Point-a-Pitre Approach Control: Three One Zero One Alpha, ready with, uh, go ahead.

N3101A: Uh Point-a-Pitre, uh, Zero One Alpha just, we just lost our oil pressure, we've got about seventeen hundred rpm and I'd say we're . . . probably about, uh, fifteen—twenty miles off your coast . . . uh . . . heading, heading into the north coast, over.

App Con: OK, uh, I understand you have any trouble, engine trouble?

(11:39:40)

N3101A: That's right. We've got an engine failure, sir.

App Con: Now you are at, uh, fifteen hundred feet and you are on a three one five and your heading for Guadeloupe now is one three five. Are, are you being obliged to land on the, on the sea?

N3101A: (Garbled) . . . five hundred and we're losing . . . we're descending about fifteen hundred feet a minute.

(11:40:05)

App Con: Fifteen hundred feet a minute, uh, you have a . . .

(11:40:55)

App Con: (to another aircraft, Seagreen Airlines DC-3, VP-LAO): We have a mayday call just, uh, one minute ago, is a Cessna one-seventy at fifteen hundred feet [sic] going to, uh, land on the sea...on the...uh...on the northwest three one five from Guadeloupe.

VP-LAO: It's a Cessna on the northwest three one five from Guadeloupe?

App Con: I think so, yes . . . uh Three One Zero Alpha, ready.

N3101A: Our heading is, uh, one four zero and we're heading right in for your ADF, sir.

App Con: Uh that's, uh, you want to go on the . . . on the direction of the Cessna going to, to land on the sea?

(11:41:45)

VP-LAO: Is that for Lima Alpha Oscar?

App Con: Affirmative.

(11:41:50)

N3101A: Point-a-Pitre, Point-a-Pitre, we're on the three one zero radial of the Point-a-Pitre VOR, three one zero radial . . . and we are leaving five thousand three hundred.

(11:42:05)

App Con: Three one zero radial from the VOR. Three one zero radial understood.

(11:42:10 to 11:42:50)

There follow some exchanges in French about our situation between the tower, a French Army helicopter and an Air France 737–AF 366–apparently on a training flight in the area. The pilot of the Air France jet asks the tower to confirm our distance from Guadeloupe.

This, then, is the real thing, the moment all good pilots prepare for every minute of their flying lives. An icy shiver runs down my body and leaves a heavy cold feeling in my guts: Guadeloupe is merely a dark mass on which not even the beaches can be distinguished, impossibly distant over the calm-appearing blue sea more than a mile below us.

I am sure we are going to have to ditch.

Despite the coolness of the cockpit at this altitude, my hands start to sweat; I hope my voice carries more self assurance than I feel as I begin to describe our situation with "mayday, mayday, mayday . . ." (Why one gets some sense of security out of knowing that another man, standing far away in a glass-enclosed elevated room called a control tower, is aware of one's predicament is not something I really comprehend. After all, the important thing in our plight is to survive the ditching–we must get out of the plane, and hopefully into the life raft, to avoid the sharks we know inhabit these waters. The man whose voice immediately acknowledges awareness of our troubles is not going to be able to soften the impact of our landing on the waves, nor will he be on hand if one or both of the doors refuse to open. Nevertheless, and for reasons that perhaps only a person who has spent many hours in a small cubicle linked to earth and his fellow man only by a stranger's voice over the earphones can understand, the calm reply "November 3101 Alpha, go ahead," steadies me at a moment when the certain realization of our danger is almost unbearable.)

For what seems an interminable length of time and at once a cruelly brief spell, I fight to keep our rate of descent–which had gotten away from me in the initial excitement–as low as possible without letting our airspeed sink below stalling speed, all the while trying futilely to get more rpm out of the engine without having it shake itself from its mounts. Larry, meanwhile, is bent over the back seat, struggling to get the fire extinguisher attached by a special pressure tube to the life raft air valve (an arrangement permitting almost instant inflation of the raft once we are in the water). He also makes sure the raft is free of any other baggage that might slow, or prevent us, from getting it out of the plane.

(11:42:55)

App Con: OK, Three One Zero Alpha [sic], your distance, can you give me the distance?

N3101A: Descending now at seven hundred feet a minute—we're running at fifteen hundred rpm but the engine has no oil so it's liable to blow at any minute, over.

(11:43:10)

App Con: OK, what distance from Guadeloupe?

(11:43:35)

N3101A: I would say . . . I'm estimating it looks to me like we're about, uh, fifteen miles off the coast. There's a— on the north coast there's an island . . . there are two islands, uh, to our left is a small one, to our right a larger one, we're heading right for that large island and I would say we're about . . . uh . . . oh, twelve miles out, over.

App Con: OK, twelve miles out, OK. Check with helicopter Thirty-One.

(11:44:00 to 11:45:40)

More exchanges in French follow between Approach Control, Air France 366, and the French Army helicopter being sent to our rescue.

(11:45:55)

N3101A: Point-a-Pitre, I'd say we're now, uh, ten miles off the coast and we're leaving, uh, three thousand six hundred.

App Con: Say again, please.

(11:46:05)

N3101A: We have an aircraft in sight at our, at our twelve o'clock position, uh, below us, uh, just passing that large island on his left . . . It's a jet aircraft, over. (The plane was Air France 366.)

App Con: It is passing the island?

N3101A: Roger, we're right now at his five-thirty position, over.

(11:46:45)

App Con: Five-thirty position. Who is speaking now?

(11:46:55)

N3101A: Roger, we've lost our engine now, and we, looks like we've got a fire in there.

App Con: You're going to land now?

(11:47:05)

N3101A: We have stopped the engine and we're gliding, it looks like we have a fire. We're heading right for that large island . . . uh . . . and I think we could probably make it to there—we're leaving three thousand feet.

App Con: (Garbled) . . . three thousand feet, OK. What's the distance from Guadeloupe?

(11:47:20)

N3101A: Uh, I would say we are now about, uh, eight miles off the coast and we're about three miles from that large island.

App Con: OK, three miles from the large island.

N3101A: That island is off the coast about two miles and there's a smaller one to the left, that is to say . . . uh . . . to the north of it, over.

(11:47:50)

App Con: OK, thank you, uh, you, you can stand by the aircraft? (This may have been directed to VP-LAO.)

The engine vibrations have by now become severe, and smoke is blowing back from the top of the engine cowling. We are both afraid that the engine will tear itself from its mounts, thus destroying the aerodynamic balance of the plane and causing it to enter a flat spin or some other uncontrollable flight attitude, carrying us to certain death. I am afraid that the smoke means we have an incipient fire in the engine, which would allow us no choice but to lose altitude as quickly as possible and ditch without regard to our distance from land. Larry turns the fuel selector to "Both Off," and seconds later the prop stops. The sudden change from violent noise and vibration to smooth, silent flight would almost seem encouraging were it not for our distance from land and our low altitude. Although my voice over the radio probably sounds calm enough, the difficulty I experience in giving our position relative to the Air France jet and my inconsistency in estimating the distance of the small island from the main coast betray the terrible stress I am feeling. Larry is not exactly suffering from boredom either—at some point he tries to jam my life vest over my head, ignoring in his excitement the headset I have put on to better understand the French-accented English of Point-a-Pitre approach control. A little more than eight minutes have passed since I began our distress call.

(11:47:55)

N3101A: We're going to have to ditch I think. I don't think this island will take us—it's too rocky. Uh . . . but we'll see what we can do, over.

(11:48:05)

App Con: There is a helicopter, a helicopter. One helicopter is going on the place now.

(11:48:10)

N3101A: We're going to try to land on the beach on the south, uh, correction, on the, uh, east side of this island . . . the southeast side.

(11:48:30)

App Con: Is landing near the beach. OK. OK.

(11:49:20)

N3101A: We're now out of one thousand five hundred and we're going to try to make this beach on the south side of the island that's on the north coast, over.

App Con: The north coast of the island, OK . . . *Alors* would you please call me, uh, tell me what (sic) who is speaking now? What aircraft is it?

As we get nearer the larger of the two small islands Larry says, "Let's try for that beach on Guadeloupe—I think we can make it." I take another look at the shortness of the beach on the small rocky island, the coral reefs jutting up on the northeast end of it, and our altimeter (which was now reading about 1,200 feet) and say, "O.K." Both of us are thinking that if we don't make it all the way to the beach, a ditching closer to the main shore will be in smoother water and nearer to help.

Air France 366, in French, says he has us in sight, and is just above us at one thousand five hundred feet.

App Con, in French, asks AF 366 if he has another aircraft in sight (the Seagreen Airlines DC-3).

Air France 366, in French, says he thinks we can reach the coast.

The beach on Guadeloupe seems much too far away to ever reach it with so little altitude between us and the almost two miles of ocean we still have to cross. Yet as we overhead the little island offshore that was our first choice, it becomes obvious that we probably would have ripped the plane apart on the rocks and coral reefs that surround the short sandy beach on which I had said we were going to try to land.

I try to advise Guadeloupe Approach that we are now attempting to glide to the main coast: I get no acknowledgment (only much later does the reason occur to me—we are now below the level of some low mountains about a mile or two inland from the coastline, and they have blocked all radio contact between us and the Guadeloupe tower). We have also lost sight of the Air France jet and do not realize in our excitement and ignorance of the French we continue to hear on the radio that the jet's pilot has us in sight. Everything, now more than ever, seems to depend on our skill and N3101A's aerodynamic resourcefulness. I remove the pencils and pens from my shirt pockets and check that the mags and master switch are off.

We seem to glide forever, silently closing the distance between us and the main coast as the altimeter measures the gradual-seeming but irreversible loss of our only remaining resource. I am amazed when we actually get over the beach with about 400 feet of altitude to spare.

Larry starts a left turn to line up with the beach to the north of a hotel, which we are by now overhead—the beach on the south side of the hotel is dotted with sun bathers (including, we learn afterwards, some nudists at the far end . . .). Just then I spot a cow pasture almost dead ahead of us, to the east of the hotel grounds, and shout, pointing, "Larry—cow pasture!" He turns the plane away from the beach and we head for the pasture.

As we start our approach I grab my camera to photograph our crash site. Larry, appalled, says loudly, "Jesus Christ, man!" Unfortunately, I have used up all the film photographing Saba and St. Kitts, but my apparent nonchalance may actually have helped calm Larry for the difficult job ahead.

I am just beginning to worry about the half dozen or so cows that we will have to weave through on the landing when I see two high tension poles on either side of the pasture on a line running diagonally to our flight path in a roughly north-south direction. I shout, "Wires!" and Larry immediately pulls on all three notches of flaps and starts a dive to get under them.

We get under the wires and miss the first few cows and hit going fast because of the airspeed we've built up in the dive. (The normal touch-down speed of the Cessna in a three-point position is 45-50 mph; we must be going 65-70). While we careen over the uneven terrain and rampage through some five-foot-high scrub, I find myself thinking, "Jesus, what a smooth ride!"

Then suddenly we see the house. An unpainted solid-looking little house built of concrete, squarely in front of us across a paved road. Somehow we managed to steer to the left and clear of the house by some 10 yards. We don't, however, clear the pile of big boulders near it: the airplane strikes the boulders resoundingly with the right landing gear, spins to the right, and comes to rest about 90° to our landing path. There is a big tree in the direction we've been going, about 15 feet from where we've stopped. Larry has done a fine job.

(11:51:35)

Air France 366 says we have crashed exactly at Fort Royale next to a road and near a house—"crashed normally" ("Il n'y a pas de probleme . . . il c'est crashe normalement . . ."—and he will fly over us to look again.

Only twelve and one-half minutes have passed since I first reached for the microphone to say the international code word for distress. We have touched down not more than 400 yards from the edge of the island, after gliding at least 15 miles over the ocean. (The glide ratio of the Cessna "dead stick"—without the drag of the windmilling prop—must be, we calculated later, at least 12 to 1.)

I yell, "Get out!" fearing fire or an explosion. We scramble clear of the plane, and are soon surrounded by native blacks talking excitedly in unintelligible island French. Air France flight 366 comes roaring over us at about 500 feet, and we stand in the road waving both arms over our heads, neither of us remembering the international airman's hand signals for "all O.K."

Fate, the Hunter, has missed its prey this time.

I return to the radio. "To any aircraft on this frequency, this is Cessna Zero One Alpha. Please advise Guadeloupe tower that we've crashed on the coast and that there are no injuries, repeat no injuries." The Air France pilot comes on and says he will relay my message; I thank him profusely for his help in looking for us.

By this time some sun-bronzed white fellows and attractive young women in bikinis or beach sarongs have approached us and say in English, with strong and delightful European accents, that they work at the "Club Mediterranee" (the hotel we flew over) and can they offer us any assistance?

With the men's help, we managed during the afternoon to answer all of the questions of the French police, the aviation authorities from Point-a-Pitre, and the customs officials. At about five p.m., the Club Mediterranee "chef de village" (manager), having learned that we'd not eaten since six that morning, hastily but ceremoniously brought us a tray containing fresh French bread, Swiss cheese, meat loaf, a pewter dish with butter 3 inches deep, bananas, apples, pears, white linen napkins and an hibiscus flower. He also brought five very cold bottles of Carlsberg beer. Squatting in the shelter of the damaged wing, we ate with the undisguished relish characteristic, I suppose, of men just pardoned from the death sentence. The elegance of the food tray, our casual posture and uninhibited appetites, and the incongruity of an aircraft at rest in that pastoral setting seemed to greatly amuse the still large native crowd.

By nightfall we had a wrecker truck from Point-a-Pitre on the scene, and with the generous help of the Club Mediterranee employees, an A&P mechanic (who happened to be vacationing at the hotel), and countless local natives, we got the wings off. At about nine p.m., soaked to the skin by a tropical evening shower, we wheeled the wings and fuselage into the safety of the Club Mediterranee compound. Shortly thereafter we entered the hotel dining room (without changing our clothes), made our way through the many tables occupied by formally dressed and curious or amused hotel guests, and sat down in the staff dining room to a delicious French fish dinner.

At the insistence of the "chef de village," Jean Claude "Ringo" Adouin, we slept and ate for free at the hotel for two days while getting the plane and its contents moved the 37 km. to the Le Raizet airport. (Every time I would ask about paying, he would say in his charming French accent. "Do not worry. Keep on truckin'!") If you ever have to crash in the Caribbean, I recommend doing so at a Club Mediterranee.

The airplane's engine was, of course, ruined by our having run it for some seven minutes without oil. The right landing gear was torn completely off the fuselage upon impact with the boulders, and the right wing got a puncture hole about a foot in diameter when it collapsed onto the rocks after the gear was torn away. But otherwise the plane was in surprisingly good shape. The entire underside of the fuselage, from the firewall back to and including the elevators, was covered with a thin film of oil.

Neither the American A&P mechanic, nor three French mechanics at the Guadeloupe airport who examined the engine, could say with any certainty where the oil had leaked from. There were no cracks or holes in the crankcase other than those made by some of the connecting rods when they blew out the top of the block just before Larry shut off the gas. Nor were there any visible cracks in the temperature gauge tube or other oil lines. The thin film of oil on the underside of the plane and our having flown three and one-half hours before anything happened both indicate a slow rather than sudden leak (roughly two quarts an hour).

Had we failed to properly close the "quick-drain" valve at the bottom of the oil sump after draining the old oil that afternoon in San Juan, a failure that went unnoticed because the new, relatively cold oil was too thick to pass through the narrow opening? Would an engine run-up after the oil change have revealed the oil leak while we could still have done something about it? Were we negligent in not having a test flight after the oil change and before taking off on a long flight over water? Had I failed to look at the engine gauges frequently enough in my instrument scan, thereby missing one or more tell-tale flickers of the oil pressure needle? If we had noticed anything even five minutes earlier, we would have been only eight minutes beyond the Monserrat airport, within easy gliding distance from 11,500 ft. Yet maybe we were lucky: Imagine an engine failure farther down the line, over the coastal swamp and jungle that characterizes the Trinidad to Georgetown to Paramaribo leg as well as the terrain between Paramaribo and Belem. Or worse: we could have lost the oil while flying above the impenetrable Amazon jungle canopy that stretches south and west from Belem towards Maraba. At no place in this desolate, hot and humid region would a commercial jet have been looking for us within three minutes of our distress call. . .nor could we have counted on the hospitality of a Club Mediterranee after we crashed.

One interesting afterthought: As a friend of mine pointed out later, our San Juan departure put us clearly within the influence of the Bermuda Triangle . . .

THE ELT MESS

by George E. Haddaway

Reprinted by Permission FLIGHT OPERATIONS MAGAZINE
(February 1976)

In researching the subject of emergency locator transmitters, which have proved anything but a boon to general aviation, we found the following report published by the National Pilots Association, which paints the best picture yet of the problem and efforts to solve it:

Incident at Indianapolis: Virgil E. Lyons was returning from Kent, Ohio, with a small cargo, to his home field, Four Winds Airport, Roachdale, Indiana. He had refueled at Kent, filling up with 100-octane fuel, which mixed in with the 80-octane already in the tank. The Grumman American TR-2 had a visual fuel gauge, and with the two grades mixed together, the fuel was clear, making it difficult to read the amount. Just 23 miles from home, Lyons glanced at the gauge and figured he had a third of a tank left. Minutes later, at 4:15 p.m., the engine sputtered and quit, and Lyons crashed into a field. For 17 hours he lay in the wrecked aircraft, most of the time in a semi-coma. Although there was no way for him to check the operation of his emergency locator transmitter, the unit in the back of the plane was working, sending out its distress signal.

Reports of the signals started flowing into the Air Force Rescue Coordination Center at Scott AFB, Illinois, and the information indicated it was a false signal coming from the Indianapolis metropolitan area. The weather was good, no aircraft were listed as missing, and the majority of the ELT reports the Center receives turn out to be false alarms. Then Indianapolis Center reported that the signals had terminated. There was a change of personnel at the AFRCC, and later in the night the ELT signals picked up again. The new shift didn't relate the signals from the Indianapolis area to the previous reports. Still no aircraft was missing. No search was initiated.

At 9:30 the next morning, a farmer spotted the wreckage of Lyons' airplane and called the state police. Lyons was taken to the hospital where it was found he had a compressed vertebra, cracked pelvis, a ruptured bladder and other injuries. The Federal Aviation Administration which is investigating the accident, hasn't determined the cause.

It was the first time that the Air Force Rescue Coordination Center had not launched a search for a downed aircraft since the ELT requirement has been in effect. The AFRCC has received an average of 500 ELT incident reports a month. Each report, says Colonel Ryland Dreibelbis, AFRCC Director of Inland Search and Rescue, must be assumed to be an emergency unless proved otherwise. But 99 percent of the signals turn out to be inadvertent transmissions.

When reports of an incident come in, the AFRCC asks the FAA to request other aircraft in the area to monitor 121.5 to try to get a better bearing on the transmitting ELT, a DF bearing if possible. Reports of missing aircraft are checked. AFRCC then calls Flight Service Stations, towers and ground stations to see if they are receiving the signal. If ground stations are picking up the signals, the search is narrowed down, as the transmissions are line of sight, and most often the transmitting unit is found in a parked aircraft. If a malfunctioning ELT has not been located within two hours, a search for a missing aircraft is launched, by air if weather permits, or by ground forces.

Since the Lyons incident, a report must be filed if a search is not initiated within two hours. Other procedures have also been changed to prevent another incident like this from happening. "We're launching searches twice as much as before," said Colonel Dreibelbis. "Prior to the Indianapolis incident, we were launching one out of 36 reports, now we're launching one out of 19."

The Air Force is also continuing its campaign to try to get better information about signals. AFRCC has asked that everyone who hears an ELT signal report the altitude of their aircraft, where and when the signal is first heard, where and when the signal is heard the loudest, and where and when the signal faded or was lost. The altitude report will give AFRCC an idea of the range of the signal; the other three points will give a line of position for plotting.

The Canadians, who are also suffering an ELT requirement and are plagued with false alerts, take a different approach to the problem. The Canadian air force director of air operations plans to "make an example of people found guilty of triggering an ELT signal when there is no emergency." Canadian search and rescue spends $80,000 a month in unnecessary air searches, and only four of the 381 alerts in a three-month period this year were actual emergencies.

The U.S. Air Force has also suggested that Jeppesen publish information on charts about how to report signals so pilots will have the procedures in the cockpit, if they should pick up a signal. NASA has proposed satellite monitoring of ELT signals that would pinpoint the transmission better.

The Air Force is handicapped, said Dreibelbis, because it is a civilian matter. The requirement for ELTs was passed by Congress, and the burden of dealing with the faulty program falls to the FAA and the AFRCC, who have no authority to change or improve the actual operation of ELTs. "We are in favor of the ELT system if the bugs can be worked out," said Dreibelbis. But, he adds, there is not enough ground monitoring equipment. For an actual search, the Air Force calls in the CAP, the local police or other military units in the search area. But the local forces often don't have the equipment to home in on an ELT signal, says Dreibelbis.

Better tracking of ELT signals is not the solution to the whole problem, of course. The transmitters must be perfected so that the number of false alarms will be greatly reduced. The Radio Technical Commission for Aeronautics is working on specifications for ELTs, which should improve the operational situation, and the FAA is conducting a study of ELT problems.

One person who is unlikely to forget the ELT problem is Virgil Lyons. "I'm speaking to my attorney," he said. "I'm not going to let the matter die." Lyons, like others concerned with the situation, wants to keep other people from going through an experience such as his.

CAS: WHAT PILOTS CAN EXPECT

by E. A. (Jerry) Jerome

Reprinted by Permission
FLIGHT OPERATIONS MAGAZINE (June 1976)

Edited for Space

For midair separation assurance, FAA has selected beacon collision avoidance systems (BCAS) using existing ATC radar beacon system elements. But this is only part of a five-point program. Here's a review to help project your future CAS involvement.

The die is C-A-S(t)—or is it?—the Federal Aviation Administration recently made an historic decision regarding the future of collision avoidance systems (CAS). It selected one system to be the national standard—BCAS (beacon collision avoidance system). For the first time, a well-integrated plan has been proposed for attacking and minimizing the midair collision problem. But a plan is not reality until it is concluded. Indeed, there are still many technical—and political—issues that the FAA must resolve before it makes a final choice of CAS hardware to solve the problems of near-midair collisions and assure safe aircraft separation. FAA's decisive selection of beacon-based BCAS still awaits determination as to whether it will be an active or semiactive basic system or perhaps a hybrid of the two.

The midair collision problem started when Orville first said to Wilbur, "Let's build another one!" Since then, midair collisions have been occuring with almost uncanny regularity. The first midair collision occurred in Austria on September 8, 1910, when two Farman biplanes piloted by brothers touched wings and crashed. Both survived but one pilot broke a leg. The next midair occurred almost a month later on October 2, 1910 over Milan, Italy. An *Antoinette* collided with a *Henry Farman*. Both pilots luckily survived the resulting double crash. The first fatal midair occurred in France on June 19, 1912, and the first collision between airliners took place on April 7, 1922 when a Farman *Goliath* operated by the French airline, Grands Express, flew into the path of a Daimler Airways DH 18 over Poix in northern France.

In the last 20 years U.S. aviation has averaged 24 midairs per year, resulting in an average of 16 fatal accidents and killing an average of 61 people each year. The top year for midairs was 1974—34 actual midair collisions. The point of all this review is to emphasize that technology has been available to produce collision avoidance systems (CAS), but there has been a noticeable lack of basic agreement upon the techniques to use the available hardware. The prime factor in this delay in using the available hardware is cost and the choice of the system to be used. There is also a problem in that big aircraft need to see the little aircraft or the system will not work. "See" meaning an electronic warning indicating to the pilot where the aircraft is, with appropriate time and means to perform an avoidance maneuver.

Defining CAS Basics

Most of the pilot community has only a vague idea of what a system for collision avoidance is all about. The avionics people have established a good definition for CAS: "A collision avoidance system is one which detects all aircraft which represent a potential collision, evaluates the hazard, and generates an appropriate warning and maneuver command that ensures safe separation of aircraft." From that definition we get the new buzzwords we often hear when discussing CAS matters—"separation assurance."

About five years ago, collision avoidance systems almost became mandatory on all aircraft using U.S. national airspace. In 1971, such legislation was introduced into the U.S. Congress but it was defeated. This legislative proposal, however, did establish and define the basic requirements of CAS as follows: (a) practical in size, weight, and cost; (b) ability to handle extremely high-density air traffic (up to 800 aircraft) within a 70-mile radius; (c) virtually a zero false alarm rate; (d) compatibility with the current ATC system; and (e) versatility—that is, the ability to meet both the cost and technical needs of high-performance jet aircraft as well as smaller general aviation propeller aircraft.

After many years of indecision, many concur that FAA's announced program is the proper direction in which to go. Insofar as hardware commitment is concerned, experts agree only in terms of generic avionics. These experts hold that there is yet much testing to be accomplished before the die is really "CASt." The current official position was taken after FAA evaluated three candidate systems, each capable of reducing the number of potential midair collisions. In simple generic terms, the three systems can be described as:

1. *Airborne collision avoidance systems (ACAS)* which are independent of the existing ATC system and operate on an aircraft-to-aircraft basis.
2. *Beacon collision avoidance systems (BCAS)*, which is an airborne system that utilizes elements of the existing ATC radar beacon system (ATCRBS).
3. *Intermittent positive control (IPC)*, which is essentially an extension of the ground-based air traffic control system.

There are many additional sophisticated modifications of these three basic systems, and to the layman the acronym dictionary that evolves with each new development is an electronic alphabet soup which would even confuse the ordinary business pilot.

Five Steps to Separation Assurance

The FAA program for aircraft separation assurance involves a five-point program, part of which is BCAS. A short discussion of each of these five points provides a small insight into the enormity of the problem and how the present FAA thinking will affect the future of *your* CAS flight operations.

First, the new program includes the installation of conflict alert (CA) system at ATC facilities, implementation of which is well under way. This automatic backup alarm for conflicting ATC traffic is a software program using ATC computers already in place. It projects what the flightpaths of transponder-equipped aircraft will be in the next two minutes—although this time envelope is flexible. It will alert controllers when a potential conflict is detected so that the controllers can take the necessary action via radio to warn the pilots. It's already working in 20 domestic air route traffic control centers for airspace above 12,500 feet. A similar capability is being developed for automated terminal systems with a planned installation at the 60-plus airports during 1977.

Second, there will be new flight-plan requirements for passenger-carrying aircraft. These aircraft will include air taxis, commuter airlines, and the executive corporate fleet. The new scheme requires these types of aircraft to file a flight plan and operate under IFR to ensure continuous monitoring by ATC. The primary benefit would be to increase the protection against midair collision involving public air carriers, and especially against catastrophic collision involving wide-bodied jets.

Third, additional requirements for carrying automatic identity and altitude reporting transponder equipment for all aircraft flying in certain controlled airspace. The altitude/identity information is displayed directly on controllers' radarscopes, giving them a more complete picture of the traffic under their control. The altitude-reporting transponder will be the key to the enhancement of both CA and the upcoming BCAS. Forthcoming discussions will determine the new requirements for protection against a midair between a general aviation (GA) aircraft and an airliner within ATC surveillance. Conceivably, this would eventually include all small private aircraft.

Fourth, development of an airborne BCAS. You might say all the plans for ACAS have now been cast aside, despite the fact that the Honeywell-tested equipment offered the best performance at the lowest cost of the three ACAS-type systems tested. (The other two ACAS types were made by RCA and McDonnell-Douglas.) The FAA points out that it was not convinced that this ACAS would provide early cost-effective complete protection compatible with the present ground-based ATC system; it requires cooperative units to be installed in *all other aircraft* to be effective. Thus, BCAS was chosen because it offers in theory collision avoidance service that is effective immediately in an aircraft in which it is installed. Its adoption is enhanced because military aircraft and GA aircraft owners would not be burdened by the need for additional avionics if they do not desire the additional protection. In the long run BCAS is believed to be the quickest and cheapest-priced way to provide an independent backup capability

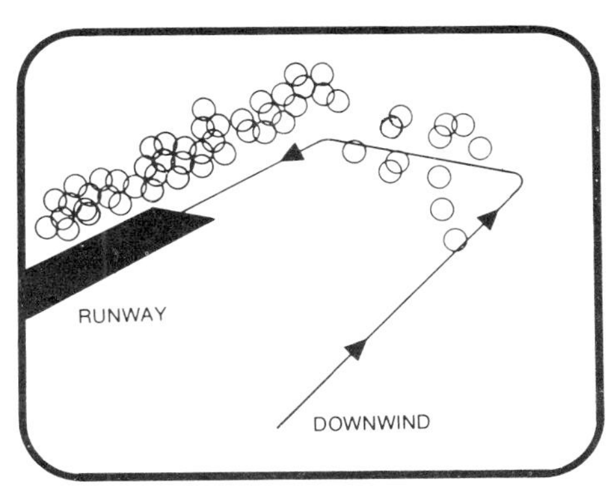

Diagram above shows approximate location of 59 midair collisions in the U.S. from 1968 to 1970 of general aviation aircraft at uncontrolled fields.

for the ground-based ATC system. The FAA considers it the wisest feasible route to extend collision avoidance services outside the present ATC system. (Honeywell disputes these assumptions.)

Fifth, development of intermittent positive control (IPC)–a system that would link ground-based ATC computers with new airborne transponder equipment to provide collision avoidance services to *normally uncontrolled aircraft*. The FAA is straightforward about the fact that such a sophisticated system will take ten years, or perhaps 15, for implementation. However, before IPC becomes operational (mid-1980s), the FAA is accelerating development of the discrete address beacon system (DABS). This interrogates aircraft on an individual basis rather than eliciting replies from all aircraft in a given area, as is the case with present radar beacon equipment. This would eliminate the problem of garbled returns and permit automatic data-link air-ground communications. Last February, Texas Instruments, Inc. was awarded an $11.9-million contract to provide three ground sensors or interrogators, 30 airborne transponders and other related equipment.

These ground sensors will be collocated with radars at FAA's National Aviation Facilities Experimental Center (NAFEC) at Atlantic City, N.J., and also at Philadelphia International Airport and at Elwood, N.J. The airborne hardware will be installed in a mix of air carrier, corporate and other GA aircraft. This will all start in late 1977 for a year or more of multi-site testing using computer-based automation equipment at NAFEC. It is hoped that this DABS-based system will provide aircraft separation assurance protection by way of a system which can cooperate with, but is not totally dependent upon, the ground-based ATC system. A more futuristic capability will be DABS in conjunction with IPC, wherein it can provide an automated three-dimensional warning and collision avoidance display in the aircraft cockpit.

CAS World of the Future

Thus we observe the short- and long-term solutions proffered by the FAA, which is determined to accomplish the task through this combination of procedural, regulatory, and hardware solutions. One must remember that the implementation scenarios of the hardware can vary between two years for optimum conflict alert operations to 15-plus years for BCAS/DABS/IPC real-world operations. If it will give you any comfort, think of what you'll be doing in the IPC aviation world of the 1990s. A few of us won't even be here.

The U.S. aviation world of the 1990s will have four times more airplanes in the sky than today. If nothing is done about the near-midair collision problem, the experts predict there will be about 16 times as many midair collisions. (That's 16 times our present average of 24 per year, or about 384 actual midairs per year statistically.) The proportionate loss of life per accident would be greater due to larger-capacity aircraft.

But most of us are not concerned with statistics predicted for one or two decades hence. Today's airspace users are more concerned with now: How will FAA's plan affect their operations, their economics and their aviation safety in the near future? If FAA's plan had been proposed in the early 1960s, it would have seemed impractical because relatively low-cost airborne avionic equipment was not available in that era. So what is the feeling about installing such equipment now? Already one can hear some of the rumblings and grumblings about the distant CAS hardware requirements. A cogent part of the problem, insofar as the users are concerned, boils down to: Do I have to put a black box in my airplane; if so, what box and when? That's an oversimplification, but below we cite a few representative positions already announced on the subject.

National Business Aircraft Association, Inc.: The NBAA position was expressed recently on two occasions: before the U.S. Congress Milford subcommittee on engineering and development in February, 1976, and at the Air Force-Industry Conference on midair collision protection held in Apple Valley, California in March, 1976. Fred B. McIntosh, director, operational services, paraphrased the NBAA position as follows:

"As a concept, we do not see a need for a collision avoidance device or a warning system in the airplane at this time—regardless of what it's called (BCAS, ACAS, IPC, etc.). In the event that a type of system is ultimately determined to be necessary, either by the FAA, Congress—or even the industry—we think it should be as *independent* of the ground-based system as is possible.

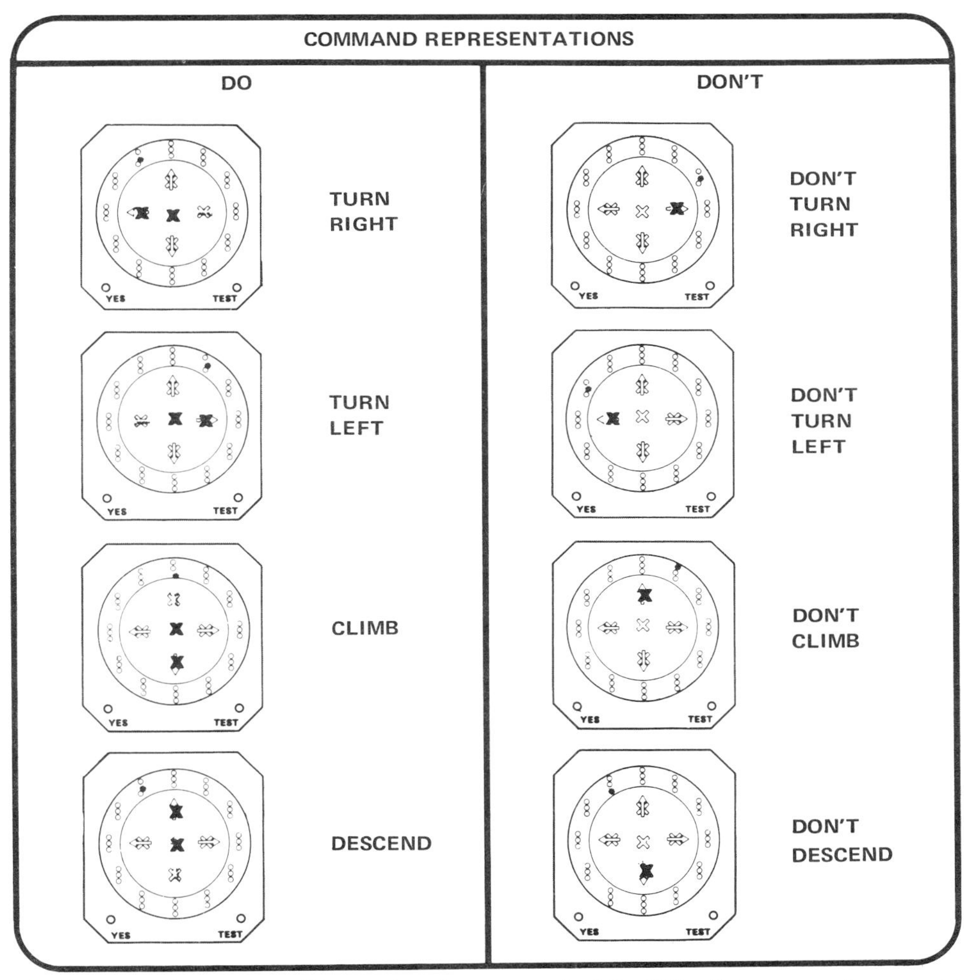

The illustration depicts all possible one-direction commands of the PWI unit by means of crosses and arrows which direct the pilot in avoiding conflict with other aircraft.

"Realizing that it may not be totally independent, we feel strongly that it should be as independent as possible because most of the near-collisions actually occur in the controlled airspace where . . . radar, transponders and the whole ATC system [is] focused on [aircraft]. And that's where the hairy near-misses are occurring—in controlled airspace.

"Another reason for our position is that the BCAS concept is being used, in part, for a justification for all the other pieces of hardware—DABS, IPC, etc. If you take any one of those blocks out, particularly the BCAS, the whole system falls under its own weight. We believe we're not alone in that feeling."

Aircraft Owners & Pilots Association: Victor J. Kayne, senior vice president, policy and technical planning, expressed AOPA's position at the Apple Valley, AF/Industry conference. These are excerpts from his presentation:

". . . The eyeball is still the basic collision prevention device under most conditions. All too often, a collision has happened because two pilots each thought that he was the only one in that particular piece of airspace and failed to keep adequate watch for other traffic. Here, a CAS or even simple proximity warning system would be useful . . . The idea of a CAS that utilizes some existing hardware in aircraft has certain attractions to general aviation, which has something like 90,000 transponders and about 40,000 encoders in operation. However, a program to require this equipment as the price of admission to large portions of the lower airspace is likely to run aground on the rocks of opposition from the many thousands of aircraft owners who do not have transponders and have no expectation of ever needing one . . .

"We do not want a mandatory program that forces everyone to carry additional equipment whether there is a need for it or not. Next, we reserve judgment on the BCAS proposal until we are convinced that it will be a viable system for GA that will offer advantages which will encourage owners to buy it, and at a price we can afford. If BCAS meets this test, then we can see no need for the DABS/IPC for further collision protection, especially in view of the large expenditures that will be required for the necessary ground system of computers, communications links between computers, ground interrogator/receiver stations, and the necessary interface with the ATC system computers and controllers with regard to IFR traffic."

Air Transport Association of America: Generally, ATA is in favor of the current plans for automatic separation assurance (ASA) endorsing ATC/IFR present methods with expansions to provide the better means for separating IFR aircraft from all other aircraft. They encourage the use of long-range (ARSR) radar equipment in the enroute environment as an extension of DABS/ASA development.

William T. Hardaker, assistant vice president, air navigation/traffic control, at the Apple Valley conference stated in part: ". . . the BCAS system must permit any user who chooses to implement it to do so. At the same time it must be possible to automatically augment BCAS with ASA/DABS service in areas where BCAS is limited. Therefore, it is essential that a BCAS system meet the test of extensive usage in high-density terminal areas without compromising itself or compromising the basic ground surveillance quality . . ."

Additionally, Hardaker stressed three points: (1) the DABS/ASA system must function satisfactorily in high-density takeoff and landing areas, whether or not BCAS plans proceed; (2) neither BCAS nor ground-based ASA systems should be permitted to adversely affect the ability of ATC to move air traffic safely and expeditiously; and (3) a single transition to a new ATCRBS transponder is essential. The FAA should advise the users as early as possible that ATC transponders purchased after a given date must have certain new capabilities and provisions for later addition of new functions.

Air line Pilots Association: ALPA's collision avoidance systems committee headed by Capt. Carl L. Smith of Western Air Lines has been monitoring the development of CAS for many years, working with both the FAA and manufacturers. In the past, this committee set forth the following CAS recommendations: (1) any warning system must be independent of ground controller operation; (2) the warning system must provide protection for adequate safe aircraft separation; and (3) it must operate equally effectively in VFR and IFR conditions for the wide range of today's aircraft speeds including the current and future Mach speeds of the SST.

The current ALPA position has been slightly modified in light of the new FAA decision to go the BCAS route. Joseph M. Schwind, deputy director, engineering and safety, summarizes: "Since the FAA has dragged us out so long—and with BCAS looking viable and on the horizon—we feel that it is worth waiting for as long as it does not present an indefinite delay. We are still not opposed to an ACAS-type of independent system, but with the recent development and decision about BCAS, ALPA is willing to allow this system to go through a normal evaluation period. Again, we sincerely hope that the FAA program and the BCAS adoption do not occasion further protracted delays in U.S. midair collision protection."

Flight Safety Foundation: This non--profit, non-government, international membership organization dedicated solely to the safety of aviation throughout the world has been concerned with the continuing midair collision threat for almost a decade. Last year FSF sponsored a special midair collision workshop at the University of Southern California. Held in Los Angeles, the workshop was attended by many high-level executives and experts of the aviation industry concerned with aircraft separation.

Jack Carroll, FSF's executive vice president and managing director, summarized the workshop's findings: "The midair collision problem is complex—it persists—and there is no single solution. The threat can be characterized into two categories. Category I is the collision of two GA aircraft having relatively low speeds, low closure rate, low angle of interception, on a bright, clear, sunny day in the vicinity of, or on final approach to, an uncontrolled airfield. The Category II threat at worst, would be involvement of two wide-bodied jets.

"Prevention measures would bear the greatest potential for reduction of the threat of a Category II midair collision in the transition altitudes between the positive control areas and the terminal control areas. In this area, generally speaking, between 4,000 feet and 14,000 feet, the threat against which to guard, primarily, is the inadvertent intrusion of a high-performance civil or military jet into the air carrier flightpath.

"The safeguard for this circumstance which would appear to offer the greatest potential would be to equip the air carrier with a relatively unsophisticated, non-cooperative, low-cost, low-range, proximity warning indicator (PWI); and to require the day and night use of high-intensity white flashing zenon (strobe) anti-collision lights in at least as much of the transition altitudes and terminal airspace as is practical."

It should be noted that several organizations are opposed to the PWI idea because the concept of "proximity warning" is not a pure form of collision avoidance. Proximity warning hardware merely indicates that something is nearby. Thus, the value of the warning depends on the timeliness and completeness of the warning information. FAA opposes PWI additionally because under certain conditions in controlled airspace, if an aircraft performs an avoiding maneuver, it may well enter another aircraft's reserved airspace, thereby creating a new separation assurance problem. Also, as an independent system, it adds little more computer capacity to make a true collision avoidance system.

On the other hand, many users of PWI have found it to be very effective. For example, the U.S. Army has been using

PWI for several years and credits this type of device for avoiding "near-misses" that could have been actual collisions. Their main success with these devices has been with helicopters.

PWI advocates realize that it is not a cure-all for the midair collision threat. They admit that it is not the ultimate solution, but they hold that it is available now, is relatively inexpensive, and has prevented many collisions. In short, PWI can fill the gap during that period between now and when CAS becomes a tangible fact of flight operations. As noted previously, this period can range from ten to 15 years.

Pleas from Avionics People

The manufacturers of avionics have become cautious, having observed the vagaries of bureaucratic indecision and aviation economics for almost a CAS decade. As a group, they know that the die has not been "CASt" when it comes to actual black-box building. Regardless of what system is ultimately chosen, the proof of their electronic pudding comes when someone specifies how the black boxes are to be wired up. Euphemistically, these manufacturers sum up their present posture with some hardnosed business philosophy:

"We have the 'smarts' to invent and make anything you want for a workable CAS, but you the users and you the feds had better make up your minds. We don't stay in business by developing equipment for which a clear need has not been identified by the potential user."

The FAA Position

Writing this type of report tends to make one empathize with the feds' head shed. For a moment, try putting yourself in the midst of the decision-making position of the FAA on the subject of CAS. Their long struggle to formulate acceptable collision-prevention rulemaking with compatible hardware for the diverse interests of U.S. and international aviation has been just short of monumental. Their responsibility is further amplified by many factors: (1) the clamor of the public and the U.S. Congress to make quantum improvements in separation assurance safety—and to make them now; (2) the total complexity of the technical problems involved; (3) the vested interests, mainly economic, of each aircraft owner and operator; and (4) the clear requirement to mandate new hardware, rules and procedures upon an aviation industry striving to remain solvent in a depressed economy additionally threatened or plagued by fuel shortages. Aggravating this is the FAA's difficulty in making everyone understand the rationale and the subtleties behind the solutions of this multi-faceted problem. In short, the FAA has been in a painful limbo about the tough decisions to be made. It's a classic case of damned if you do and damned if you don't.

The FAA is straightforward about one point without batting an eye: Collision avoidance will not be cheap. Like any safety innovation, a price must be paid for CAS. The price exacted will be in terms of money spent and aviation system efficiency, that is, the cost of hardware—airborne and ground—plus the cost of the additional effort required under new ATC procedures. The long-term benefits expected are manifestly safer flight operations for all forms of aviation and a public confidence in the U.S. aviation system and in those who use and regulate the system. All well and good, you say, but what about the price?

CAS 'Membership' Costs

In discussing aviation costs, the term "relative" is significant. Another point

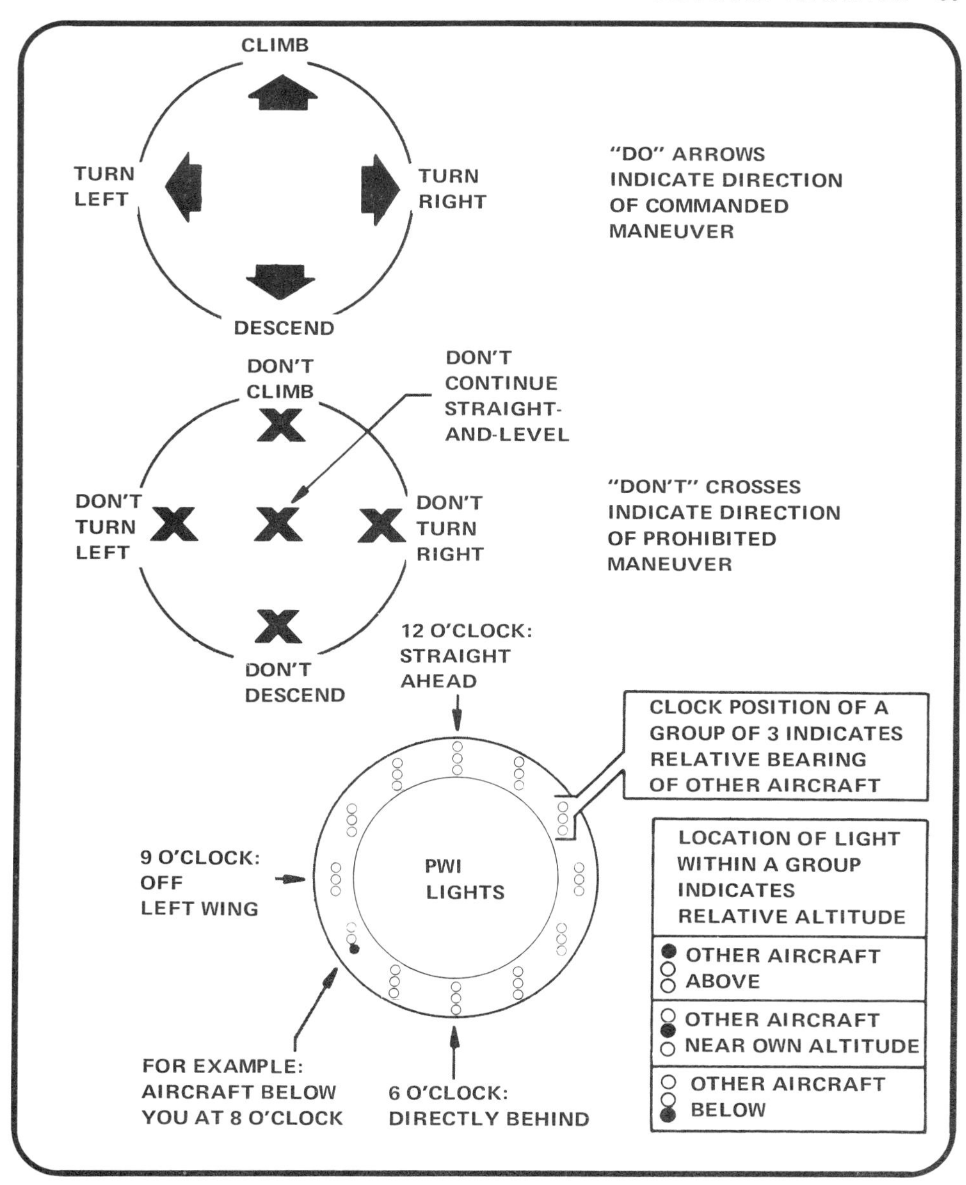

Command presentations are displayed to pilot by lights and symbols that permit rapid reaction to danger from intruding aircraft.

questions whether the costs are gradual and voluntary rather than mandatory. These impact most heavily on smaller general aviation aircraft when it comes to CAS, because all airliners, larger GA aircraft and military aircraft now carry airborne Mode C transponders, an essential part of the planned BCAS system.

Let's say that your aircraft is not equipped with a Mode C automatic altitude reporting transponder. The balm that makes the thought of paying about $1,500 for such new equipment less painful is the comforting assurance that equipped aircraft will obtain greatly increased protection against possible midair collision. At present, nearly 60 percent of the total GA fleet is equipped with transponders but not all of these are outfitted for Mode C automatic altitude reporting.

The present cost of a GA ATCRBS transponder is about $750. Compared to this, a futuristic DABS transponder with IPC cockpit display is estimated to sell for about $2,000, excluding the cost of the altitude encoder. For the additional cost, a GA aircraft will obtain protection against other GA aircraft which cannot afford a BCAS. And the price of a DABS/IPC at $2,000 (1980 prices?) will be no more than a conventional transponder, plus an airborne CAS.

As mentioned, part of FAA's new plan is the use of airborne CAS relying on the widespread use of ATCRBS transponders for operation. Current FAA thinking is that it will seek mandatory installation of BCAS on all aircraft capable of transporting ten passengers or more. This would exclude most of the GA fleet as well as much of the military fleet. Current estimates are that an *active* BCAS, in production quantities, would sell for about $20,000, with the semiactive version costing slightly more. An active BCAS would be designed to transmit, at one-second intervals, a Mode C interrogation similar to that emitted by ATCRBS ground stations, and listen for replies from all transponder-equipped aircraft within a radius of about 30 miles. Thus, the active BCAS-equipped aircraft can determine the relative altitude and distance separation, and closing rates, of potential threats. Regardless of how super-sophisticated such a system may be, many operators may balk at the near $20,000 price tag that it will have.

Honeywell's Position

On February 9, 1976, the FAA administrator wrote to Senator Howard W. Cannon, chairman of the Subcommittee on Aviation, stating, in part, "We have concluded that the Honeywell system offers superior performance and lowest cost of the three systems tested (the other two being RCA and McDonnell-Douglas), but concerns still remain in the area of false alarms and compatibility with the present ATC system . . . . We have concluded that the desired increase in separation assurance protection can best be achieved by alternatives other than ACAS in a manner which provides protection in an equivalent time, at reduced cost, and minimizes the problems of false alarms, ATC compatibility, and transition to future ATC services . . . ."

These phrases by Dr. John L. McLucas were considered to be the "coup de grace" for ACAS, since Honeywell was the principal advocate for the ACAS concept in recent months. Honeywell disputes this FAA adversary position and does not intend to give up the ghost on the ACAS facts as they see them, to wit: Comparing ACAS to BCAS/IPC, there are five areas in which ACAS wins out.

1. Implementation Timing—ACAS implementation will begin in 1978, thus having a two-year start-up advantage over BCAS. Military and GA protection will be five to seven years behind the Honeywell system, and then only a small percentage will be protected by BCAS. The sole BCAS advantage, providing air carriers more protection in the early 1980s against GA and the military, this will disappear if BCAS development slips just one year, a likely possibility.

2. Cost: The following figures indicate a more than two-to-one advantage for ACAS, with specific advantages being enjoyed also by all users:

3. False Alarm/ATC Compatibility: In three separate government reports by respected technical agencies, the Honeywell ACAS system (called AVOID for avionic observation of intruder danger) was found to have no false alarms or a very low probability of false alarms. It was also found to be compatible with present and future ATC capabilities, including high-density-area use.

4. Development Status: Three years of government-sponsored AVOID technical evaluation have produced positive results demonstrating they meet all performance requirements for CAS and system safety enhancement. In contrast, Honeywell claims that BCAS is now where ACAS was five years ago; i.e., feasibility stage of development, unproven technical capability, unanswered questions of performance, costs and related data. And because DABS, an unproven system, must be implemented first, Honeywell says it may well take until the 1990s before any IPC can be operational.

5. Level of Safety: Again, Honeywell claims that AVOID provides greater protection than the combined BCAS and IPC systems because it protects in all airspace for the entire aviation community. In contrast, they say BCAS will be limited to the air carriers and a small percentage of military and GA aircraft. In addition, Honeywell claims IPC will be restricted to airspace covered by DABS/IPC ground surveillance.

Honeywell concludes that the current FAA-proposed systems put an inordinate burden on commercial airlines in terms of individual installations (about $20,000 per airliner). As for general aviation, total costs will surpass those of the airlines, simply because there are more eligible aircraft.

The FAA decision to use beacon-based collision prevention has engendered several dissenting reactions, as we have observed. Generally, the operators have reacted in two basic ways. One segment seems to be saying: "We think the BCAS decision is okay, although it's thoroughly distasteful because of the cost and complexity involved. We'll just wait and see what happens." The other segment comments: "We don't believe any drastic measures should be taken in cockpit hardware. The present ATC system is sufficiently safe for the foreseeable needs. And cockpit gadgets supplementing the present system are not needed for improvement in the immediate future."

And then we hear a faint, bleak and cold voice from the energy and fuel experts. Cynically, they warn: "Stop worrying too much about CAS. In the 1980s, aviation fuel shortages will be so acute that there won't be a midair collision problem because of a lack of airborne aircraft."

What we have not yet heard is a unified concerted voice from the aviation community saying: "Federal Aviation Administration, your beacon collision avoidance systems approach to aircraft separation assurance is all wrong!" Consultative talks scheduled for the upcoming months will give aviation some answers as to whether or not the die has finally been "CASt."

THE CONTINUING AVGAS CONTROVERSY

by David A. Shugarts, Editor
FLIGHT LINE TIMES

Written especially for AVIATION YEARBOOK 1977

Trading blue for red made some Americans turn white during our bicentennial year.

They were the folks in airplanes with 80 octane engines, and they were very worried. In what has become a national embarrassment for the aviation industry, thousands of airplanes were forced to burn the new blue 100 octane "low-lead" avgas as a substitute for the familiar red 80 octane.

Blue gas gunks up the valves and spark plugs. This costs people money, and it may jeopardize their safety.

It took all of 1976 before anyone in aviation could make those statements without fear of argument. Though at year's end many "solutions" to the problem were on hand, it should stand as a lesson that it took so long to admit the problem in the first place.

A quick history: In 1967, the oil companies warned they would soon have to stop making two grades of avgas. In 1971, Shell brought out a form of the new single-grade fuel, but it was considered only a nice option for the 100 octane airplanes.

The new blue gas has four times the lead of 80 octane, and by 1972, Lycoming had approved such a fuel for "continuous use" in all its engines. Late in 1974, Embry-Riddle Aeronautical University sounded the aviation equivalent of a silent alarm.

Formed under FAA's aegis was a committee drawn from a half-dozen aero firms to "study" the problem. They worked through the summer of 1975, while around them the oil companies were switching to blue gas in most places east of the Mississippi.

The Embry-Riddle report, done by late Fall, said blue gas causes four times as much valve sticking, 10 times the spark plug fouling, and 32 cents more per operating hour when burned in an 80 octane airplane engine. The industry did its best to keep this report under wraps, but it got out by early 1976.

This set the stage. By June, Cessna had announced it was changing engines in the venerable Skyhawk line. Grumman-American made moves to follow suit. Piper held out with a wait-and-see attitude, and Beech, with few low-compression mounts in its stable, had almost no reaction.

In the meantime, FAA in April began a Directed Safety Investigation of the fuel problem, though the agency refused to label it a problem. By November, the investigation was still not finished.

(Until mid-summer, blue gas was "cause for concern." Later, it became "a clear maintenance problem." FAA never would admit a "safety problem" existed, so it didn't!)

Other changes took place during the year. Lycoming and Continental worked up "fixes" on their valves, but couldn't promise they would solve the fouling problem.

The companies started designing a new line of 100 octane engines, as exemplified by the 0-320 model that went into the 1977 Skyhawk.

In Texas, a company began marketing an anti-fouling additive called tri-cresyl phosphate (TCP). The firm said it would sell TCP even though FAA and the engine makers hadn't approved it.

At least two other companies designed conversion kits so some owners could put in 100 octane powerplants when their old ones failed (anywhere from 400 to 1,200 hours). More kits were sure to come.

The Experimental Aircraft Association injected real enthusiasm into the fray when it started flying a homebuilt on car gas. A shudder went through the industry when the little bird clambered to 20,000 feet without vapor-locking.

Engine makers hollered like stuck pigs about the dangers of car gas, but EAA got dozens of letters from airplane owners who used it for years without a problem.

As a side issue, the auto gas controversy should last for years to come. Airplane owners are best advised, however, to wait before filling up at the corner garage, because use of auto gas could give one legal and insurance trouble in the event of an accident, even if it doesn't seem to bother the engine.

Back in the mainstream, by the Fall of 1976 there were two strong efforts to find a national solution to blue gas.

One was a congressional subcommittee's probe. Whether this would develop into full-blown hearings remained to be seen, but it did put the aviation community on notice that it had better find an answer.

Second, and perhaps the most promising, was an effort to hasten decontrol of avgas. FAA Administrator John L. McLucas teamed up with Chairman Webster B. Todd of the National Transportation Safety Board to urge prompt decontrol.

Over at the Federal Energy Administration, officials said decontrol was coming anyway, but it was nice to have the Todd-McLucas letters to keep them on the track.

The decontrol theory states that if some oil companies were allowed, they would offer 80 octane avgas to others, so that everyone would get a chance to buy it. The rub is, the gas will probably cost more, maybe even more than blue gas.

But there are an estimated 130,000 airplanes out there with engines just

pining for 80 octane, and lots of owners say they would rather pay more for the gas than have to shell out for an early overhaul, or worse, have the engine quit some dark night.

So, depending upon the mood of the aviation community, the whole fuss could end up like one of those fencing matches where two sabre-wielding fanatics hack at each other for five minutes and the judge stops them to say, "Nothing done."

The 80 octane folks will have their gas back and all will be right with the world.

Except the gas will cost more. And some people will be out the cost of an overhaul. And others will have a collection of $9 spark plugs in their closets. And a lot of people will have worried about flying safely.

That's if the owners accept the judge's decision. The year 1977 will tell us if there really was "nothing done."

BFR: A TOWER OF JELLO

by Keith Connes

Reprinted by Permission
AIR PROGRESS (February 1976)

"And now for today's Grand Prize Question! What is it that flies . . . you can pass, but you can't fail . . . is an FAA reg, but has no specific requirements...and must be taken by most pilots, but not *all* pilots, prior to their acting as pilot-in-command of an aircraft? You have 10 seconds."

"Um, a chameleon?" "No. Eight seconds left." "Bowl of Jello?" "Nope." "Ooohh—wait, a Picasso drawing?" "No, I'm awfully sorry, time's up. It's a Biennial Flight Review!"

"A Biennial Flight Review? What's that?" "Who knows?" (Audience laughter, increasing to hysteria as contestant is hit in the face with a large custard pie.)

Personally, I am in favor of the BFR concept, and was one of its early proponents. It's possible to drift into a sloppy habit or two without even realizing it. Also, we tend not to practice such procedures as stall recovery or full flap go-around, and then if the real thing comes along, we're vulnerable. So it doesn't hurt to keep up the franchise by doing our stuff before a professional critic every so often.

But a funny thing happened on the way to Part 61. FAA decided, if you could call it a decision, that the BFR would be anything any instructor wanted it to be. Or, as stated in an FAA Advisory Circular, "There is no minimum amount of time required for a biennial flight review nor are there specific requirements for particular procedures or maneuvers that must be reviewed. We intend to leave these matters to the discretion of the person giving the flight review."

Isn't that nice? It means you don't get any assembly line BFR that looks like the Joneses next door. No sir, your BFR is as individual as your instructor's thumbprint. And that can be pretty individual, as I found out when I took my two BFRs recently. (You need to pass only one, but I was so fascinated with the experience, I took two.)

I got things started by calling my local airport and booking an instructor and a Cherokee 180. Then I got busy poring over the Cherokee numbers, since I had not flown for exactly one year. I also checked the FARs Part 91, which is just about the only specific requirement mentioned in the BFR Advisory Circular. I needn't have bothered.

At the scheduled hour, I was directed to the aircraft and told to pre-flight it while awaiting my instructor whom I will call Jack.

I had no sooner completed inspecting the plane when the instructor appeared and instantly established the rapport that was to cement our relationship. Me: "Hi, are you Jack?" Jack: "Uh." Me: "I'm Keith Connes." Jack: "Untie the airplane and let's get going."

I had somehow expected Jack to sit back and observe my flying, especially as I was checked out to rent that airplane at that airport, including, in their own notation, "day, night and strong crosswinds." But no, Jack told me how to start the airplane, how to taxi it and how to take it off. He also knocked my right hand down every time it strayed up to rest on the control wheel, an apparently lethal habit I'd picked up some 3000 hours ago.

The ride included a variety of stalls, 720s, hood work, etc. During the 1-1/2 hours we were up, I think I did some things right and some things badly. I know I was clumsy in tracking a VOR in a strong wind, but I don't know why. The only comments I got were terse put-downs, which didn't do much to improve my performance. By the time we landed, I was convinced I'd busted my first check ride ever. When we got back to the office, to my real surprise, he signed me off. Then he spoke, for the first and only time since touch down. "Cash or charge?"

Jack wrote up the charge, and even as he handed me the slip, looking elsewhere, I knew I had passed completely out of his consciousness forever. I shrugged, thanked the back of a head, and left.

Okay, I had paid my $43.65 and would be legal for the next 24 months. Fair exchange, right? Dammit, no! I had given up a chunk of my valuable time and money and hadn't learned a thing except to stay away from Jack. Legal or no, I'd been had. So I made an appointment to see the airport owner and told him of my experience. The owner seemed very sorry about the whole thing and apologized for the instructor's rudeness. I explained as well as I could that rudeness was just the tip of the iceberg. We'd had a total communications failure.

The result of this discussion was the offer of another BFR with another instructor, on the house. I accepted.

This ride was with a pleasant young man. He communicated very well, made some suggestions in a constructive way. But in spite of the contrast with Jack, I was surprised to find that he, too, started right out *instructing*. Like, "Okay, let's go over the checklist," and "Just crack the throttle." I still expected the man to sit back and see what I did with the airplane. Wasn't that what a BFR was supposed to be? Or was it?

I asked the airport's chief pilot. He didn't seem too sure. "We asked FAA for guidance on this and they said, 'just do whatever you think best.' Personally I think the pilot should demonstrate the basic maneuvers he took on his flight test. If he's a private pilot, he should demonstrate the maneuvers for the private. If he has a commercial rating, he should do an acceptable job on the commercial maneuvers. Also, I think the pilot should be quizzed on Part 61 and Part 91."

I left, somewhat puzzled. Neither instructor who rode with me had reviewed any commercial maneuvers (I hold a commercial license) and neither instructor had asked me anything about Part 61 or 91, which was probably just as well.

Finally, I called FAA in Washington and asked what *they* conceived the BFR to be. Answer: "First of all, it's not a flight check at all. If it were, it would have to be conducted by an FAA employee or designee. The instructor does not have the power to withhold the pilot's certificate. He simply acts for us as a knowledgeable evaluator of the pilot's proficiency, and therefore is given broad discretion."

All of which means the BFR you get depends on the instructor you get. It'll probably take between 1 and 1-1/2 hours, but I've been told they have run as much as 3 hours. The instructor can pass you but he cannot fail you. If he is satisfied with your proficiency, he notes in your log that your biennial was satisfactorily completed as of that date. If he is not satisfied, he makes no negative record in your log. You are then free to take another ride with that instructor (presumably after some brush-up) or go find another instructor.

Under these circumstances, it would seem like a good idea to shop before you buy. When you're arranging for your BFR, ask the instructor what he expects of you. You should be satisfied with the answer and you should make sure you get what you bargained for.

After all, the BFR can be whatever you and your instructor make of it.

Either a pain in the neck to stay legal or a fresh look at your flying to stay sharp.

You can still fly without a BFR . . . if . . . within the preceding 24 months you have logged one of the following flight checks:

a. Certification flight test for a pilot certificate or added rating.

b. Any proficiency flight check required under the FARs.

c. Pilot Examiner annual flight checks.

d. Proficiency flight checks given by a U.S. Armed Forces check pilot.

Or if you do not fly as pilot in command.

FOLLOW-UP

Spurred largely by the disastrous crash of TWA 514 at Dulles, the DOT Special Task Force recently released their report aimed at increasing aviation safety. Among other things, the Biennial Safety Review program came under fire, and the various organizations of the general-aviation lobby were quick to respond in their turn with a White Paper detailing their suggestions to improve the BFR program. In an effort to prevent *Air Progress* from looking like the *Congressional Record*, we locked Senior Editor Peter Lert into a windowless office cubicle with our copy of the White Paper and a typewriter; herewith, his condensation of its suggestions:

By now, discontent with the current state of the BFR program is manifest even at the highest levels, and the recent DOT Task Force report recommended that the BFR be standardized, at least to some extent, and that compliance with the requirement for the BFR be checked somewhat more stringently. As could be expected, the various noodles afloat in the Washington alphabet soup (i.e. AOPA, NPA, EAA, GAMA, NAFI *et. al.*) rose nobly to the occasion and came up with the following underwhelming suggestions, imposingly armored with obfuscating rhetoric:

1.) The instructor and student should discuss the BFR beforehand, in order to familiarize the instructor with the BFRee's experience and background, allow both parties to decide what is expected of each, and possibly find "... discernible, or potentially irreconcilable, conflicts of philosophies, personalities, etc.," at which time the supplicant can presumably depart in search of a more amenable mentor.

2.) FAR Part 91 should be reviewed orally. It is suggested that this review "... should not be conducted on a strictly test basis, rather it is an opportunity for the reviewing instructor to assist the pilot through a discussion of the regulations ..." Obviously, this can still range from a catechism-like session to the other extreme in which the BFRee sits mute while the instructor recites Part 91, perhaps with appropriate gestures.

3.) Preflight. "... an assessment of all those activities which the pilot would normally be expected to engage in prior to actually starting the engine of the aircraft." This, I suppose, could include everything from getting up and brushing one's teeth prior to driving to the airport; the Brain Trust mentioned above modestly suggest only that "... this phase of the Review could include, but not necessarily be limited to, an analysis of current and forecast weather, flight planning procedures, requesting and receiving reservations at high density airports *(truly germane for someone taking his BFR at, say, Bismarck, North Dakota–ED.!)*, and preflight of the aircraft in general and for the particular flight anticipated ..." They go on to extol the virtues of inclusion of the complete planning of an entire cross-country flight, but enjoin the instructor to "... render assistance by questioning, correcting, and instructing rather than testing."

4.) The BFR itself. Ah— we finally get to fly the airplane. Here, the suggestions given are valid: Emphasis on safety of flight should be stressed rather than pure training maneuvers; the ride should be based on the needs of each individual pilot, and should include the types of operation that he'll be performing; and some attention should be paid to emergency procedures, including inadvertent weather penetration, which is the official way of saying "flying into a cloud." Moreover, they suggest that particular attention be paid to ". . . those operations that seem to have the greatest tendency to get pilots into trouble." Exactly *which* operations these are has been left, alas, an exercise for the reader.

Mention is also made of the BFR's potential as an instructional, as well as evaluation, tool. After noting that the possibility of this instruction can be discussed before the flight by the instructor and student, the suggestions go on to say, ". . . it is generally accepted that one reviewer-performed 'demonstration' of proper technique is a realistic expectation *following* an initial marginal or unsatisfactory pilot-performed flight operation." (Italics mine.) This may be the most valuable suggestion to emerge from the whole seven close-spaced pages, and may go far to eliminate the current spread of instructor attitudes, which ranges from "demonstrating" the entire flight to sitting silent unless things get to the "Jesus—I've got it!" stage.

5.) Postflight discussion. ". . . This phase of the BFR is potentially the most important part of the entire evaluation process because it is here that the reviewing flight instructor gets a chance to discuss with the pilot what the entire review has revealed." Hear, hear. In fact, this holds true of *all* flight instruction, and far too many instructors and students share in the error of thinking that the lesson's over when the mixture is pulled to idle cutoff and the Hobbs meter stops ticking.

All in all, then, the BFR remains—at least for the moment—about where it's been since the program started. Assuming that these suggestions are accepted by those who run things at the Puzzle Palace on the Potomac, it *may* shape up to be the safety aid it was designed to be. In the meantime, if you've got one coming up, you're on your own with the instructor you select or are assigned. Good luck.

PITCH VS. POWER: WHAT'S RIGHT?

by JESSE F. ADAMS, CFII

Reprinted by Permission PRIVATE PILOT (October 1976)

It's been fun as well as highly informative, in the few years I've been in the flight instructing game, to sit through several FAA Flight Instructor Revalidation courses. There was a time not long ago when we avoided full stalls in flight training and on the flight test – required spins on the Private Pilot check ride were before my time–and we were exhorted to teach power for stall recovery. Since then, the "newer" principle is to reduce angle of attack with elevator together with power (if you have it). Now, when trainers approved for intentional spins are less common and so full of gyros the FBO doesn't want them banged about, we're encouraged to teach full stalls and even spins. Some of us always did anyway.

A couple of years ago, the official pronouncement was made at revalidation sessions that pitch is primary for altitude and power is primary for airspeed and the boys from OKC wouldn't even argue it, just made the flat statement, held our little heads, and tried to shove it down our throats without ketchup. The Word was that the Stark, 1-2-3 system, pitching to airspeed, won't work but 70% of the time. Just when or how or in what it won't work wasn't explained, but the message was, we should teach power to airspeed and pitch to altitude, always and forever so far as any of the team explained, and now we had a new official discovery. Ho hum.

Lately, there is grudging admission we should perhaps explain to the student the equivalence, in a way, between power, altitude and airspeed and how sometimes we can trade some of one for another, but we must still teach power to airspeed and pitch to altitude. Engines and gyros don't fail much anymore, it seems.

Through the twists and turns and shifts in policy, there has continued to be a lot of official anguish over the stall-spin accident rate.

Fellow instructors, students and sport fans, we are being had again. We are getting another half-explanation instead of the whole truth. Neither the old, gray flight instructor who nags pitch to airspeed *always*, nor the brand-new breed preaching the newest FAA "discovery" of power to airspeed, pitch to altitude is telling the truth–not all of it.

The truth is, both pitch and power control airspeed and altitude. If the rate of change of altitude remains constant, power determines airspeed, but if power is constant, pitch controls airspeed at the expense of altitude. Anyone who insists on one to the exclusion of the other without carefully qualifying the statement is, as they say in legalese, not making full disclosure. When someone who is, or should be, in full possession of all the facts tells only enough to convince you of his view, that's deception–to put the kindest construction on it–right?

The horns on this dilemma are that both statements are correct, IF, and both are wrong, IF . . .

An engineer, or one who pretends to it, will whip on you the airspeed vs. power diagram to prove power controls airspeed but nowhere does the label say that altitude, or rate of change of altitude, is constant, which it has to be for the diagram to be valid.

I'm not aware of any pitch vs. airspeed graphs, but let's make one up. To be valid, this one has to have a "Constant Power" stipulation. The shape of the curve, but not the trend, might vary with aircraft weight and airfoil. We haven't specified that altitude remains constant, and obviously it won't. An increase in pitch gains altitude, but at the sacrifice of airspeed until the airspeed account is overdrawn and then there is a rather abrupt loss of altitude. Pitch controls airspeed until either staff or terminal velocity is reached and then further pitch change in the same direction has little immediate effect on airspeed. The general principle can be seen: With power constant, pitch controls airspeed until we reach stalling speed or terminal dive velocity and then things come unglued.

Not being a mathematician or even an engineer, I can't quite formulate a law of equivalence with everything in a glib package, but a rough principle can be stated. With a constant value of any one of pitch, power, altitude or airspeed, changing another value will produce a change in the remaining pair, directly in some, indirectly for others, but not necessarily in linear proportion.

With power constant, airspeed varies inversely with pitch and altitude varies directly.

With airspeed constant, altitude and pitch vary directly with power.

With a constant rate of change of altitude (from zero rate to any other constant rate of change), airspeed is proportional to power and inverse pitch is necessary to maintain the rate of altitude change.

With constant pitch, both airspeed and altitude vary directly with power.

Pitch plus power equals performance, constrained by the limit of power available *and* where you are on the power vs. airspeed curve.

What does all this sum up to? Con-

fusion, no doubt, but conviction, I hope, that there is no simple, one line answer. There is such interdependence that change in any one factor affects the others. We are subjected to a bombardment of arguments on modern life and technology from environment vs. commerce to politics vs. ethics, but rarely do we get all the facts, only what it takes to huckster successfully, playing on the universal eagerness to grasp for oversimple explanations of complex problems. Sadly, few complexities can be expressed simply. If you don't believe it, try arguing the downwind turn question.

To oversimplify the pitch-power/airspeed-altitude interaction is as silly as to rely only on power for stall recovery. Sure, it works up to a point, but power isn't unlimited and unless the angle of attack is reduced somehow, whether with elevator or power or decreasing the load factor or any combination, nothing will get that wing unstalled and the airplane will come down when airspeed and power are insufficient. We prove this on every flight, one way or another.

When we fly on instruments, it's more convenient to use small increments of pitch to control rate of altitude change and we use power to effect small changes in airspeed, but we also have to accept the result of these small borrowings of airspeed in the first case and changing altitude in the second and a reckoning will occur just as surely as a loan company wants to be paid back. The payment may be delayed by the inertia of the machine, but ultimately, payment has to be made.

We should continue to require primary students in light airplanes to control airspeed with pitch in the landing glide and to adjust the rate of change of altitude with power, but then to prove we're dealing with two sides of the same energy coin, at the flare we check descent with pitch and trade airspeed for the maintenance of altitude until touchdown. Anything else will lead the student to believe he can dive at a field and land on it and we set the stage for overshoot or wheelbarrowing. In light airplanes, the acceleration is fairly prompt in response to pitch changes and the horsepower relatively small. In large, heavy aircraft with high wing loading and mucho horsepressure, it's awkward and probably dangerous to try to pitch down to gain airspeed when there's not much altitude to spare and the sink rate is already high, but then, these behemoths are also equipped with enough thrust to snatch them along the axis of thrust. When the student changes to that category of flying machine is time enough to change the emphasis.

There is nothing incompatible in using one method one time and the other for a different purpose or condition IF we make sure the interaction is well understood. We must not teach the use of either method exclusively for all situations. That 150-pound, self-powered computer, mass produced by unskilled labor, when programmed to be a pilot is still the best thing to control the airplane in all regimes of flight. He's probably cheaper, too, considering the price of avionics these days. The flight instructor who programs a student with only one method for all situations leaves the job unfinished and the computer axiom: "Garbage in, garbage out" will shortly be demonstrated. I'll tell you something else. The bloodless machine spewing out predictions from Weather Central isn't as good at it either as the old-fashioned weather man who was in the business because he understood and loved it.

Meantime, flight instructors, hang in there. Tell 'em the whole story and strive for full understanding by your students. Someday the FAA will "discover" another new (old) principle and the party line will change and they'll rewrite the books and try to pretend it was never any different, just like before.

Once, when the FAA evangelist said there was no such thing as "getting an airplane on the step," his handout—would you believe?—had a graph proving there was a step. The fancy footwork went: "Well, it's so small a difference it's hardly there and, anyway, that graph will darn well be redrawn before we go on the road again." C'mon, fellas, we're fond of you and we have a lot to learn together, and everything Mother said isn't necessarily so either.

DEBATE RAGES OVER QUESTION OF "CONSTRAINTS ON GENERAL AVIATION"

Reprinted by Permission
FEDERAL AVIATION AGENCY
and
AIRCRAFT OWNERS AND PILOTS ASSOC.

One of the more significant controversies in the area of General Aviation concerned the question of Federal constraints on the growth of General Aviation. The points of discussion can best be seen in two speeches by Mr. F.A. Meister, FAA Acting Associate Administrator for Policy Development and Review, and an open letter written by Victor J. Kayne, Senior Vice President, Policy and Technical Planning, Aircraft Owners and Pilots Association. Portions of that discussion follow.

THE NEXT 20 YEARS

A presentation by F.A. Meister
at
20th Annual Meeting and Technical Program
Air Traffic Control Association
Washington, D.C.
October 29, 1975

I have been asked to expand upon certain comments made at the FAA's Aviation Review Conference in Washington on May 19. Specifically, on those comments which I made relating to possible future constraints on aviation growth—especially General Aviation growth.

I shall be happy to expand on that topic, but first, let me state that my overriding concern then was, and still is, our ability and willingness to respond to change.

Aviation, along with the rest of our society, is experiencing rapid, unforeseen and often misunderstood change. To respond to these events, aviation must pool its considerable talents and expertise to reach decisions and to attack in a forthright and comprehensive fashion the issues of today. Above all, we must face the fact that avoidance of issues is decision making—but the worst kind—decision by default. The problems and issues facing us are fundamental, both near and long term. Critical self-evaluation combined with positive action is necessary for effective response to change.

Now, that is the context in which I posed in May the question about aviation growth.

I want to talk about the specific factors that have steered me to the conclusion that we should even have to address the possibility of future constraints.

Today's discussion will focus on three major points. First, that the earlier reference to possible future constraint was understated. We should be talking instead about possible continued and increasing constraint. Second, that no obvious panacea exists for today's forecasted demands let alone the increased demands of tomorrow. And finally, without the benefit of our total involvement with this problem we may be forfeiting our part in framing the ultimate solution.

Current and Continued Constraints

In discussing current and projected system constraints, I want to draw your attention to some very interesting statistics related to General Aviation activity.

In a recent study of the 25 largest air carrier airports, G.A. accounted for only 24 percent of all operations. On a national basis, however, G.A. accounted for 75 percent of total operations at all FAA towered airports. It is presently forecast that by 1985 national G.A. activity will climb to 81 percent of total operations at towered airports while at the 25 largest air carrier airports, G.A. activity will drop to 21 percent of total operations.

But let's look at the situation from a more microscopic view. In the New York area, for example, La Guardia Airport has displayed the following trend in G.A. activity as measured in terms of the percentage of total activity:

Mr. F. A. Meister Photograph courtesy FAA

Meanwhile, the large, adjacent, all-G.A. Teterboro, New Jersey Airport was similarly showing declines in activity from 265,679 operations in 1965, to 209,772 operations in 1973.

But if G.A. activity is growing on a national basis, where is it going? —And why is it going there? Closer inspection shows that G.A. activity is shifting in two ways. First, away from the larger TCA restricted air carrier airports (as was evidenced by the La Guardia Airport statistics) and second, beyond that, "personal use" G.A. activity is being pushed out of major G.A. airports by rising costs of operations and maintenance which, generally speaking, only the business and corporate G.A. interests can absorb. This hypothesis is confirmed by looking at a further breakdown of Teterboro activity. "Local" G.A. activity dropped from 60 percent of all G.A. activity in 1965 to 22 percent in 1973. These figures do not represent isolated instances but instead depict the situation we face across the country.

The same trend is evident at other large hubs. For instance, at Chicago O'Hare in 1955 G.A. activity accounted for 33 percent of total activity and in 1973, only 6 percent; at Atlanta, G.A. activity was 48 percent in 1955 and only 11 percent in 1973; and at Miami, the figures show G.A. at 30 percent in 1955 increasing to 48 percent in 1965 but dropping back again to 28 percent in 1973.

In effect, the system, either by virtue of operational (safety related) requirements or by basic economics, has constrained the nature of market demand by causing a shift of "personal use" G.A. interests from the major air carrier and major G.A. airports to the less costly, less operationally restricted rural areas. This trend in turn has influenced the near doubling of U.S. airports from 6,426 in

1960 to 12,700 in 1975. In view of heightening environmental concerns and spiralling costs of new airport construction, one can only wonder how long this source—or any source—of relief can be maintained.

To see where G.A. is going, we need only to reflect on a few more statistics for medium and small hubs. In the medium category, we find, for instance, at Omaha, G.A. activity amounted to 60 percent of total traffic in 1955 and 75 percent in 1973; at Salt Lake City, 52 percent in 1955 and 64 percent in 1973; at Milwaukee, 38 percent in 1955 and 64 percent in 1973; at Birmingham, the percentages for the same years are 37 percent and 68 percent; at Des Moines, 67 percent and 82 percent; and at Tucson, 59 percent and 69 percent. Similar comparisons can be drawn for small hubs: at Fresno, California, the 1955/1973 G.A. percentage comparisons are 43 percent and 84 percent; at Grand Rapids, Michigan, 60 percent and 84 percent; at Daytona Beach, Florida, 62 percent and 92 percent; and at Youngstown, Ohio, 27 percent and 82 percent.

Faced with this problem, FAA and the aviation community must explore sensibly and without emotion the twin forces of cost-consciousness and constrained or shifting growth. Until we have been assured that marginal system improvements have been carried to their cost-effective limit, we must live within the confines of the basic system. As a result of this cost-conscious approach, we have come to explore the possibility that in the long term we may not be able to continue increasing system capacity to simultaneously accommodate unconstrained growth in demand, maintain reasonable prices of entry into the national aviation system and retain the absolutely necessary high degree of safety.

Relief from Future System Improvements

But what can we expect from future system improvements to relieve constraints (measured in delay) after the current system has taken us as far as it can?

In its report of June 15, 1973, the House Appropriation Committee directed the FAA to investigate the effectiveness of its R,E&D and F&E programs in producing capacity and productivity increases. As a result, a special study was conducted of the following eight major airports:

Atlanta
Chicago O'Hare
Denver
Los Angeles
Miami
JFK
Philadelphia
San Francisco.

Those eight airports were chosen because they had consistently been responsible for well over 60 percent of all aircraft delays.

Among the assumptions of the study are included:

- A modest growth in traffic over the next decade back to or above the 1968-1970 level.
- Extensive use of wide-body aircraft.
- In the 1976-1978 time period, variable IFR longitudinal separation standards on final approach based on wake vortex conditions (2 n.m. minimum).

In the 1979-1982 time period, ATC improvements developed in the UG3RD ATC system including:

- advanced metering and spacing,
- MLS,
- 2,500 ft. simultaneous parallel IFR approaches, and

- airport surface traffic control systems.

Based on the above and other additional assumptions and improvement, it was concluded in the study that:

> ". . . (delays) remain undesirable and substantial increases may be expected in the future if the improvements to the system are not instituted."

But since that study was developed, a couple of key events have taken place. First, that "modest" growth now looks more like a national increase of almost 50 percent in operations from 1970 to 1980. Second, considerations of fuel and operating costs have caused air carrier operators to favor other than the wide-body aircraft. Third, just a couple of weeks ago the FAA announced that on November 1 it was instituting a 6- and 4-mile separation standard—for small aircraft landing behind "heavy" and "large" aircraft, respectively. This is a far cry from the 2 n.m. minimum separation referenced as an assumed feature of the 1976-1978 ATC system. The 6/4-mile standard is being instituted because of findings of recent studies on the potential impact of wake turbulence. The capacity/delay impact of this action is not yet fully determined.

In view of these facts, is it unrealistic or excessively controversial to evaluate the implications of continued or increased demand constraints of one kind or another? I think not! Under the circumstances we would be delinquent if we did not evaluate them.

But, if unconstrained demand cannot be met in a cost-effective, safe manner, what kind of constraints should be applied?

Regulatory: Exclusive airspace usage?

Economic: Peak hour pricing of landing fees?

Operational: Curfews? or quotas?

These are the questions we must come to grips with if we are to be ready with the answers when needed.

External System Constraints

And let us not remain oblivious to elements external to the aviation system which may also impact on the nature of future demand.

As an example, the Federal Energy Administration's Office of Transportation research is about midway through a three-year, $3.7 million study program to explore transportation energy use patterns and alternative government policies to implement transportation energy conservation measures. To summarize the FEA approach to air transportation, we see a search for alternative means to induce shifts away from air travel in general with a particular focus on short-haul air transportation. Specific policy strategies presently being studied include:

- rail and bus fare subsidies,
- air fare surcharges, and
- automobile fuel pricing and speed limits.

Should any of these specific policy strategies advance to the point of implementation, aviation-interested groups had better be ready to address the potential impacts of those strategies on the system.

But while we're keeping one eye on those "external elements," we must also be addressing the basic internal policy questions facing aviation. Specifically:

- Should the federal government continue an active policy of promotion of aviation, especially General Aviation?
- Should more effective policies be

adopted which would shift demand of airlines to off-peak hours and General Aviation away from major hub airports?

- How far should cost-consciousness go and at what point does it conflict with safety, capacity and efficiency?

I do not claim to have all the answers to these questions, nor do I suggest knowledge of all the questions.

What is obvious, however, is that resolution of these problems requires everyone's input: the users, state and local governments, the system operators, and just plain John Q. Public.

Together, I'm sure, we will find the right path.

AN OPEN LETTER TO MR. FRED MEISTER, FAA ACTING ASSOCIATE ADMINISTRATOR FOR POLICY DEVELOPMENT AND REVIEW, CONCERNING POSSIBLE FUTURE CONSTRAINTS ON GENERAL AVIATION GROWTH

from
Victor J. Kayne, Senior Vice President
Policy and Technical Planning
Aircraft Owners and Pilots Association
April 1976

In speeches to the FAA Aviation Review Conference on May 19, 1975 and the Air Traffic Control Association Annual Meeting on October 29, 1975, you, Mr. Meister, spoke at length on possible future constraints on aviation growth, and especially on the growth of general aviation.

While you carefully did not advocate such constraints, you nevertheless treated the subject in such a manner as to leave no doubt in the minds of your listeners that in fact, you were advocating such constraints on behalf of your employer, the Federal Aviation Administration and the Department of Transportation. The Department of Transportation has denied that you were advocating constraints and classified your talk as posing a hypothetical question of how constraints should be imposed if unconstrainted demand cannot be met in a cost-effective and safe manner.

The first speech on May 19 drew a considerable amount of criticism to the point that you felt obliged to expand on those comments in your speech on October 29. In the latter, you cited a number of statistics to support your theory that certain changes were taking place with regard to general aviation and air carrier use of many of our major and smaller airports. We have examined up-to-date statistics for the airports that you cited and get an entirely different reading. We will expand on that later in this letter.

There are a number of points that you failed to mention in your speeches that have a bearing on this subject.

First, the Administrator is directed by the Congress (Section 305 of the Federal Aviation Act of 1958) "to encourage and foster the development of civil aeronautics and air commerce in the United States and abroad." Section 306 becomes even more specific by stating "In exercising the authority granted in, and discharging the duties imposed by, this Act, the Administrator shall give full consideration to the requirements of national defense, and of commercial and general aviation, and to the public right of freedom of transit through the navigable airspace."

FAA budgets have been notable for their lack of any provision "to encourage and foster" the development of general aviation. The FAA's main effort has been to regulate rather than foster and encourage. Further, many of the FAA require-

ments imposed on general aviation come dangerously close to infringing on the public right of freedom of transit through the navigable airspace.

In your speech of May 19, 1975, Mr. Meister, you stated that "the FAA is guided by the Administration's intent" in relation to federal/state functions, and you then couple this with a question regarding the proper role of the federal government to promote and protect healthy interstate commerce. Since when does the "intent of the Administration" contravene a legislative mandate of the Congress?

Next, civil aviation, and especially general aviation, has been operating under an ever-increasing burden of constraints imposed by the Federal Aviation Administration. These constraints are both operational and economical. They include items such as the requirement for various pieces of electronic equipment as the "price of admission" to certain areas of airspace, imposition of time-consuming air traffic control procedures that destroy the utility of the aircraft and consume inordinate amounts of precious fuel, and imposition of unduly restrictive requirements concerning licensing and current competency of flight personnel. Thus, to infer that we are operating in an atmosphere of unconstrained demand is fallacious. We have been operating under rather severe constraints for some time.

One of the constraints that is becoming more troublesome as time goes on is the philosophy of FAA that the users must serve the ATC system rather than the system serving the users. Everyone who flies will agree that reasonable rules must prevail, but the dictatorial air traffic control decisions are imposing increasing constraints on general aviation operations.

Those who have written letters of protest to their Congressmen about your speeches, Mr. Meister, have received letters in return from the FAA through the Congressmen referring to the ever-increasing traffic at the major airports as a reason for imposing constraints. Yet, FAA's own traffic reports indicate quite the contrary. In your October 29, 1975, speech you cited traffic figures comparing 1955 with 1973 for some strange and unknown reason. You might just as well have gone back to the days of the Wright brothers, when there was no aviation except general aviation.

In 1955, the airlines were still operating more propeller aircraft than jets. Many cities that had airline service from the major carriers were just beginning to get the first inkling that the big jets were best at long haul flights and that intermediate stops were going to be dropped by the trunk carriers. As smaller cities were dropped from the large carrier schedules, scheduled air taxi began to develop in order to provide the needs of these communities for air transportation. It should be noted that the air taxi services, later blooming into commuter airlines, had to operate into the major hubs in order to give the small community passenger access to the large city and to the long haul trunk carriers.

Those communities that lost their service from the major carriers can be thankful that the Meister philosophy of constraining the growth of general aviation was not in effect in those days, otherwise they would have been completely cut off from air transportation.

The advent of the wide-body and jumbo jets further reduced the number of flights by the major carriers. Then, the fuel embargo and the subsequent rise in fuel prices reduced both air carrier and some types of general aviation operations. Business aviation became more attractive as convenient schedules disappeared. While commuter and scheduled air taxi filled the gap in providing service to those communities where the larger carriers had dropped schedules or cut off service completely, the general ratio of air carrier to general aviation operations had not changed significantly in recent years. The significant fact is that many airports have lost a considerable amount of their total traffic. Some of these airports are having tough sledding economically. The last thing that they need is further reduction of their traffic through constraints on civil aviation.

In looking at the FAA traffic reports for times more modern than 1955, we find that the peak year for air carrier traffic was 1969. It also was the last full year before imposition of user taxes under the Airport-Airway program. Comparing that year with 1974 (the latest full year the FAA has published statistics) tells an entirely different story than that you attempted to show by using the 1955 base. For example, the ratio of general aviation to air carrier does not change dramatically. Further, of the 25 busiest airports in terms of air carrier traffic, 23 of them had a decrease in air carrier traffic in the six-year period. The overall decrease was 13.8%, with some airports losing more than 40% and even 50% of their airline traffic. Only two of the top 25 gained in air carrier traffic. These were Atlanta and Denver.

Considering the total number of FAA control towers, the average annual number of air carrier operations per tower fell from 33,320 in 1969 to 22,890 in 1974.

You spoke of the great decrease in "local" general aviation activity at Teterboro, but you failed to mention that much of that decrease was due to the prohibition by airport management of touch-and-go landings. Increased airport fees and the inconvenience imposed by the New York TCA have also taken their toll. However, there is another very interesting aspect of the Teterboro operations. In 1969, general aviation comprised 99% of the total operations, with air carrier being .05%. In 1974, general aviation showed a loss of 3% from 1969 and comprised 96% of the total traffic. Air carrier traffic gained 696% during the same period and increased to 4% of the total. The overall gain of traffic for the airport was .05%.

Taking a look at a few of the top airports from the viewpoint of air carrier operations and the relationship to general aviation, we find the ratios between air carrier and general aviation quite different than described in your speech. In comparing figures for 1974 with 1969, we must keep in mind that air taxi and commuter operations are now counted separately from general aviation, whereas in 1969 they were included in the general aviation count. Thus, they are added in our 1974 figures to provide a comparable baseline.

For O'Hare in 1969, the ratio of air carrier to general aviation operations was air carrier 93% and general aviation 6%. (In cases where the percentages do not add to 100%, the difference is in military operations.) In 1974, the ratio was air carrier 79% and general aviation 20%. Overall, in the six-year period, the air carriers lost 17.1% of their traffic and general aviation gained 31%. The total airport traffic decreased by 1.6%.

The story for Los Angeles shows the ratios remaining nearly static, but with a dramatic 25% overall loss of traffic for the airport. Air carrier was 72% in 1969 and 74% in 1974. General aviation was 26% in 1969 and 25% in 1974. The air carriers lost 22.7% of their traffic and general aviation lost 29%.

John F. Kennedy in New York was much the same as Los Angeles, with the ratio between air carrier and general aviation from 1969 to 1974 remaining about the same, but with the airport suffering an overall 21.7% loss of traffic. Air carrier was 86% in 1969 and 83% in 1974. General aviation was 14% and 17% for the same years. There was a 24.2% decline in air carrier traffic during the period and a 4.6% decline in general aviation.

Atlanta was one of the top 25 air carrier used airports where a gain was registered, although the air carrier/general aviation ratio remained static with air carrier remaining at 84% and general aviation going from 16% to 15%. Air carrier traffic increased 13.1% and general aviation gained 8.3%. In the same period, Fulton County (Charlie Brown Airport) lost 12.5% of its general aviation traffic and Peachtree DeKalb gained 4.7% in its general aviation count.

Miami was another of the top 25 that suffered a significant loss in total traffic, going down by 21% from 1969 to 1974. The air carrier ratio was 68% in 1969 and 70% in 1974. General aviation was 32% and 30% respectively. The air carriers lost 7% while general aviation lost 27%. In the same period Tamiami gained .2% and Opa Locka .5% in general aviation traffic.

At LaGuardia, the air carrier ratio was 78% for both 1969 and 1974, with general aviation being 21% in 1969 and 22% in 1974. The air carrier loss was .4% and general aviation gained 6% during the period. Overall loss of traffic for the airport was 2%.

It may be seen from the above that the ratios of air carrier to general aviation at the major airports are not changing greatly contrary to your out-of-date statistics. A significant factor of note is the overall decrease in traffic at these airports, which also refutes the FAA bromide of "ever-increasing traffic" at the major air carrier airports.

You stated that there was a shift of "personal use" general aviation from the major air carrier and major general aviation airports to the less costly, less operationally restricted rural areas. This, of course, is fallacious. Some traffic has shifted to more convenient urban airports, but you cannot classify Opa Locka, Tamiami, Addison and Peachtree DeKalb as "rural" airports. The person who lives in a large city will keep his airplane at the most convenient airport that offers some

measure of service without too much economic and operational penalty. But he won't go so far as to base it 100 miles away in a rural area.

You cited this fallacious argument as the cause for the number of airports doubling from 6,426 in 1960 to 12,700 in 1975. (FAA figures show 13,062 for 1974.) Of the 13,062 airports recorded by FAA for 1974, 8,487 are privately owned. Of these, 2,888 are public use, which, when added to the 4,211 public use airports of the 4,575 publicly owned airports, gives 7,099 for public use. The great increase had been in privately owned airports. This could be caused by many factors, including lack of publicly owned airports (due in part to FAA's own policies), poor management and bad pricing policies at some publicly owned airports, or not having an airport convenient to an aircraft owner who has the land available to build his own.

You also brought out statistics to show the trend for general aviation at the medium and small hubs. Again you used 1955 figures as a base.

In the medium hub category, Omaha's air carrier ratio in 1969 was 20% and 26% in 1974. General aviation was 78% in 1969 and 74% in 1974. This contrasts with your figures that showed the general aviation ratio going from 60% in 1955 to 75% in 1973. During the 1969-1974 period, Omaha's air carrier traffic dropped off 14%, general aviation dropped 36% and the overall drop in total traffic for the airport was 32%.

Salt Lake City was much the same with a loss in total traffic of 27%. The air carrier ratio was 22% in 1969 and 26% in 1974. General aviation was 71% and 69% respectively. The air carrier loss in traffic was 17% and general aviation lost 29%.

Another of your examples was Milwaukee. The air carrier ratio was 29% in 1969 and 33% in 1974. General aviation was 65% and 69% respectively, as contrasted with your ratios of 38% and 64%. The airport lost 11% of its total traffic during the period, with air carrier losing 1%, general aviation losing 15% and military losing 20%.

Birmingham was another big loser with 25% of its total traffic disappearing between 1969 and 1974. The ratios were air carrier 21% and 23%, general aviation 66% and 69% and military 13% and 7%. Loss in each category was air carrier 18%, general aviation 21%, and military 58%. Again, your ratios for general aviation were 37% and 68% using your obsolete figures.

The next medium hub example that you used, Mr. Meister, was Des Moines, which was one of the few to show an overall gain. Its total traffic went up 5%. The ratios for air carrier in 1969 and 1974 were 19% and 15%. General aviation was 73% and 82% while military was 8% and 3%. The air carriers lost 19% of their traffic, general aviation gained 18% and the military lost 55%.

Tucson was another gainer with an increase in total traffic of 17%. The ratios did not change much. Air carrier was 18% and 14%, general aviation was 73% and 71%, and military 9% and 14%. The air carrier traffic declined by 10%, general aviation gained by 14% and the military gained by 86%.

In the small hub examples that you gave, Fresno was another loser with 19% of its total traffic disappearing between 1969 and 1974. The ratios were 12% and 9% for air carrier, 77% and 84% for general aviation and 11% and 7% for military. The loss in each category was air carrier 36%, general aviation 12% and military 51%. Again, the change in the general aviation ratios is quite different than your 43% and 84%.

Grand Rapids was another gainer in total traffic—6%. The ratios were 19% and 13% for air carrier and 81% and 86% for general aviation. The air carrier loss was 26% and the general aviation gain was 13%.

Daytona Beach was another example of the ratios not changing greatly. For air carrier they were 8% and 6%. General aviation was 92% and 94%. Air carrier traffic declined by 26% and general aviation gained by 1%, with the airport having a loss in total traffic of 1%.

Youngstown was another big loser with a decline of 25% in total traffic. The ratios were for air carrier 13% and 8%, 79% and 83% for general aviation and 8% and 8% for military. Air carrier lost a whopping 52% of its traffic, general aviation lost 20% and the military lost 22%. The ratios for general aviation remained rather constant in contrast to your example of a 27%-82% spread using the 1955 figures.

So much for the "trends" that you were conjuring up by using 1955 figures instead of more up-to-date statistics that reflect the real world today.

Up to this point, we have been dealing at face value with the air carrier and general aviation traffic counts reported by FAA control towers. However, a disturbing factor was revealed in our research of your statistics, which appears to negate some of your argument regarding ratios between general aviation and the air carriers at selected airports. It also completely destroys your arguments on cost allocation and your charges that "general aviation is not paying its fair share" of the costs of services and facilities provided by the FAA. The cost allocation process is based on traffic counts made by FAA facilities in large measure.

In our pursuit of comparable traffic figures, we discovered the traffic count for the Miami Dade Collier airport. This is the big airport that is used almost exclusively by the air carriers for training purposes. In 1974, your tower recorded 2 air carrier operations and 50,101 "general aviation" operations. According to your cost allocation thinking, this airport and tower should be paid for entirely by general aviation. However, those "general aviation" operations are almost all by two-, three-, and four-engine jets carrying the name of various airlines on their side. On questioning FAA field people in several regions, we were told that all air carrier operations such as training, crew qualification, ferry, flight test and other nonrevenue flights were put in the general aviation count and not counted as air carrier.

Miami International Airport has a large volume of air carrier training activities, in addition to the airline planes that come and go in the process of using Dade Collier. All that traffic is "credited" to general aviation.

Thus, you have inflated the general aviation count with untold thousands of operations by the air carriers and expect general aviation to pay their "fair share" to support these operations.

Another question that is a natural consequence of the rather large traffic drops shown for many of the airports is whether the FAA has reduced the staffing and grades accordingly. We have read where tower complements have been upgraded when they reached certain volumes of traffic, but we have seen nothing with regard to towers that have suffered a significant loss in traffic. Are the taxpayers being charged for tower complements that are out of date?

In summary, Mr. Meister, we commend to you a careful reading of the statutory responsibilities imposed on the Administrator by the Congress. We also suggest that you and your staff start planning how to handle a healthy and growing national asset rather than trying to strangle it with constraints. The FAA has spent billions on the air traffic control system with no real improvement in capacity. In fact, most of the "improvements" result in either increased cost or reduction in capacity. General aviation has grown without your help and despite the constraints imposed by the FAA. If more constraints are needed, the place to impose them is on the FAA and its staff.

GENERAL AVIATION – THE OUTLOOK

Opening remarks only of presentation made by
Mr. F. A. Meister at
Aviation/Space Writers Association
National News Conference
Denver, Colorado
May 17, 1976

It's a pleasure to have the opportunity to expand on certain remarks I have made in the past regarding the future of General Aviation. It is particularly important for me to do this in order that certain distortions of those past remarks be corrected once and for all. To illustrate the general impression these distortions have created, let me read you the opening line of the invitation I received which brought me here today.

> "Dear Mr. Meister: Your thoughts on the necessity to constrain the growth of General Aviation, as reported in the aviation media, have generated considerable interest to say the least."

At the very outset I want to state that neither I nor any other responsible FAA representative advocates the constraint of any element of aviation. In order to lend supportive evidence to that statement, I want to review for you some of the comments and observations concerning General Aviation that I have made in the past. Then I will discuss the related FAA activities which have gone forward since that time. And finally, the associated needs of the future will be explored as we at FAA now perceive them.

Prior Observations Regarding General Aviation

At last year's Aviation Review Conference in Washington, D.C., the subject of "Major New FAA Policy Directions" was presented. In that presentation, a number of "driving policy forces" were discussed including:

- The national aviation system as an open system.
- This force is caused by the "fish bowl" type of existence the FAA finds itself in today. Thus, every action taken, be it programmatic or regulatory, is viewed critically from many viewpoints. Although much good has come from this new force, the by-product too often has been delay in actions taken.

Other similar policy-impacting forces were treated, such as:

- New federalism, or the relinquishment of federal control to state/local authorities whenever possible.
- The philosophy of cost allocation.
- Increased system capacity demand.
- Diminishing returns from technology breakthroughs (thereby limiting opportunities for system capacity expansions).
- Cost consciousness in government.
- The concern for the environment.
- The continued demand for system safety.

It was pointed out at that time that certain of these "forces" do not always pull in the same direction. As an example, the issue was raised concerning the relinquishment of certain federal airport responsibilities to state or local authorities. The tenets of *new federalism* would act to support such relinquishment while proponents of the federal policy commitment to safeguard our environment might see an abrogation of that commitment by such action.

A further comparison was made regarding the forces of cost-consciousness and increased system capacity demand. The notion of these two interacting and opposing forces was the basis for the question regarding the proper federal role in the promotion of aviation, especially General Aviation. The aviation community was urged to think about these opposing forces and provide their thinking as to the tradeoffs between them.

Following the Aviation Review Conference, little response was received from the General Aviation community on these very serious issues. So, in an October presentation before the Air Traffic Control Association I repeated the message regarding these opposing forces and their future impacts on General Aviation. Here, however, evidence was presented which indicated that, "in effect, the system, either by virtue of operational (safety related) requirements or by basic economics, has (already) constrained the nature of market demand by causing a shift of personal use G.A. interests from the major air carrier and major G.A. airports to the less costly, less operationally restricted rural areas." In light of this evidence, the question was again asked as to what the appropriate FAA role should be regarding aviation promotion.

After that presentation, we started to get some of the response we had sought. Unfortunately, some of this was accomplished in a reactive, emotional manner by certain of the General Aviation respondents. From AOPA, for example, came the following:

> "While you carefully did not advocate such constraints, you nevertheless treated the subject in such a manner as to leave no doubt in the minds of your listeners that, in fact,

> you were advocating . . . constraints . . ."

While it is difficult to argue with such subjective remarks, it must be done. I again repeat, I did not explicitly or implicitly advocate constraints. What was suggested, however, was the possibility that unbridled demand might one day exceed FAA's ability to meet that demand in a *cost-effective, safe* manner. Accordingly, a shift in, or constraint on that demand might be one viable solution.

Again from AOPA:

> "One of the constraints that is becoming more troublesome as time goes on is the philosophy of FAA that the users must serve the ATC system rather than the system serving the users."

This has never been a philosophy of FAA. What *is* an FAA position, however, is that the users should equitably pay for the system service they receive.

Another quote from AOPA:

> ". . . the Administrator is directed by the Congress (Section 305 of the Federal Aviation Act of 1958) 'to encourage and foster the development of civil aeronautics and air commerce in the United States and abroad.'"

The Administrator is charged with other concerns as well, e.g., the safety of aircraft and the efficient utilization of airspace. As has already been pointed out, the "promotion" aspects do not necessarily complement these other concerns at all times.

All the comments received were not as negative but instead provided *constructive criticism* while agreeing with the FAA observations. From EAA, for example, came the following:

> "Mr. Meister said recently that government constraints on General Aviation have a significant effect on its growth. We here at EAA headquarters have recognized this for many years. He further states—and we certainly agree with him—that General Aviation is being pushed out of major airports by rising costs of operation and maintenance, and in addition, the air traffic control system has constrained the nature of the demand by causing a shift of personal use aircraft from the major air carrier and General Aviation airports, to the less costly and less restricted rural airports."

This commentor continued from this point to express this organization's strong objections to the process that is causing this to happen, but as can be seen, he does not claim that the problem does not exist or that the FAA observations are wrong.

So far, I have only expanded upon my past statements and have responded to specific reactions by General Aviation industry representatives. Accordingly, these comments have "plowed no new ground" on the very serious subject of General Aviation. What has been reconfirmed, however, is something that all sides agreed to in the first place. Namely, that General Aviation has *special concerns*. Beyond that, the causes and solutions to those concerns remain hotly disputed.

WHIPPING BOY

by Robert J. Serling

Reprinted by Permission
FLIGHT OPERATIONS MAGAZINE
(June 1976)

(Robert J. Serling is one of the universally recognized, competent and respected aviation writers of our generation. In a speech before a recent Flight Safety Foundation seminar for pilots, he pinned down some meaty issues that should be the concern of all U.S. civil aviation, specifically the business flying community. Here are excerpts that deserve wide distribution.)

In all the years I've been proudly associated with aviation, I can't remember seeing so much garbage, so many phony accusations, such an over-supply of glib over-generalizations, all directed against every phase of aviation from the airlines down to the private pilot.

Headline-hunting, vote-scrounging politicians, so-called experts weaned in ivory towers and suddenly ensconsed in positions of bureaucratic power, short-sighted exponents of ecology, make up the breed who keep insisting aviation isn't really important . . . that the airlines are a plaything for the jet set . . . that business/corporate aviation is for fat-cat industrialists who regard a company airplane as their own expensive toy . . . that general aviation consists largely of weekend hot-rodders with wings instead of motorcycles and drag races.

These are the real enemies of aviation. Some of them are well-meaning, perhaps, but the majority feed from the trough of their own blind prejudices, from an insatiable appetite for both publicity and power, and finally, from just plain ignorance. Those in the latter category aren't helped by what they so frequently hear and read from the communications media—the news media of which I was a part for more than 30 years.

I worked for two daily newspapers and was with the United Press for almost three decades. From my first job as a cub reporter in 1936 to the day I resigned from UPI in 1966, I was weaned on the very cornerstones of a free press—objectivity, fairness, accuracy, truth. In simplest terms, keep your own opinions out of your stories and leave the editorializing up to the editorial page.

Do we in aviation get a fair shake from the news media of today?

We most assuredly do not!

Not when one of the nation's supposedly great newspapers, the *Washington Post*, devoted many thousands of words against the proposed U.S. supersonic transport, quoting in full every anti-SST statement uttered by anyone, regardless of their qualifications as aviation experts, running almost daily anti-SST cartoons, printing columnist after columnist with an anti-SST axe to grind, publishing every anti-SST letter that came in to the editor (and not printing, incidentally, any of the 10 or 15 I wrote myself defending the project). All this avalanche, this flood of anti-SST propaganda. And yet when Bill Magruder, who once headed the SST program, held a news conference to defend the SST against some of the charges being made against it, the *Post* didn't bother to send a reporter around to cover it.

The *Washington Post* is hell-bent for deregulating the airline industry. Fine. That's its privilege. But when Eddie Carlson of United Airlines made a speech in Washington explaining what was harmful in deregulation, the *Post* didn't print a single word of what he said.

These incidents are par for the course these days when it's so fashionable to be either indifferent toward aviation or outright hostile. Why? Partly because of the new breed of media journalists, those enamored of so-called in-depth reporting, but who dig only as deep as their own preconceived beliefs. The bulk of them don't know the first thing about aviation, but they know, or think they know, a lot about those gold dust twins of emotionalism—ecology and consumerism. You could count on your hands the number of fulltime aviation-writing specialists employed by large metropolitan newspapers today.

Guys like myself who understand aviation, who respect it and who feel it only fair to write about its virtues as well as its faults, are a dying breed. We've been replaced by the would-be crusaders, the followers of the Ralph Nader cult, the not-too-far-removed-from-journalism-school reporters who cover virtually every aviation story as if it were a criminal trial. Only 30 percent of the membership of the Aviation Writers Association is composed of the working press. AWA has never been able to get younger men and women interested in joining, which is easy to understand: most of them cover aviation only sporadically. And when I call the true aviation reporter a dying breed, I'm not referring to the state of our blood pressure and clinical health. I'm talking about an attitude that's dying out—one of fairness toward aviation and its problems.

In this kind of an atmosphere it is only too easy for the William Proxmires of politics to flourish. It was Bill Magruder who once said of this leading exponent of hair transplants: "If God had meant man to fly, he would never have invented William Proxmire." Amen.

If I had a dollar for every word of untruth, half-truth, or outright deliberate falsehood Proxmire has uttered about aviation, I could retire.

But Proxmire isn't the only offender. He's more of a symptom of the anti-aviation disease, the virus that causes it. It seems to me that aviation began getting the shaft with the rise of the ecology and consumerism issues. Being against either is like coming out in favor of sin. It's a simple matter of arithmetic. There are more votes to be gained from being pro-ecology or pro-consumer than from being pro-aviation. The fact that aviation is not really an enemy of either makes no difference.

I submit to you that the issue of ecology itself has been perverted by half-baked, unproven theories, falsehoods that are substituted for facts, emotion that has pushed logic aside, and the replacement of reason by a flock of mumbo-jumbo, pseudo-scientific claims that would look ridiculous if you put them in "Star Trek." The U.S. supersonic transport program died in precisely that kind of atmosphere. You may not agree that the SST project should have been saved, but I'll tell you one thing: it sure wasn't defeated by any legitimate scientific evidence. Ecology and technology are not natural enemies. They're natural allies if the excesses on both sides can be avoided. Ecological advancements are based largely on technological achievements. It's just as simple as that.

What have we in aviation come to when the Federal Energy Administration awards a $100,000 contract to a California firm for a study involving commercial aviation? The purpose of the study? You wouldn't believe it: to develop ways of discouraging flying; $100,000 in taxpayer money to sabotage a mode of transportation that consumes less than one percent of all the fuel refined—not just the airlines—but all civil aviation. When I talk about the enemies of aviation, I'm not referring merely to those in Congress who'd sell their mother down the river for a fat headline and a pocketful of votes. I am referring to the bureaucrats, too.

You in corporate aviation got a taste of the bureaucrat's indifference to commercial aviation, his ignorance of aviation's importance. Just think back to the early days of the fuel crisis, when the original fuel allocations for business flying apparently stemmed from a belief that all corporate aircraft were used largely for flying millionaires to weekend fishing spots.

All we have to do is remember this one simple fact. Every mistake those of us in aviation make, plays into the hands of our enemies. If a pilot, a company, an airline, a manufacturer, a mechanic, a writer—if any one of us goofs—we supply the opposition with ammunition.

Some way, somehow, we must unify against the forces that would hurt, hamper or even partially destroy civil aviation. For, in a sense, if we are to list the enemies of aviation, we must include ourselves. When we vacillate, when we get careless with planes or policies or facts, when we let lies and slander go unchallenged, we are handing the enemy the weapons he needs. We no longer have room for feuds within aviation's own family. We can no longer afford the luxury of intramural quarrels. We either work together or we crash together. If we share a common heritage of progress, remember that our enemies share a common failure to recognize or appreciate what aviation has accomplished. For every step backward, aviation has taken a thousand steps forward. What we occasionally forget is that there are those who only look at the one step backward. They are the enemy. We are their whipping boy.

CORPORATE TRANSPORTATION: THE NEED FOR CITY-CENTER PUBLIC-USE HELIPORTS

by John E. Meehan
Manager, Metroport East 60th

Reprinted by Permission
ROTOR & WING (November/December 1975)

While the onus is on public officials to include heliports in the master plans for city centers, it's also up to helicopter operators and manufacturers to achieve a much greater degree of cooperation between industry and government.

On Aug. 1 of this year, the West 30th Street Heliport in New York City temporarily ceased to operate for a variety of economic reasons. West 30th Street was one of the first city-center heliports built in the United States, having started operations in 1956. Its closing will have very little effect on the overall transportation picture in New York City, but it represents another setback in our industry's efforts to establish the city-center heliport as a vital link in a national air transportation plan.

On Nov. 4, 1968, Pan American World Airways Inc. opened Metroport East 60th, located at E. 60th St. and the East River in New York City. This strategically located heliport serves the massive business community concentrated in the mid-Manhattan area. It is the first U.S. city-center heliport with hangar facilities to be operated as a public facility by other than a city, state, or bi-state agency.

The name "Metroport" itself denotes the airline's interest in improving metropolitan air transportation and helping corporate aircraft owners and travelers move rapidly in and out of the city.

Many corporations owning their own helicopters come and go from the Metroport daily. It is also used extensively by commuting businessmen, the news media, city and state officials, and by individuals who charter helicopters for quick business trips to locations within 150 miles of New York City. More and more, the business community is depending on helicopters not only for transportation from one point to another, but also for such important functions as property inspection, new site surveys, and quick delivery of vital parts and personnel to job locations.

Corporate fixed-wing travelers also regularly pass through the Metroport, arriving at reliever airports, such as Teterboro in New Jersey, and helicoptering to Metroport East 60th within six to eight minutes after stepping from their planes.

The activity at the Metroport is generated by approximately 150 helicopters operating within 100 miles of New York City. This figure is expected to increase.

At present, Metroport East 60th is

approximately 300 feet long by 90 feet wide, and has four landing pads. It operates seven days a week and is equipped with a temporary trailer terminal facility, turbine and aviation-gasoline fuel pumps, night lighting and protected overnight hangar facilities. Limousine and taxi service can be arranged by pilots radioing their requests 10 or 15 minutes before landing.

In planning the helicopter approach and departure paths, considerable effort was made to minimize the disturbance to the surrounding area. Flight paths were laid out to keep 99 percent of aerial activity over the East River. The absence of complaints has proven that helicopters can indeed be good neighbors and that they can readily fit into any congested area if their operations are properly planned.

Metroport East 60th demonstrates how a modern city and a progressive airline can join together in facilitating and expediting the movements of an important segment of its traveling public. It also represents a forward step in helping to relieve congestion at the major jetports, in the airways, and on the highways.

Metroport's four landing pads handle municipal helicopters as well as corporate ones.

The success of Metroport East 60th is clearly manifested in the statistics compiled since its inception through December 1974. After a relatively slow start in 1969, movements have risen from 11,500 in 1970 to 15,800 in 1974. Figures for the first half of 1975 indicate a continued growth pattern.

The example set by Pan Am and its cooperation with the city of New York in operating a public-use heliport should have set a pattern for other cities to follow. Unfortunately, it didn't. As we approach our nation's 200th birthday, neither Philadelphia, Washington, D.C., nor Chicago have public-use city-center heliports. Those in Los Angeles are being challenged on the basis of noise pollution. This trend must be reversed. And it can be if attention is focused upon the importance of the helicopter in urban society.

The most widely recognized role of helicopters today is in public-safety applications, such as law enforcement, fire fighting, and emergency rescue. One recent example of this recognition is New York City's announcement that the Cornell Medical Center of New York Hospital would build a $25-million burn-treatment center on Manhattan's East Side and use Metroport East 60th as its aerial receiving point. This burn-treatment facility will be the largest of its kind in the United States.

A less understood but equally vital role of the helicopter is that of providing transportation to the city center for business and industry leaders who, in turn, provide the jobs that support that community. Through the future development of intra/intercity helicopter transport, the city can enjoy renewed economic vitality and growth. However, this growth in helicopter transportation and the future economy can take place only by first establishing centrally located public-use heliports.

City-center public-use heliports must be included in the master plans. The onus for making available strategically located heliport sites in city centers is primarily on public officials. However, the helicopter industry, working through its manufacturers–should strive for greater cooperation with local governments. The DOT and the FAA should be petitioned to take a more active role, as well, in establishing public-use heliports. In the same way, state aeronautical departments should be made fully aware of the importance of helicopters and heliports in insuring the economic vitality of major cities.

Ideally, the city-center heliport should be constructed by the city on property removed from the tax rolls. It should be placed as close as possible to the heart of the business community–not in the fringe or slum areas. Through application to the FAA, Airport Development Aid Program (ADAP) funds could be made available to cover 75 percent of the construction costs. The construction itself can vary from the unmanned, single-pad facility to an all-weather, multipad heliport with all the extras currently found at today's modern airports. The facility can be operated by either a local heliport authority or a private operator.

If the operation is to be conducted by a private operator, steps must be taken to assure a reasonable profit for that operator: reasonable rent, tax abatement, etc. Landing fees, parking and hangar charges, and fuel prices should be kept compatible with local airports.

Local communities should be more aware of how vital a tool the helicopter has become to modern American industry. In 1973, IBM announced plans to build a multimillion-dollar training center in Ridgefield, Conn., on 200 acres of company-owned land. IBM proposed a college campus type of environment with no pollution, but the key to making the school project successful was the corporation's need for helicopter transport of key executives who would teach many of the classes at the training center. Ground transportation to and from the training center and IBM's headquarters would be too time consuming.

The lack of community support, represented by a small but vocal minority of citizens, prevented the issuance of a heliport permit. For this reason alone, IBM was forced to change its plans and is now in the process of establishing the training center in another area. Perhaps without even realizing it, that small suburban community lost an extremely good ratable.

The problem is already clearly defined. The proper solutions require greater cooperation between industry representatives and local governments. Special emphasis should be placed on the capability of the helicopter to handle emergency situations, but we should not overlook the great economic benefits helicopters can bring to the community.

Flight paths at Metroport were laid out to keep aerial activity over the East River.

READING SHOW IS BEST EVER

Reprinted by Permission FLIGHT LINE TIMES (June 1976) FLT Photos by Bill Baker

Perhaps the only way to measure the Reading Show is by its own yardstick—and this year, the yardstick was too short.

The 27th annual Maintenance and Operations Seminar, as general aviation's summer festival properly is called, was the biggest ever in attendance, displays, traffic and any other quantity one could measure.

The four hazy but rainless days commencing June 8 drew upwards of 175,000 people to watch the Navy Blue Angels, the Canadian Snowbirds, the great solo aerobats and the 100-plus ramp displays packed into Reading (Pa.) Municipal Airport. Last year, a day of rain kept the general attendance down to about 125,000, according to Paul Doelp, show director.

For those who paid the extra dollars to register, there were 184 indoor exhibitions representing easily the widest cross-section of general aviation products to be found in the world. Last year, there were only 167 exhibitors. Registration was 12,802, up from 10,317 last year.

More sheer numbers: At one point, there were 700 airplanes parked on the field—"more airplanes than I've ever seen," said one experienced control tower official. Traffic movements for the week totalled 10,410, up from 7,837 the year before.

But the Reading Show is more than numbers. Where else, for instance, can a weekend pilot actually be invited to sit at the controls of a King Air and discuss the value of a heads-up display?

How many ever even see a 10-waypoint RNAV, let alone get to push the buttons and pretend their favorite VOR has moved

to their favorite runway threshold?

Where else can an aviation history buff turn around in the crowd to see Max Conrad, the Flying Grandfather?

At Reading, one can take a walk of 100 yards and go from aviation's earliest commercial craft (the Swallow biplane) to one of its most strikingly advanced (the Hustler, an avant garde machine with a turbo-prop in its nose and a jet in its tail).

At Reading, there is always something going on. On opening day, a troop of Air Explorers from Topeka, Kan., left for a bicentennial tour of the 48 contiguous states, aiming to end in Philadelphia on July 4.

For hundreds, a highlight of the week was the Reading Air Service awards banquet, held on the grass during the June 10 air show performance.

All during the day, the Goodyear blimp Mayflower could be seen chugging its way around the Pennsylvania country-side, stopping briefly at the ramp to change passengers every hour and charg-ing its way uphill at 25 mph in what a blimp calls a takeoff.

At showtime each day, the air crackled with the sound of jet engines as the Blue Angels and the noticeably quieter Snow-birds executed flawless performances.

Other sounds were not as loud, but caught the attention just as firmly. When the Great Lakes biplane was taken through its paces, it was a time to listen. The plane's effortless loops and rolls testified to the smooth power of an Allison 250 turbine engine teamed with a three-bladed controllable-pitch prop.

There was a chilling lack of sound when R.A. "Bob" Hoover, master of the understated, shut down both engines on his stock Shrike Commander, pulled into a loop and landed, rolling on one main wheel and then the other.

For those who like only the best, there was the country's champion aerobat, Leo Loudenslager of Sussex, N.J., flying the mid-wing modified Stephens Akro which took him to the winner's circle earlier this year. He was joined by Henry Haigh, second place in the national standings, flying a Pitts. The two were raising money to send the U.S. Aerobatic Team to the international tourney in Kiev, Russia.

A superb demonstration was turned in by aerobatic great Art Scholl in his Super Chipmunk, an airplane whose relatively large size and slowness helped to make even a lomcevak understandable for the crowd. Scholl made the viewing easier by performing at such a low altitude that the Angels, Snowbirds and other aerobats gawked and held their breath like novice airshow spectators.

But by any standards, the week was a safe one. Only two incidents occurred in all 10,000 traffic movements. The first

came June 7 when the American Jet Industries Super Pinto arrived and pilot Dick Hunt could not get the nose gear down. The single-engine jet was smoothed onto the runway on its main gear and came to a stop with minimal damage. Fully 16 emergency vehicles rushed to the plane, according to accounts, but there were no injuries. The Pinto was repaired in time for the next day's performances. Later in the week, a Mooney dinged a prop while taxiing on the grass.

An unrelenting heat wave made beer and soft drink concessionaires happy, and when more than 50 lightplanes lined up on the ramp to leave after each evening's airshow, it was hard to tell whether the resultant cloud was smoke or perspiration.

For the serious pilots, there were informative seminars on topics ranging from propeller care to reading charts. For those with endurance, it was easy to amass five pounds of brochures in a single circuit of the booths.

Few genuinely new products were exhibited, however. The major airplane manufacturers displayed models that have been in production for months, and none was ready to talk specifically about the coming year's planes. Cessna was somewhat of an exception, since only days before it had announced that the Skyhawk line will convert to engines burning 100 octane gas, thus averting the problems of the new blue 100 "low-lead," which has been found to cause fouling and contamination problems when burned in engines rated for 80 octane. But Cessna salesmen were only recently made aware of the change.

The blue gas controversy formed an undercurrent throughout the show. Two oil companies had booths and drew a steady stream of pilots who voiced widely varying concerns about the blue gas. Some said they are able to "get along," other spoke of changing spark plugs and cylinders on their aircraft and still others angrily demanded that the oil companies return to producing 80 octane.

At the booths for Lycoming and Continental, the same atmosphere was apparent, although many pilots' fears were assuaged when the engine manufacturers' representatives assured them that proper operation will prevent much of the lead fouling blamed on the new gas. One engine maker challenged each pilot to prove that his engine problems stemmed from the fuel switch, and at last count said he could find only three out of more than a dozen who could stick to their guns.

HIGH ROLLERS IN THE MILE-HIGH CITY

by William Garvey

NBAA meets in Denver

Reprinted by Permission
THE AOPA PILOT (November 1976)

A smallish man, looking silly in a giveaway cowboy hat, listened calmly as the airplane salesman finished his spiel.

". . . we're guaranteeing 1979 or 1980 delivery at $4,375,000. Either Canadian or American. Your choice. Now that's complete with full avionics, but not long-range navigation. And, of course, you'd have the interior installed yourself. You can get a nice one for about $375,000."

The convention cowboy nodded, apparently satisfied with the information. The salesman had told him for four million bucks he could have a LearStar executive jet but one without so much as an orange crate for its passengers to sit upon. And the inquirer seemed not at all surprised by such an offer.

Across the cavernous Currigan Exhibition Hall was stationed a representative of Delco Electronics, makers of the Carousel INS, one of those long-range navigational units not included in the LearStar's price. The Carousel goes for $107,000, the man explained, almost bored.

Excepting the quarter Cokes and $2 sandwiches on sale at the Hall's eatery, the Carousel was one of the less expensive items on display at the National Business Aircraft Assn. convention held in mid September in Denver.

Photo Courtesy Combs Gates

For those who gee-whizz a lot over a brand new Bonanza or Arrow, the NBAA show can be a numbing experience. The buyers at this show take Mooneys as change. This is Gucci country where people discuss the shortcomings of others' multi-million dollar jets. Their air conditioners could outpull a Cessna 150.

What was surprising about the NBAA convention was to discover just how many such people there are. A record 5,382 of them registered during the three-day affair.

On hand to enlighten and entertain them was an army of aviation salesmen, showmen and leggy ladies representing 165 companies. The importance of the event was reflected in the company exhibits—all very expensive.

Gates Learjet, for example, rented hall space for 30 booths and then proceeded to erect a small wild west town—complete with Opera House and Buffalo Gals. Sikorsky Aircraft blocked off another two dozen booth spaces, constructed a pedestal upon which it perched a full-sized mockup of its jet-powered business helicopter, the $875,000 S-76. Beside the gleaming white whirly bird was a two-story edifice, the lower part containing a theater featuring continuous showings of S-76 flicks, while free coffee and rolls were served in the open cafe above.

Conventions have become arcades for adults and NBAA's was no exception. Teledyne Continental employed a magician to levitate hot pantsed girls, Mitsubishi Aircraft had a ventriloquist's dummy hawking its MU-2s, you could play Matt Dillon to Exec-Air's electronic badman, take a chance on Exxon stuffed tigers, win a Hangar One T-shirt, a B-25 painting from Dee Howard Co., a $500 gift certificate at Executive Jet Aviation, French champagne at Aerospatiale, even a vacation in Vail from Denver Beechcraft.

The idea of luring heavy rollers with penny gimcracks and sleight of hand seems a little ludicrous, but it did lend a gay air to what was otherwise a serious business. Unlike other aviation gatherings, the NBAA convention displayed complete disinterest in the aura of flight. No one used their hands to describe a certain maneuver; there was none of that ". . . there I was." Not once was the word "fun" used in relation to any product. The hardware's just too darn expensive.

The men selling the tri-jet Falcon 50, the Augusta 109 business chopper, the re-engined Century Jetstream III turbo-prop or the Swearingen Merlin IVA prop-jet are not selling good times. Those craft cost $4.25 million, $695,000, $875,000 and $940,000, respectively. Their customers are no nonsense corporations; the airplanes are tools that just happen to fly.

The buzz words in Denver during NBAA were "dispatch reliability," "seat miles," and "utilization." Dull stuff, perhaps, but the stuff that keeps corporate aviation growing. And growing it is.

When NBAA last convened in Denver six years earlier, it rented just half of Currigan Hall. This year it rented the whole block-square building. But the vitality of this heady segment of general aviation was best underscored by the new products shown there. And when the question was, "What's new?" the first answer was inevitably "Cessna."

The Wichita giant unveiled three new Citation jets during the affair. Citation I, to be available in December, will have upgraded engines and modified wings improving its performance and will sell for $945,000. Citation II is a "stretched" version of the I, with seating for 10. It will be available in 1978 at $1,295,000 per copy. Citation III, now Cessna's piece de resistance, features swept wings, trans-continental range, seats for a dozen and a sticker of about $2.4 million.

Board chairman Russ Meyer said development of the three new jets set his company back about $80 million. After mentioning that figure, he paused and added, "Damn, that's a lot of 150s, isn't it?"

Rockwell International said it had been tinkering with its Sabreliners and was well pleased with the results, henceforth to be known as the 65A and the 80A. The two craft, to appear in 1978, are the product of aerodynamic meddling by Rockwell in conjunction with the Raisbeck Group of Seattle (the latter having scored good marks for its Howard/Raisbeck Mark II Learjet modifications.

Lockheed made sure there was no misunderstanding about its commitment to corporate aviation. It took over a Stapleton nightspot, packed it with conventioneers and, while whiskey and roast beef coursed down several hundred throats, the new JetStar II put on a taxi, takeoff and landing performance outside the picture windows.

Beech, Piper, Grumman American, Ted Smith Aerostar were there as well, with big displays inside the hall and their planes among the 85 strong gaggle of magnificent muscle birds covering about 2,000 square feet of the Stapleton ramp. One major change was announced by Ted Smith Aerostar. The wings of the 290-hp turbocharged Model 601 have been extended 15 inches per side, resulting in a gross weight increase to 6,000 pounds (270 pounds more useful load) and a new designation—the 601B.

While the airplanes were quite naturally the stars of the show, the gear that goes within them and without them took the lion's share of booths.

RCA Avionics, Collins, Champion, King Radio, Narco, Telex, Marathon Battery, Garrett Corp., B.F. Goodrich, Avco Lycoming, Shell Oil, Grimes Manufacturing—they were all there with products as diverse and remarkable as Collins' WVR 250 digital radar and Bendix' BX 2000 avionics system.

It was particularly interesting to note the growing presence of helicopters striving to make it in corporate flying.

As already mentioned, Sikorsky, Aerospatiale and Augusta choppers were present but so too were Bell, Boeing Vertol, Hughes and Enstrom. And the interest evident in those vertical flight machines was probably to be expected. For if the NBAA convention is any kind of gauge, there's only one direction for corporation aviation: Up.

Next year's 30th NBAA gathering will convene Sept. 27th in Houston's Albert Thomas Convention Center.

SHORT NOTES

The items that follow are subjects of significance that could not be included at article length in this year's General Aviation section.

AMERICAN JET INDUSTRIES' HUSTLER

AMERICAN JET INDUSTRIES' HUSTLER is moving ahead in development. With the first flight now scheduled for March 15, 1977, the Hustler now has initial deposits on record for 38 aircraft.

The seven-passenger Hustler will be the only business aircraft using both prop and jet propulsion. The primary power plant is a Pratt & Whitney PT6A-41 turboprop, derated to 850 SHP. The standby turbojet engine, located in the aircraft's tail, is the Teledyne CAE J402-CA-700, of 660 lb. thrust.

American Jet Industries' Hustler Performance & Specifications
Preliminary Specifications Weight & Loading

Weight & Loading	
Max. Take-off Weight	6500 lbs.
Empty Weight	3700 lbs.
Useful Load	2800 lbs.
Fuel Cap. — 1440 lbs. pass. load	1360 lbs./200 gal.
— 760 lbs. pass. load	2040 lbs./300 gal.
Wing Area	181.15 sq. ft.
Max. Wing Loading	35.88 PSF
Max. Landing Weight	6500 lbs.
Take-Off Performance	
Minimum S/E Control Speed	68 MPH (CAS)
Take-off Prop RPM	2000 RPM
Lift-off Speed	70 MPH (CAS)
Take-off Flap Setting	25 Degrees
Take-off Ground Run S/E	670 ft.
Total Take-off to 50 ft.	1270 ft.
Climb Performance	
Best S/E Climb Speed	170 MPH
Best S/E Rate-of-Climb	3000 FPM
Cruise Performance	
Max. Cruise Speed — Single Engine (front)	380 MPH (TAS) at 20,000 ft.
Range at Max. S/E Cruise Speed, Reserves 30 Min.	2076 Miles at 20,000ft.
Econ. Cruise Speed S/E	330 MPH (TAS) at 35,000 ft.
Max. Range at Econ. Cruise Speed S/E, Reserves 30 Min.	2970 Miles
Fuel Mileage at Econ. Cruise Speed	11 MPG
Forward Engine Oper. Ceiling	36,200 ft.
Aft Engine Operational Ceiling	13,500 ft.
Pressure Differential	8 PSI
Landing Performance	
Approach Speed	85 MPH (IAS)
Ground Roll with Prop Reverse	500 ft.
Total Landing Distance with Prop Reverse	1000 ft. over 50 ft.

Note: All performance quoted at ISA Standard Day conditions with an airplane in good repair at gross weights of 6500 lbs. for all data except landing which is quoted at 6175.

MOONEY 201

The MOONEY 201 is heralded as the light single that provides 200 mph from 200 horsepower. The product of an extensive—and effective—drag reduction program based on the Mooney Executive, the 201 uses such devices as reworked intake and exhaust ducting, a new cowling, flush riveting on the wings, fairing-in or flush-mounting antennas and drains, streamlining essential brackets, and improving gear door coverage to achieve its impressive performance.

Mooney 201 Performance & Specifications Highlights

Engine	Lycoming 1O-360-A1B6D
Horsepower	200
Gross weight	2740 lbs. (1243 kg)
Empty weight	1640 lbs. (744 kg)
Useful load	1100 lbs. (499 kg)
Wing span	35 ft. (10.7m)
Length	24 ft. 8" (7.5m)
Height	8ft. 4" (2.5m)
Power loading	13.7 lb./HP (6.2 kg/HP)
Wing loading	16.4 lb./sq. ft (80.2 kg/ca)
Luggage capacity	130 lb. (59 kg)
Fuel capacity, usable	64 gal./384 lb. (242.1/174 kg)
Wheel tread	9' ¾" (2.8m)
Wing area	167 sq. ft. (15.5ca)
Aspect ratio	7.338
Landing gear	Retractable
Number of seats	4
Top speed	201mph/174k (324km/h)
Cruise speed, 75% power	195mph/169k (315km/h)
Range, 75% power, no reserve	1156sm/ 1004nm (1865km)
Fuel flow/ mpg. 75% power	10.8gph/ 18.1mpg (40.9 l/h/ 7.7km/l)
Cruise speed, 55% power	174mph/151k (281km/h)
Range, 55% power, no reserve	1.295sm/ 1.125nm (2089km)
Fuel flow/ mpg, 55% power	8.6gph/ 20.2mpg (32.5 l/h) 8.7km/l)
Rate of climb, sea level	1.030fpm (5.23m/sec)
Stall speed (gear & flaps down, power off)	61 mph/53k (98km/h)
Service ceiling	18,800 ft. (5730m)

Performance figures ± 3%.

BENDIX BX 2000

The BENDIX BX 2000 series introduces new technology to General Aviation avionics. Microprocessor computers, large-scale integrated circuits, and digitalized display systems have been combined into a system the manufacturer claims surpasses "many systems in airline service" for accuracy and performance.

Equipment included in the BX 2000 line includes NAV/COM systems, DME, ADF, digital course direction indicators, and transponders. System prices range from about $3,000 to about $18,000, based on the features included.

LOCKHEED JETSTAR II

The LOCKHEED JETSTAR II is a follow-on development of the original Jetstar model. New fanjet engines improve the fuel efficiency of the aircraft. Work on the final assembly line at the Lockheed Marietta plant is shown.

BEECH'S MODEL 77

BEECH'S MODEL 77, a two-place, single-engine trainer, has been approved for certification and production. First deliveries are scheduled for 1978.

Featuring a low wing with a new-technology (GAW-1) airfoil, the new T-tail trainer is powered by a Lycoming 0-235 engine of 115 horsepower. The canopy design of the cabin provides wide visibility, while left and right doors simplify access for instructor and student.

CESSNA'S 172 SKYHAWK

CESSNA'S 172 SKYHAWK received major improvements in the powerplant area for 1977. The "basic" Skyhawk 100 now uses a 160 hp Lycoming 0-320-H engine designed to run on 100 octane gas. The "extra performance" HAWK XP makes a significant advance over any previous Skyhawk model with the installation of a 195 hp fuel-injected engine and a constant speed propeller.

Both aircraft provide the pilot with pre-select flap control (detents at 10°, 20°, and 40°), vernier mixture control, and controllable rudder trim.

1977 Hawk XP Performance & Specifications

SPEED:	
80% power at 5500 feet	131 knots (243 kph)
maximum range w/49 gal.	485 nm (898km)
reserve.	3.7hr@80%
RATE OF CLIMB AT SEA LEVEL	870fpm (265mpm)
SERVICE CEILING	17,000 ft. (5,182m)
TAKEOFF (over 50 ft. obstacle)	1,360 ft. (415m)
LANDING (over 50 ft. obstacle)	1,270 ft. (387m)
STALL SPEED, CAS:	
Flaps up, power off	53 knots (98kph)
Flaps down, power off	46 knots (85kph)
MAXIMUM WEIGHT	2,550 lb. (1,157kg)
STANDARD EMPTY WEIGHT:	
HAWK XP	1,549 lb. (703 kg)
MAXIMUM USEFUL LOAD	
HAWK XP	1,001 lb. (454 kg)
BAGGAGE ALLOWANCE	200 lb. (91kg)
FUEL CAPACITY, TOTAL	52 gal. (197 liters)

1977 Skyhawk 100 Performance & Specifications

SPEED	
75% power at 8,000 feet	122 knots (226 kph)
maximum range 2 w/40 gal.	485 nm. (898km)
reserve.	4.1 hr @ 75%
RATE OF CLIMB AT SEA LEVEL	770 fpm (235mpm)
SERVICE CEILING	14,200 ft. (4,328m)
TAKEOFF (over 50 ft. obstacle)	1,440 ft. (439m)
LANDING (over 50 ft. obstacle).	1,250 ft. (381m)
STALL SPEED, CAS:	
Flaps up, power off	50 knots (93 kph)
Flaps down, power off	44 knots (81 kph)
MAXIMUM WEIGHT	2,300 lb. (1,043 kg)
STANDARD EMPTY WEIGHT	1,379 lb. (625kg)
MAXIMUM USEFUL LOAD	921 lb. (418kg)
BAGGAGE ALLOWANCE	120 lb. (54kg)
FUEL CAPACITY	
Standard Tanks	43 gal. (163 liters)

ISRAEL AIRCRAFT INDUSTRIES 1124 WESTWIND

ISRAEL AIRCRAFT INDUSTRIES 1124 WESTWIND was certificated by both the United States and Canada during 1976.

Using twin Garrett TFE-731-3 turbofan engines, the Westwind meets all FAR 36 requirements. Maximum cruise speed is 552 mph; maximum range is 2,960 statute miles. The standard interior seats seven plus a crew.

LEARSTAR 600 CONSTRUCTION SET

The *LEARSTAR 600*, Learjet designer William P. Lear's latest effort, will be manufactured by Canadair Ltd. of Montreal. Construction of the prototype will begin in January, 1977, with the first flight scheduled for January, 1978. Certification and initial deliveries are anticipated by the summer of 1979.

Canadair currently has 63 orders for the corporate/commuter/cargo jet—38 executive versions, and 23 cargo versions. The cargo versions are a "conditional agreement" with Federal Express, the Memphis-based small-package cargo firm. The conditions of the agreement are expected to be resolved in early 1978.

The LearStar design incorporates ideas first used in the Learjet to keep the aircraft simple and light. One new idea used in the LearStar 600 is the NASA-designed supercritical wing. Use of this wing allows a larger maximum fuel load, since the LearStar's fuel will be carried in the wings.

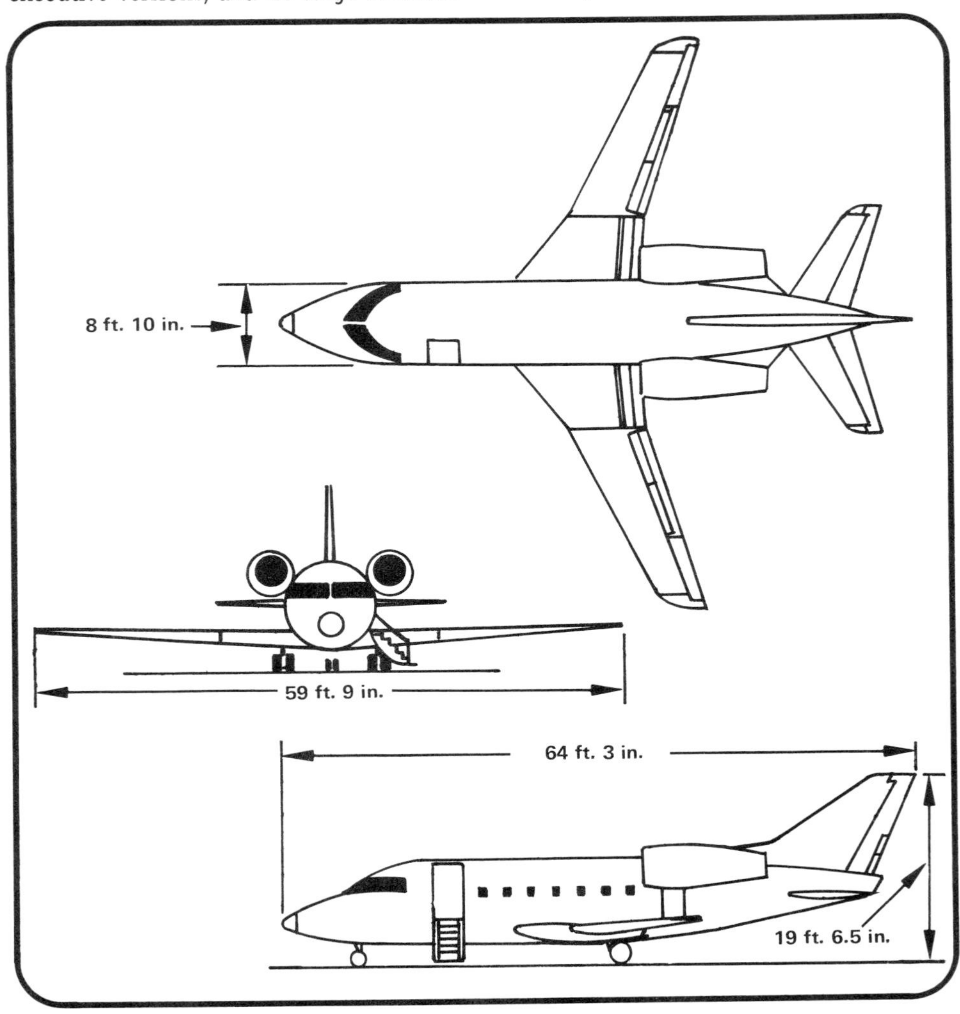

LearStar 600 Specifications

WEIGHTS

Max. Ramp	32,650 lb.
Max. Take-Off	32,500 lb.
Max. Landing	31,000 lb.
Max. Zero Fuel	27,000 lb.
Max. Fuel	14,810 lb.
Basic Operating	16,900 lb.
Payload with Full Fuel	940 lb.

SPEEDS

High-Speed Cruise	505 kts/581 mph
Normal Cruise Speed	488 kts/561 mph
Long-Range Cruise	459 kts/528 mph

RANGE

4,020 n.mi. (4,623 s.mi.) with NBAA IFR Reserves for 100 n.mi. alternate remaining.

ENGINES

Two Avco Lycoming ALF 502 high bypass turbofans (5:1 ratio)
7,500 pounds thrust each
Exceptionally low fuel consumption

NOISE

Take-Off	78 EPdB
Approach	90 EPdB
Sideline	87 EPdB

NTSB MEMBERS ARE NAMED

WEBSTER B. TODD, Jr.
Chairman

Webster B. Todd, Jr. was appointed by President Ford in February 1976 to a five-year term as a Member of the Safety Board and to a two-year term as its Chairman.

He had served since April 1974 as Inspector General of Foreign Assistance in the Department of State with the rank of Assistant Secretary. His Federal service had begun in 1969 as Special Assistant to the Chairman of the Civil Aeronautics Board. In 1971 he became Executive Director of the White House Conference on Aging; in 1973, Deputy Special Assistant to the President.

Mr. Todd was owner and President of Princeton Aviation Corporation, a commuter airline and fixed-base operator, from 1963 to 1969. In 1967-68 he served a two-year term as an Assemblyman in the New Jersey State Legislature.

Born in New York City, Mr. Todd received an A.B. degree from Princeton University in 1961. He holds a commercial pilot certificate with single and multi-engine land and instrument ratings, and is both a Certificated Flight Instructor and an Instrument Flight Instructor.

KAY BAILEY
Vice Chairman

Kay Bailey was appointed by President Ford in June 1976 to a Safety Board term expiring December 31, 1979, and to a two-year term as Vice Chairman.

An attorney, Miss Bailey was elected to the Texas House of Representatives in 1972. She was re-elected in 1974, and was seeking a third term at the time of her nomination to the Board.

As a legislator, Miss Bailey concentrated on transportation problems. She co-sponsored the bill to re-organize the Texas Highway Department and to create a mass transit authority for Houston, Texas. She also was appointed to the Transportation Task Force of the National Legislative Conference, which studied the problems of railroads, highways, airports and the trucking industry.

In 1974, Miss Bailey was a delegate to the White House Summit Conference on Inflation and also was a delegate to the Texas Constitutional Convention. In 1971, she was press secretary for Ambassador Anne Armstrong, who was then the co-chairman of the Republican National Committee.

Born in Galveston, Texas, Miss Bailey attended the University of Texas, and the University of Texas School of Law, receiving an LL.B in 1967.

FRANCIS H. McADAMS
Member

Francis H. McAdams was appointed a Member of the Safety Board by President Johnson in April 1967, and was reappointed by President Johnson on September 20, 1967 and by President Nixon on June 13, 1974.

From 1958 to 1967 he served as Assistant to a Member of the Civil Aeronautics Board and during 1960 as Assistant to the Chairman of that Board. His duties involved advising and counseling

in legal, economic, and safety matters within the jurisdiction of the CAB. From 1954 to 1958 he practiced law, specializing in aviation matters.

For six years prior to practice of law, he served in the CAB, from 1948 to 1954, as Attorney-Trial Examiner, where he participated in field investigations of aircraft accidents, presided at subsequent public hearings to help determine cause, and drafted the Board final report. Later, he served as Senior Trial Attorney concerned with the preparation and trial of the Government's case in all types of aviation regulatory matters. Prior to this he was a general corporate attorney for Capital Airlines from 1946 through 1948.

Certificated as a commercial pilot in 1939 he entered military service with the U.S. Navy in 1942 serving for four years as a pilot and executive officer of an aircraft carrier fighter squadron. From 1946 to 1954 he continued his military flying as Commanding Officer of a U.S. Naval Reserve Fighter Squadron.

He holds an A.B., LL.B., and LL.M. from Georgetown University and Georgetown University Law School.

WILLIAM R. HALEY
Member

William R. Haley of Washington, D.C. was appointed a Member of the Safety Board by President Nixon in June 1972.

Prior to his appointment he served as Legislative Counsel to Senator John Sherman Cooper (R. Kentucky) for eight years. During 1966 he was Minority Counsel to the Republican members of the U.S. Senate Public Works Committee. He is a graduate of Princeton University, A.B. 1945, and of the Harvard Law School, LL.B. 1948, and is a member of the Bar of the District of Columbia.

Prior to serving with Senator Cooper, Mr. Haley, was in private law practice and

an attorney with the International Division of the Mobil Oil Company, New York City. During the period 1955 to 1959, he was staff attorney with the Federal Trade Commission, and then with the Securities and Exchange Commission. He has lectured on Anti-trust law and S.E.C. practice at St. John's University Law School in New York.

PHILIP A. HOGUE
Member

Philip A. Hogue of Virginia was appointed by President Ford in April 1976 to fill the balance of a Safety Board term which expires December 31, 1978. A career Coast Guard pilot, Captain Hogue retired from the Coast Guard at the time of his appointment; he had been serving in Washington Headquarters as Deputy Chief of the Office of Coast Guard Operations since 1974.

Captain Hogue began his 27-year Coast Guard career in 1949 after seven years as

a Naval Aviator. He was a Coast Guard search and rescue pilot in Massachusetts, Newfoundland, Florida and Puerto Rico. From 1962 to 1966 he was Chief of the Training and Logistics Branch of the Aviation Division in Washington. Subsequent assignments included the post of Coast Guard Liaison Officer on the staff of the Chief of Naval Air Training in Pensacola, Florida; Chief of Search and Rescue in New Orleans; Commanding Officer of the Coast Guard Air Station and Chief of the Kodiak Search and Rescue Sector in Kodiak, Alaska; and Chief of Aviation, Senior Member of the Commandant's Aviation Safety Board, and Chief of Search and Rescue, in Coast Guard Headquarters in Washington.

Among Captain Hogue's numerous awards are the Coast Guard Commendation Medal, which he received for development of wide-area, high-intensity lighting systems for search and rescue helicopters, and a commendation for his search and rescue work in New Orleans during Hurricane Camille. He also holds two Meritorious Service Medals.

A native of Kansas City, Missouri, Captain Hogue attended George Washington University and the University of Virginia.

AVIATION SAFETY REPORTING SYSTEM PROVIDES ANONYMOUS WAY TO IDENTIFY AVIATION HAZARD

by YEARBOOK Staff

The National Aviation System has an outstanding safety record, but situations and conditions can develop that pose a threat to safety. Pilots, controllers, and others using and working in the system have the best opportunity to recognize and report these problems before someone has an accident.

To continue this safety record, a non-punitive safety hazard reporting system has been designed by NASA for the FAA. Run by NASA, the Aviation Safety Reporting System (ASRS) acts as an early warning system. Safety reports describing problems are submitted by pilots, controllers, and others in the system. Safety information is extracted and sent to those who can do something about the problem—hopefully in time to prevent an accident.

By being the central point for collecting safety reports, ASRS can also detect trends which alert the aviation community to hidden problems. For this reason, each ASRS safety report is considered not only for the information it contains alone, but also for what it contains when compared to all others.

ASRS OPERATION

The operation of ASRS is straightforward. As each safety report is received, it is promptly given to an expert safety analyst. He examines the report and decides whether or not he needs to call the person who submitted it for more information.

The analyst who calls is an expert in the subject described in the report. NASA has employed experts in all phases of aviation, including air traffic control, general aviation operations, and airline operations to operate the system.

When the analyst has obtained the information he believes necessary, he removes the identification portion of the report and mails it back to the source. He also removes all other information that could be used later to trace the report back to the source or to any other person. NASA calls this step "de-identification."

The analyst then codes the de-identified information and enters it into the ASRS computer. If the problem encountered poses an immediate threat to aviation safety, pertinent de-identified information is quickly relayed to the FAA so they can take action to remove the danger.

The ASRS computer is designed to continuously examine the coded information from all reports. This allows NASA and the aviation community to discover subtle changes and uncover hidden problems in the aviation system. Also, the effectiveness of corrective actions can be evaluated.

NASA will routinely provide the results of its study of the coded ASRS information to all segments of the aviation community. Additionally, the results of these studies will be reported from time-to-time in aviation magazines and publications.

SAFEGUARDS

The ASRS is a voluntary reporting program. NASA has designed it so that members of the aviation community can report in confidence without being concerned that the information provided will be used against the reporter or anyone else. There are only two exceptions to this: (1) criminal activities like hijacking, sabotage, or smuggling; or (2) actual aircraft accidents. NASA has to pass those reports to the proper government officials with all identifying information.

The FAA has taken a number of steps to make the ASRS a meaningful safety reporting system. The three steps of most concern to pilots relate to what enforcement action FAA might take if there is a possible violation of the Federal Air Regulations (FAR).

First, FAA will never request and NASA will never report if alleged violations of FARs are revealed in a safety report. Second, FAA will never ask NASA the identity of an individual submitting a safety report. And third, for unintentional FAR violations, FAA will not take disciplinary action when a timely safety report has been sent to ASRS.

The FAA will continue to enforce the FAR for intentional acts that threaten safety whether or not an ASRS safety report has been submitted. However, the FAA would have to learn of these acts through means other than the ASRS.

In the event of possible enforcement action by FAA, there are two safeguards available to show that the pilot has reported to ASRS. The first is the identity slip that NASA returned. It will be date stamped by NASA.

The second is a separate computer file kept by NASA that notes only the date, time, location, and type of incident of each safety report received. An entry will be kept in this file for 45 days after the incident. If FAA believes enforcement action may be appropriate, it asks NASA to check the "45-Day File" to see if a safety report has been submitted. If it has or if FAA does not ask NASA within 45 days of the incident, enforcement action will not be taken except for those intentional acts that threaten safety.

As a final safeguard, NASA has organized a committee of aviation safety experts to advise NASA on the design and operation of ASRS. Within the committee there is a security group that examines ASRS periodically to assure that individual confidentiality is being protected. Members of the security group are associated with AOPA, ALPA, and PATCO.

AIR CARRIER

AVIATION YEARBOOK

INTRODUCTION

It's incredible.

Fish are not dying in unusual numbers.

The incidence of skin cancer has not increased.

The ozone layer is not dissolving.

All of this despite the regularity of trans-Atlantic flights by the Anglo-French, supersonic "Concorde." After a year of scheduled service, the absurd, doomsday prognostications of the anti-SST establishment finally can be laid to rest.

The "Concorde" may not be economically viable (production, we have learned, will halt after the 16th model is manufactured), but at least the concept of hauling humanity supersonically has proven to be both safe and feasible. And within the following pages are detailed not only the creation of the first supersonic transport, but some of NASA's ingenious technological developments that eventually may lead to an American TST (transonic transport) and, hopefully, an advanced, quiet, more efficient SST. Surely the progress of civil aviation will not be aborted at Mach One.

But to keep the critically needed ozone layer intact, we are told, it is necessary to avoid the use of aerosol spray cans. Fortunately for aviation, aerosol propellents have not been regarded seriously as substitutes for turbine engines. Otherwise, airframe and powerplant manufacturers would have to launch an aggressive campaign to overcome the socio-political bias against aerosol propelled airliners.

But don't laugh. At least one application of aerosol spray is becoming more commonplace on the flight deck: hairspray. When Emily Howell became America's first female airline pilot (for Frontier Airlines), this was regarded as somewhat of a novelty. But now Ms. Howell is Capt. Howell, a four-striper flying a DeHavilland "Otter." And several other U.S. carriers (Braniff, Western, et al) have opted to follow Frontier's lead and have furthered the cause of women's liberation by introducing co-ed cockpits. And this is only the beginning. Just the other day, for example, I was discussing the feminist movement with my first officer and flight engineer, Louise and Judy . . .

But not to be outdone, the men are striving to settle the score. They are, in ever-increasing numbers, infiltrating the ranks of hostesses and stewardesses (who are now regarded non-sexually as "flight attendants"). Somehow it seems a little strange to see Bruce hustling toward the cockpit of a 727 with a cup of coffee for Valerie. But I guess that's progress.

Fortunately, 1976 was financially rewarding to the airline industry and it seems as though the recession can be regarded historically. As a result, most carriers are (or soon will be) recalling furloughed pilots or hiring new ones. So this may be the time for an aspiring airline jockey to bring his logbook up-to-date, get another rating or two and shotgun applications to every airline in the country (or out of it).

The long-term forecast, however, is more uncertain. Much of it depends on the Carter Administration and its view toward deregulation. Most airline experts agree that an attempt to dramatically deregulate the industry will result in widespread chaos, some bankruptcies and a reduced demand for pilots.

The oil-rich sheikdoms and kingdoms also may have additional future impact on the airlines. Because of OPEC, fuel has become an expensive commodity and further price hikes could make it unprofitable to fly an airliner even when it is completely loaded with paying passengers.

To reduce the cost of operation as much as possible, air carriers are working feverishly to reduce the fuel burn on any given flight. One possible and fascinating technique that you'll read about in this 1977 edition of the JEPPESEN SANDERSON AVIATION YEARBOOK is the NASA-developed scheme based on energy conversion and conservation. The idea is to descend toward an airport with all engines idling (the aircraft is in a glide). The pilot monitors an electronic computer that tells him when to lower the flaps and add power, actions that are delayed until the last possible second.

Safe? Time and experimentation will tell.

A reduced-power approach also softens the noise footprint beneath the aircraft, something that airport neighbors have been complaining about for years. And herein lies another distressful fact of economic life for the airlines. According to current Federal environmental regulations, the older jets (primarily the DC-8's and 707's) are too noisy and must be phased out during the next several years in deference to the quieter, wide-bodied aircraft. Now that's a bitter pill to swallow. Not many air carriers can afford to junk large portions of their fleets and then dig up the billions of greenies necessary to buy new machines. Presently, the airlines are playing against a stacked deck. Unless the Feds are prepared to help out in some way, the long-range airline forecast is far from optimistic.

And have ya' heard about the guy who flicked his Bic in the aft john of a Varig Airlines' 707? Apparently, a passenger (who–for obvious reasons–prefers to remain anonymous) tossed a lit cigarette in the trash bin. Moments later, smoke filled the cabin and finally entered the cockpit. The smoke was so black and dense that the captain elected not to continue the approach to Orly Airport (Paris, France). Instead he . . . well, you'll read about the tragic conclusion.

Fortunately, no such difficulty occurred aboard Pan American's 747SP (Special Performance) "Clipper Liberty Bell" betwen May 1 and May 3, 1976. Otherwise, this long-legged jumbo would not have shattered the around-the-world speed record (previously held by the Flying Tiger Lines' 707, "The Pole Cat"). After taking off from New York's John F. Kennedy International Airport, 71 paying passengers (and a contingency of the world press) were treated to a globe-girdling ride that lasted only 46 hours and 50 seconds and that included ground time at the only two stops made enroute (new Delhi, India and Tokyo, Japan).

True, this was a long time for any passenger to voluntarily remain aloft in one sitting, but how many people can claim to having flown around the world in less than two days? The only ones to have done it any faster are, of course, the astronauts (one orbit requires less than two hours) and a group of Strategic Air Command pilots who flew a formation of Boeing B-52 bombers around the world NONSTOP, a feat made possible by mid-air refuelling.

Unfortunately, the military brass never bothered to have this flight observed and certificated as an official world speed record. Pan Am's flight in a 747SP, therefore, has captured the honor of being–officially–the fastest flight around the world.

But 1976 was not all fun and games–especially for the pilots of an American Airlines DC-10 and a Trans World Airlines' L1011. Because of a human error by Air Traffic Control, these wide-bodied aircraft came within 100 feet of having a high altitude, head-on, mid-air collision over Southern Michigan. Poor judgment almost snuffed 306 lives. (The errant controller has been relieved of ATC duties and requalified for a less demanding job in the Federal bureaucracy).

While struggling to improve methods of controlling air traffic, the FAA also asked a group of six retired airline captains to form a Special Air Safety Advisory Group (SASAG). These experts rode in the cockpits of hundreds of airline flights while objectively observing the delicate interface between pilots doing their jobs and FAA controllers doing theirs.

The results were collated into a fascinating report that we are proud to be able to publish for you. What makes this report particularly unique is that each operational aspect of flight that attracted critical attention is followed up with a suggested method of improvement *and* FAA's reaction to each suggestion.

As this is written, Federal regulatory agencies are contemplating whether to ban smoking entirely on all U.S. airliners. Surely these noble ones can find something more meaningful with which to become involved. After all, non-smoking passengers are already immunized from the polluters by being provided with their own smoke-free zones. Gee, don't smokers have rights too?

An attempt was made in early 1976 by Ralph Nader and a group of 76 (in 1976, get it?) airline pilots to ban smoking in airline cockpits. The Federal Air Surgeon's reply was swift and final: "Nicotine withdrawal symptoms can affect an airline pilot's performance more than the effects caused by smoking cigarettes. Petition denied."

If airline pilots should not be forced to suffer from nicotine withdrawal, then why should such suffering be imposed on passengers, especially those about to embark on a 12-hour, nonstop flight from London to Los Angeles?

In the meantime, whether you smoke or not, we hope you'll enjoy the following review of the past year. My only frustration is that space did not permit more articles to have been included.

Capt. Barry Schiff
Fall 1976

AIRCRAFT ACCIDENT REPORT EASTERN AIR LINES, INC. BOEING 727-225, JOHN F. KENNEDY INTERNATIONAL AIRPORT JAMAICA, NEW YORK JUNE 24, 1975

Adopted: March 12, 1976

Edited for Space

SYNOPSIS

About 1605 e.d.t. on June 24, 1975, Eastern Air Lines Flight 66, a Boeing 727-225, crashed into the approach lights to runway 22L at the John F. Kennedy International Airport, Jamaica, New York. The aircraft was on the ILS approach to the runway through a very strong thunderstorm that was located astride the ILS localizer course. Of the 124 persons aboard, 113 died of injuries received in the crash. The aircraft was destroyed by impact and fire.

The National Transportation Safety Board determines that the probable cause of this accident was the aircraft's encounter with adverse winds associated with a very strong thunderstorm located astride the ILS localizer course, which resulted in a high descent rate into the nonfrangible approach light towers. The flightcrew's delayed recognition and correction of the high descent rate were probably associated with their reliance upon visual cues rather than on flight instrument references. However, the adverse winds might have been too severe for a successful approach and landing even had they relied upon and responded rapidly to the indications of the flight instruments.

Contributing to the accident was the continued use of runway 22L when it should have become evident to both air traffic control personnel and the flightcrew that a severe weather hazard existed along the approach path.

1. INVESTIGATION

1.1 History of the Flight

On June 24, 1975, Eastern Air Lines Flight 66, a Boeing 727-225, operated as a scheduled passenger flight from New Orleans, Louisiana, to New York, New York. The flight departed New Orleans about 1319 e.d.t.[1] with 116 passengers and 8 crewmembers aboard.

Eastern 66 arrived in the New York City terminal area without reported dif-

[1] All times herein are eastern daylight based on a 24-hour clock.

ficulty, and, beginning at 1535:11, Kennedy approach control (Southgate arrival controller) provided radar vectors to sequence the flight with other traffic and to position it for an instrument landing system (ILS) approach to runway 22L at the Kennedy airport. The flight had received a broadcast on the automatic terminal information service (ATIS), which gave in part the 1251 Kennedy weather observation and other data as follows: "Kennedy weather, VFR, sky partially obscured, estimated ceiling 4,000 broken, 5 miles with haze . . . wind 210° at 10, altimeter 30.15. Expect vectors to an ILS runway 22L, landing runway 22L, departures are off 22R . . ."

At 1551:54, the Southgate arrival controller broadcast to all aircraft on his frequency, " . . . we're VFR with a 5-mile, light, very light rain shower with haze, altimeter check 30.13 . . . It's ILS 22L, also." At 1552:43, the controller transmitted, "All aircraft this frequency, we just went IFR with 2 miles very light rain showers and haze. The runway visual range is–not available, and Eastern 66 descend and maintain four thousand, Kennedy radar one three two four." Eastern 66 acknowledged this transmission.

Eastern 66 was one of a number of aircraft that were being vectored to intercept the ILS localizer course for runway 22L. At 1553:22, the flight contacted the Kennedy final vector controller, who continued to provide radar vectors around thunderstorms in the area, to sequence the flight with other traffic, and to position the flight on the localizer course. About 1557:21, the flightcrew discussed the problems associated with carrying minimum fuel loads when confronted with delays in terminal areas. One of the crewmembers stated that he was going to check the weather at the alternate airport, which was LaGuardia Airport, Flushing, New York. Less than a minute later, one of the crewmembers remarked, ". . . one more hour and we'd come down whether we wanted to or not." At 1559:19, the final vector controller transmitted a message to all aircraft on his frequency that "a severe wind shift" had been reported on the final approach and that he would report more information shortly.

Eastern Air Lines Flight 902, a Lockheed 1011, had abandoned its approach to runway 22L at 1557:30. At 1559:40, Eastern 902 reestablished radio communications with the Kennedy final vector controller, and the flightcrew reported, ". . . we had . . . a pretty good shear pulling us to the right and . . . down and visibility was nil, nil out over the marker . . . correction . . . at 200 feet it was . . . nothing." The final vector controller responded, "Okay, the shear you say pulled you right and down?" Eastern 902 replied, "Yeah, we were on course and down to about 250 feet. The airspeed dropped to about 10 kn below the bug and our rate of descent was up to 1,500 feet a minute, so we put takeoff power on and we went around at a hundred feet."

Eastern 902's wind shear report to the final vector controller was recorded on Eastern 66's cockpit voice recorder (CVR). While Eastern 902 was making this report, the captain of Eastern 66, at 1600:33, said, "You know this is asinine." An unidentified crewmember responded, "I wonder if they're covering for themselves."

The final vector controller asked Eastern 66 if they had heard Eastern 902's report. Eastern 66 replied, ". . . affirmative." The controller then established the flight's position as being 5 miles from the outer marker (OM) and cleared the flight for an ILS approach to runway 22L.

Eastern 66 acknowledged the clearance at 1600:54.5, "Okay, we'll let you know about the conditions." At 1601:49.5, the first officer, who was flying the aircraft, called for completion of the final checklist. While the final checklist items were being completed, the captain stated that the radar was, "Up and off . . . standby." At 1602:20, the captain said, ". . . . I have the radar on standby in case I need it, I can get it off later."

At 1602:42, the final vector controller asked Eastern 902, ". . . would you classify that as severe wind shift, correction, shear?" The flight responded, "Affirmative."

At 1602:50.5, the first officer of Eastern 66 said, "Gonna keep a pretty healthy margin on this one." An unidentified crewmember said, "I . . . would suggest that you do;" the first officer responded, "In case he's right."

At 1602:58.7, Eastern 66 reported over the OM, and the final vector controller cleared the flight to contact the Kennedy tower. At 1603:12.4, the flight established communications with Kennedy tower local controller and reported that they were, "outer marker, inbound." At 1603:44, the Kennedy tower local controller cleared Eastern 66 to land. The captain acknowledged the clearance and asked, "Got any reports on braking action . . . ?" The local controller did not respond until the query was repeated. At 1604:14.1, the local controller replied, "No, none, approach end of runway is wet . . . but I'd say about the first half is wet–we've had no adverse reports."

At 1604:45.8, National Air Lines Flight 1004 reported to Kennedy tower, "By the outer marker" and asked the local controller, ". . . everyone else . . . having a good ride through?" At 1604:58.0, the local controller responded, "Eastern 66 and National 1004, the only adverse reports we've had about the approach is a wind shear on short final . . ." National 1004 acknowledged that transmission–Eastern 66 did not.

Both flight attendants who were seated in the aft portion of the passenger cabin, described Eastern 66's approach as normal–there was little or no turbulence. According to one of the attendants, the aircraft rolled to the left, and she heard engine power increase significantly. The aircraft then rolled upright and rocked back and forth. She was thrown forward and then upright; several seconds later she saw the cabin emergency lights illuminate and oxygen masks drop from their retainers. Her next recollection was her escape from the wreckage.

Witnesses near the middle marker (MM) for runway 22L saw the aircraft at a low altitude and in heavy rain. It first struck an approach light tower which was located about 1,200 feet southwest of the MM; it then struck several more towers, caught fire, and came to rest on Rockaway Boulevard. Initial impact was recorded on the CVR at 1605:11.4. The accident occurred during daylight hours at 40°39' N. latitude and 73°45' W. longitude.

Five witnesses located along the localizer course, from about 1.6 miles from the threshold of runway 22L to near the MM, described the weather conditions when Eastern 66 passed overhead as follows: Heavy rain was falling and there was lightning and thunder; the wind was blowing hard from directions ranging from north through east.

Persons driving on Rockaway Boulevard stated that a driving rainstorm was in progress when they saw the aircraft hit the approach light towers and skid to a stop on the Boulevard. Persons located about 0.6 miles south of the accident site stated that no rain was falling at their location when they saw the crash. They stated that the visibility to the northeast was good, but that visibility to the north was reduced. Persons who were in the north and northwest areas of the airport between 1555 and 1600 stated that heavy rain was falling; one stated that a violent wind was blowing from the northwest.

Flying Tiger Line Flight 161, a DC-8, had preceded Eastern 902 on the approach and had landed on runway 22L about 1556:15. After clearing the runway, at 1557:30, the captain reported to the local controller: "I just highly recommend that you change the runways and . . . land northwest, you have such a tremendous wind shear down near . . . the ground on final." The local controller responded, "Okay, we're indicating wind right down the runway at 15 kn when you landed." At 1557:50, the captain of Flight 161 said, "I don't care what you're indicating; I'm just telling you that there's such a wind shear on the final on that runway you should change it to the northwest." The local controller did not respond. At 1557:55, he transmitted missed approach directions to Eastern 902 and asked ". . . was wind a problem?" Eastern 902 answered, " Affirmative."

The captain of Flying Tiger 161 stated that during his approach to runway 22L he entered precipitation at about 1,000 feet[2], and he experienced severe changes of wind direction, turbulence, and downdrafts between the OM and the airport. He observed airspeed fluctuations of 15 to 30 kn and at 300 feet he had to apply almost maximum thrust to arrest his descent and to strive to maintain 140 kn on his inertial navigation system groundspeed indicator. The aircraft began to drift rapidly to the left, and he eventually had to apply 25° to 30° of heading correction to overcome the drift. He believed that the conditions were so severe that he would not have been able to abandon the approach after he had applied near maximum thrust, and therefore he landed.

The captain of Eastern 902 stated that on his approach to runway 22L he flew into heavy rain near 400 feet. The indicated airspeed dropped from about 150 kn to 120 kn in seconds and his rate of descent increased significantly. The aircraft moved to the right of the localizer course, and he abandoned the approach. He was unable to arrest the aircraft's descent until he had established a high noseup attitude and had applied near maximum thrust. He thought the aircraft had descended to about 100 feet before it began to climb.

Two aircraft, Finnair Flight 105, a DC-8, and N240V, a Beechcraft Baron, followed Eastern 902 on the approach. Their pilots stated that they also experienced significant airspeed losses and increased rates of descent. However, they were able to cope with the problem because they had been warned of the wind shear condition and had increased their airspeeds substantially to account for the condition. Neither pilot reported the wind shear conditions; one pilot stated that he did not report the wind shear because it had already been reported and he believed that the controllers were aware of the situation.

1.7 Meteorological Information

The weather in the New York City area at the time of the accident included scattered thunderstorm activity. Weather radar observations established that the thunderstorms near the Kennedy Airport were very strong with associated heavy precipitation.

[2] All altitudes herein are mean sea level.

The anemometer, which provides the official wind information on the Kennedy Airport, is located about midway between runways 22L and 22R and about a mile from the threshold of runway 22L. Remote indicating equipment is located in the control tower and the NWS office on the airport.

At 1526, the National Weather Service Forecast Office (NWS), located in midtown Manhattan, issued a strong wind warning which was valid from 1600 to 2000. The warning called for gusty surface winds to 50 kn from the west in thunderstorms in the New York City terminal area. The NWS distributed the warning to various facilities in the area, including the Kennedy control tower and approach control and Eastern Air Lines operations at the Kennedy Airport. There was no evidence that the warning was disseminated to flightcrews operating in the area.

About 8 minutes before the accident, the NWS weather radar located at Atlantic City, New Jersey, showed that an area of thunderstorm activity was centered along the northern edge of Kennedy Airport. The area was oriented west-northwest to east-southeast and was 30 to 35 miles long and about 15 miles wide. Several groups of thunderstorm cells in the area had tops which exceeded 50,000 feet. The tropopause was reported at 46,500 feet. About the time of the accident, the largest group of cells, moving east-southeast at a speed of 30 to 35 kn, merged with a smaller group of cells, moving east-northeast at a speed of about 20 to 25 kn; the cells merged over the approach course to runway 22L. There is no evidence that this information was available to either air traffic control (ATC) agencies or flightcrews who were operating in the New York City terminal area.

The NWS terminal forecast for Kennedy Airport, which was valid before Eastern 66 departed New Orleans, called for thunderstorms and moderate rain showers after 1800. The forecast was amended at 1430 to include thunderstorms and moderate rain showers after 1515. At 1545, the forecast was further amended to call for thunderstorms, heavy rain showers with visibilities as low as 1/2 mile, and winds from 270° at 30 kn with gusts to 50 kn after 1615. There was no evidence that the flightcrew of Eastern 66 received any of these forecasts.

At the time of the accident, there was no SIGMET in effect for the New York City terminal area.

The Eastern Air Lines forecast, which was issued at 1208 and which was valid from 1215 to 2000, predicted widely scattered thunderstorms with tops from 30,000 to 40,000 feet in New York and eastern New Jersey. The terminal forecast for New York City predicted scattered clouds until 2000; thereafter, thunderstorms were possible with light rain showers. The flightcrew of Eastern 66 received this forecast before departing New Orleans.

1.10 Aerodrome and Ground Facilities

Kennedy Airport, located in Queens County, New York, is about 12 miles southeast of midtown Manhattan, about 9 miles south-southeast of LaGuardia Airport, and about 18 miles east-southeast of Newark International Airport in New Jersey. Two sets of parallel runways are available—4-22 and 13-31, left and right. These runways are equipped with ILS facilities; however, under IFR weather conditions, only one runway at a time can be used for instrument approaches.

Runway 22L is 8,400 feet long and 150 feet wide. The elevation at the touchdown zone is 12 feet. The runway is equipped with high intensity runway lights and a high intensity approach lighting system with sequence flashing lights. There were no visual approach slope indicators (VASI) on runway 22L. According to the local controller, the runway and approach lights were on when Eastern 66 crashed, and they were set one step below maximum intensity.

The approach light towers struck by the aircraft were spaced 100 feet apart and constructed of nonfrangible material.

1.11 Flight Recorders

The correlation of CVR, FDR, and radar data shows that N8845E intercepted the glideslope at an altitude of about 3,000 feet at 1601:20. At that time, the captain commented, "Just fly the localizer and glideslope," and the first officer replied, "Yeah, you save noise that way and get a little more stability." The flaps were extended to 15° and the landing gear were lowered. The flightcrew was engaged in final checklist duties for the next 30 seconds, and the aircraft was bracketing the glideslope. The airspeed varied between 160 and 170 kn.

At 1603:05.5, the first officer requested 30° of flaps. The aircraft continued to bracket the glideslope and the airspeed oscillated between 140 and 145 kn. At 1603:57.7, the flight engineer called, "1,000 feet," and at 1604:25, the sound of rain was recorded.

At 1604:38.3, N8845E was nearly centered on the glideslope when the flight engineer called, "500 feet." The airspeed was oscillating between 140 and 148 kn. The sound of heavy rain could be heard as the aircraft descended below 500 feet, and the windshield wipers were switched to high speed.

At 1604:40.5, the captain said, "Stay on the gauges." the first officer responded, "Oh, yes. I'm right with it." At 1604:48.0, the flight engineer said, "Three greens, 30 degrees, final checklist," and the captain responded, "Right."

At 1604:52.6, the captain said, "I have approach lights," and the first officer said, "Okay." At 1604:54.7, the captain again said, "Stay on the gauges," and the first officer replied, "I'm with it." N8845E then was passing through 400 feet, and its rate of descent increased from an average of about 675 feet per minute (fpm) to 1,500 fpm. The aircraft rapidly began to deviate below the glideslope, and 4 seconds later, the airspeed decreased from 138 kn to 123 kn in 2.5 seconds.

N8845E continued to deviate further below the glideslope, and at 1605:06.2, when the aircraft was at 150 feet, the captain said, "runway in sight." Less than a second later, the first officer said, "I got it." The captain replied, "got it?" and a second later, at 1605:10.2, an unintelligible exclamation was recorded, and the first officer commanded, "Takeoff thrust." The sound of impact was recorded at 1605:11.4.

Because of the landing problems reported by the pilots of Flying Tiger 161 and Eastern 902, the Safety Board obtained their FDR's and examined them. Also, the FDR from Finnair 105 was examined. The NAS Stage-A radar data provided a basis for determining the time intervals between the flights. Flying Tiger 161, Eastern 902, and Finnair 105 preceded Eastern 66 on the approach by 8 minutes 59 seconds, 7 minutes 28 seconds, and 6 minutes 45 seconds, respectively.

Flying Tiger 161 recorder traces showed that after the flight had descended through 500 feet, its airspeed decreased from 154 to 137 kn within 10 seconds. During the same period, the aircraft's rate of descent increased from

750 fpm to 1,650 fpm.

Eastern 902, a Lockheed 1011, data showed altitude and airspeed deviations similar to those encountered by Eastern 66.

After Eastern 902 had descended below 400 feet, its rate of descent increased from 750 fpm to 1,215 fpm, and its airspeed decreased from 145 to 121 kn in 10 seconds. When the airspeed reached 121 kn, the engine pressure ratios increased from 1.1 to 1.5. The airspeed remained at 121 kn for about 6 seconds and then began to increase. The aircraft continued to deviate below the glideslope, however, until it reached 75 feet. At that time, Eastern 902 was about 120 feet below the ILS glideslope, and a positive rate of climb was established to execute the missed approach procedure.

Finnair 105 traces showed that the flight was maintaining about 160 kn while it descended to 750 feet. Between 750 and 500 feet, the airspeed oscillated between 148 and 154 kn. After Finnair 105 descended through 500 feet, the airspeed began to decrease to 122 kn within the following 20-second period. The rate of descent increased momentarily; however, it decreased when the aircraft descended through 250 feet. The airspeed increased slightly and continued to oscillate until touchdown.

1.12 Wreckage

Eastern 66 first contacted the top of the No. 7 approach light tower at an elevation of 27 feet above the mean low-water level and 2,400 feet from the threshold of runway 22L. The aft end of the jackscrew fairing for the left, outboard trailing edge flap lodged in the tower. The aircraft continued and struck towers 8 and 9. The aircraft's left wing was damaged severely by impact with these towers—the outboard section was severed. The aircraft then rolled into a steep left bank (well in excess of 90°) between towers 9 and 10, where it first contacted the ground. Its descent angle between the No. 7 tower and the beginning of the ground mark was 4.5°. It missed towers 10, 11, and 12; a gouge in the earth, about 340 feet long, paralleled the approach light towers on the northwest side from near tower No. 10 to tower No. 13. Three large outboard sections of the left wing were located near the beginning of the gouge.

Near the No. 13 tower, the aircraft's direction of travel changed from a magnetic heading of 220° to 205°; the fuselage struck towers 13, 14, 15, 16, and 17. The aircraft then continued to Rockaway Boulevard, where it came to rest. The approach light towers and large boulders along the latter portion of the path caused the fuselage to collapse and disintegrate.

The stabilizer trim setting was 8.25 units airplane noseup. The wing leading edge devices were extended fully and the trailing edge flaps were extended 30°. The landing gears were fully extended.

1.14 Fire

Fire erupted after the left wing failed and released fuel as the aircraft skidded through the approach light towers. There were numerous ignition sources--hot engine components, electrical wiring in the aircraft, the approach light system, and the street light system—and many friction sources. Destruction of the fuselage caused more fuel to be released, and the fire continued to burn after the aircraft came to rest.

The assistant chief of the Kennedy tower activated the fire alarm about 1606 and the Port Authority of New York and New Jersey's fire department, which is located at Kennedy Airport, responded immediately. The first firetruck arrived at the scene about 2 minutes later. The New York City Fire Department was notified about 1609, and its first units arrived about 4 minutes later.

The main fire was under control in about 2 minutes and was extinguished about 3 minutes later. The firemen extinguished a number of small fires with portable fire extinguishers.

The fire department's rapid response prevented fatal burns to the 9 passengers who ultimately survived; some were found lying in pools of fuel and fire-extinguisher foam.

1.15 Survival Aspects

The accident generally was not survivable because of the near complete destruction of the aircraft's fuselage. The cockpit seats, the forward flight attendants' seats, and the passengers' seats were torn from their supporting structures. The seats were mangled and twisted and were scattered throughout the area along the last 500 to 600 feet that the aircraft traveled. Only the aft flight attendants' seats remained attached to their supporting structure. Almost all passenger seatbelts remained attached to their seat structures and remained fastened.

When the fuselage disintegrated and the cabin floors and seat anchors failed, the aircraft's occupants became unrestrained and unconfined. They collided with each other and their surroundings, causing multiple extreme impact injuries.

The 14 survivors were seated in the inverted rear portion of the passenger cabin. Although their seat support structures (except the aft flight attendants') also failed, they were less severely injured because the rear portion of the passenger cabin and the empennage section remained relatively intact. The aft flight attendants were able to escape unaided because their restraint systems did not fail, and they were protected from flying debris.

Personnel from the Port Authority Medical Clinic arrived at the scene promptly, and they administered first aid to the survivors. Only one ambulance was available and it was used to transport six survivors to the Jamaica Hospital. Firemen transported the remaining survivors to the hospital in a firetruck.

Two of the 14 survivors died shortly after they arrived at the hospital. Two passengers died within 5 days after the accident and one passenger died 9 days after the accident.

1.17.4 Runway Use at Kennedy

The Chief of the Air Traffic Division in the FAA's Eastern Region established the procedures for runway use at Kennedy, LaGuardia, Newark, and Teterboro Airports. The tower supervisors were responsible for selecting runways in accordance with their respective runway-use programs; the following considerations were paramount: (1) Safety, (2) aircraft noise abatement, and (3) operational advantages. The tower supervisors then coordinated with the assistant chief of the CIFRR before making the runway assignments. The latter was responsible for determining that the selected runways created the least adverse impact on the traffic flow to all of the airports, and he was the final authority for determining the runway configurations to be used.

The runway-use program at Kennedy Airport provided for a computer to assist the tower supervisor in making runway selections. The objective of the program was to optimize noise abatement throughout the airport community without derogating the safe, orderly, and expeditious flow of traffic.

The program applied to all turbojet aircraft when the wind speed was 15 kn or less and when there was no ice, slush, water, or any other conditions which would render the selected runway unsuitable for the intended operation. If the wind changed from one direction/ quadrant or velocity category to another or if a runway combination had been in use for 6 hours, a new runway configuration would be selected. Runways could be used with crosswinds up to 15 kn. The computer's first selection of runways could be rejected and another runway configuration could be selected if: (1) The computer's selection would have an unacceptable impact on adjacent airports, and (2) one set of parallel runways was closed and traffic delays of 30 minutes or more were likely.

In the event of computer failure, criteria were established to alter runway use providing the surface winds did not exceed 15 kn. Runways could be selected for use even though crosswinds of 15 kn existed. If the surface winds exceeded 15 kn, the runway use program was not to be used.

On June 24, 1975, runways 31L/R at Kennedy Airport had been in use from 0718 to 1347. At 1347, operations were changed to runways 22L/R. From 1500 to 1900 was a peak traffic period and shortly after 1500, inbound traffic was being delayed. According to the approach control logs, about 1510 the watch supervisor of the CIFRR requested that the Kennedy tower permit some of the arriving traffic to use runway 13L. The control tower advised that a flight check was in progress on runway 31R (reciprocal runway) and that they would accept traffic spaced 10 miles apart. At 1539, the tower advised that they could not allow any landing traffic on runway 13L because the visibility was too low.

About 1543, Kennedy approach control began to hold inbound traffic at Southgate.[4] Five minutes later, all low-level traffic inbound from Philadelphia was suspended. About 1550, Kennedy departure control began to delay all traffic departing Kennedy via the Oakwood[5] departure routes. About 1554, Kennedy approach control began to hold all inbound traffic, and at 1602 Kennedy approach control anticipated arrival delays of 15 minutes at Southgate and 12 minutes at Bohemia.[6] The reason for the delays was the thunderstorm activity in the area.

At the Safety Board's public hearing, the assistant chief of the Kennedy tower, who was in charge of the control tower cab personnel, testified that the 1500-to-2300 duty period generally was very busy. Shortly after 1500, he observed thunderstorms to the northwest of Kennedy on the tower radar. Thereafter, he was busy coordinating various activities and did not notice the rain and lightning northeast of the airport. He was aware that Eastern 902 had abandoned its approach to runway 22L but did not know why; the local controller was too busy to be interrupted for an explanation. Also, he did not know that Flying Tiger 161 had reported the wind shear and had recommended that the runway be changed. He stated, however, that had he known of Flight 161's report, he would not have changed the runway

---

[4] A navigation fix about 30 miles south of the Kennedy airport defined by the intersection of the 131° radial of the Colts Neck VOR and the 221° radial of the Deer Park VOR.

[5] Routes toward the northwest to the Huguenot VOR.

[6] A navigation fix about 32 miles east-northeast of the Kennedy Airport defined by the intersection of the 083° radial of the Deer Park VOR and the 191° radial of the Bridgeport VOR.

because the surface wind was most nearly aligned with runway 22L.

The local controller testified that he was aware of thunderstorms to the north of Kennedy about 15 minutes before the accident, but he considered them to be weak. He was very busy with his duties and did not have time to pass either Flying Tiger 161's report or Eastern 902's report to the assistant chief. He stated that he did not consider a change of runway either before Flight 161's and 902's problems or in response to Flight 161's recommendation because the official wind instrument was indicating that the surface wind was most nearly aligned with runway 22L. He further stated that it would take anywhere from a few minutes to 30 minutes to change the runway.

The local control coordinator testified that shortly after 1500 he saw dark clouds to the west and northwest of Kennedy. On radar, he confirmed that there was a large thunderstorm to the west and that it was moving east. He was concerned about the weather situation and he expected it to deteriorate. About 1551, he observed the official prevailing visibility to be 2 miles. He stated that a thunderstorm with considerable lightning activity was north of the airport and that during the 10 to 15 minutes before the accident there was heavy rain just off the approach end of runway 22L. He described the rain as forming a solid wall beyond which he could not see. He said that throughout this period he and the local controller were very busy controlling the inbound and outbound traffic.

The Kennedy approach control final vector controller stated that on his radar screen he saw a small thunderstorm cell centered on the localizer course about the time he cleared Eastern 66 for the ILS approach. The cell was located about midway between the OM and the airport. He said that he was very busy with his duties, and that he had received no report that wind shear had affected Flying Tiger 161. The only report he had received was from Eastern 902.

A number of airline pilots stated that when they conduct instrument approaches to airports affected by weather hazards they rely substantially on the experiences of pilots who precede them when they decide whether to make the approach themselves or to choose a different course of action.

The manager of B-727 training for Eastern testified that under IFR conditions at high density traffic airports such as Kennedy, Miami, and others, a pilot could expect substantial delays (about 30 minutes) if he chose to land on a runway other than the one which ATC had established as the runway for instrument approaches. These delays could be anticipated because ATC could not provide simultaneous instrument approaches to different runways. Therefore, the pilot would have to wait for ATC to resequence the traffic and provide separation from the normal flow. Most pilots are familiar with these delays, and their fuel supply becomes a significant factor in their decisions whether to accept the delays, to continue in the flow of traffic that ATC has established, or to proceed to their alternate airport.

1.17.6 Installation of Frangible Approach Light Towers

The nonfrangible approach light towers were responsible for much of the severe destruction of the aircraft. The need for frangible approach light towers on the approach paths to runways has been recognized. On April 15, 1975, the FAA issued Order No. 6850.9 on revised approach lighting criteria. Among other things, the order provided that frangible structures would be used for the full length of all future approach light installations. Additionally, a retrofit program would be considered if funds were available.

The Chief, NAVAID/Radar Facility Branch, Airway Facility Service, FAA, testified that funding for part of the retrofit program was expected in the fiscal year 1977 budget. He stated that the towers currently being installed were designed to fracture at impact speeds of 80 kn or higher and that the towers would probably fracture at speeds well below 80 kn, depending on the type of aircraft involved.

2. ANALYSIS AND CONCLUSIONS

2.1 Analysis

It is clear from surface weather reports, weather radar data, and witness and pilot statements that a large area of very strong thunderstorms accompanied by strong, variable, and gusty surface winds was moving rapidly along the northern perimeter of Kennedy Airport between 1540 and 1620. The storm area was moving east-southeasterly, and about 1550 it began to seriously affect safe approach operations to runway 22L. Although the weather along the final approach course to that runway deteriorated rapidly from about 1550 to the time of the accident, the approach paths to the northwest runways remained relatively unaffected by the storms. Significant clues (both visual and radar) were available to air traffic controllers and flightcrews alike to indicate the existence of these conditions on and near Kennedy Airport.

Given the above circumstances, two causal aspects of this accident require discussion and analysis: (1) The weather hazards that existed along the approach path to runway 22L and how they affected Eastern 66, and (2) the reason or reasons why approach operations to runway 22L were continued even though the thunderstorms along the final approach course were evident and hazardous wind conditions had been reported.

How Thunderstorms Affected Eastern 66

Air flow is disturbed significantly within a mature thunderstorm cell and in the air mass surrounding the cell. These disturbances are dominated generally by vertical drafts, both up and down, which are created when the relatively cold and more dense air formed at higher altitudes displaces the warmer and less dense air near the surface. The downdrafts, which are frequently accompanied by heavy rain, can reach vertical speeds exceeding 30 fps. The interaction between the descending air and the earth's surface causes the flow to change from the vertical direction to the horizontal direction and creates a horizontal outflow of air in all directions beneath the cell and near the surface. The speeds of the vertical drafts and horizontal outflows depend on the severity of the storm. An aircraft passing through, below, or near a thunderstorm cell at low altitude may encounter these rapidly changing vertical and horizontal winds.

Passage through either a downdraft or a decreasing headwind can be singularly hazardous; however, when combined, the two conditions produce an even more critical situation. A mature thunderstorm cell contains both. As the airplane approaches the storm, it encounters the influence of the horizontal outflow in the opposite direction of flight as an increasing headwind; as the flight continues, it passes below the storm and through the peak downdraft. Almost immediately, the change in direction of the horizontal outflow will affect the aircraft as an abrupt decrease or loss or headwind. The sequence of the wind change can be particularly dangerous since the pilot might reduce power when he senses the positive performance effect caused by the initially increasing headwind. Therefore, the airplane may already be power deficient when it encounters the downdraft and loss of headwind; thus, their negative effect on the airplane's performance is compounded.

The Safety Board concludes from the evidence that Eastern 66 and at least four of the flights which preceded it encountered abrupt changes in the vertical and horizontal winds on the approach path to runway 22L.

When Eastern 66 was tracking the glideslope near the OM, the airplane was affected by a slight headwind and little or no vertical winds. While the airplane descended and approached the strongest cells of the thunderstorm, it was influenced by the vertical winds and the horizontal outflow. The increase in headwind of about 15 kn and possibly an updraft produced a reduction in the rate of descent and the airplane moved slightly above the glidepath as it descended between 600 feet and 500 feet. When the flight descended through 500 feet, about 8,000 feet from the runway threshold, the airplane was passing into the most severe part of the storm. The vertical draft changed to a downdraft of about 16 fps and the headwind diminished about 5 kn. As the airplane descended through 400 feet, the downdraft velocity increased to about 21 fps and the airplane began to descend rapidly below the glideslope. Almost simultaneously, the change in the direction of the horizontal outflow produced a 15-kn decrease in the airplane's headwind component, which caused the airplane to lose more lift and to pitch nose down. Consequently, the descent rate increased.

The wind conditions encountered by Flying Tiger 161, Eastern 902, Finnair 105, and N240V were similar but possi-

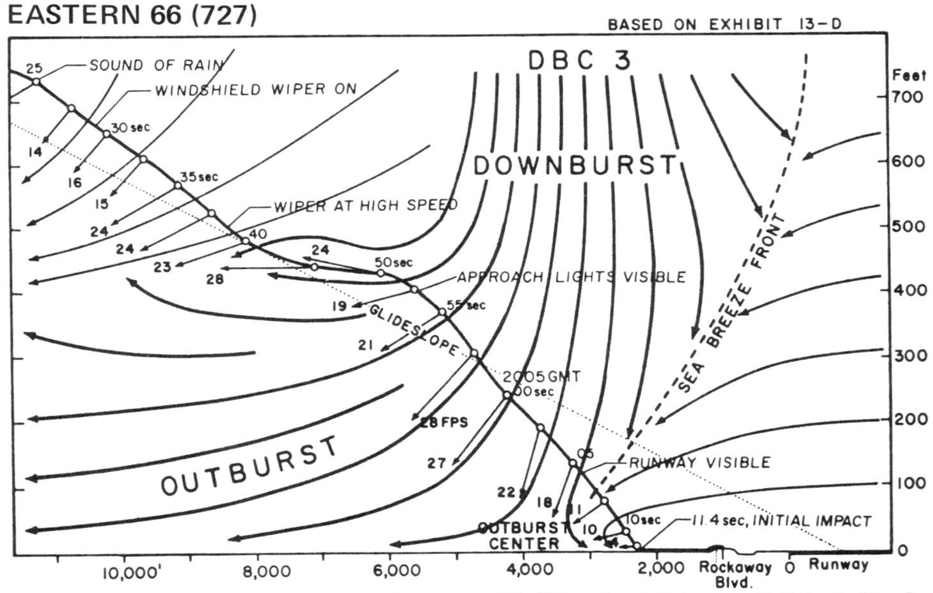

Chart from SMRP Research Paper 137 (March 1976) "Spearhead Echo . . ." T.T. Fujita. See following article for discussion of report contents.

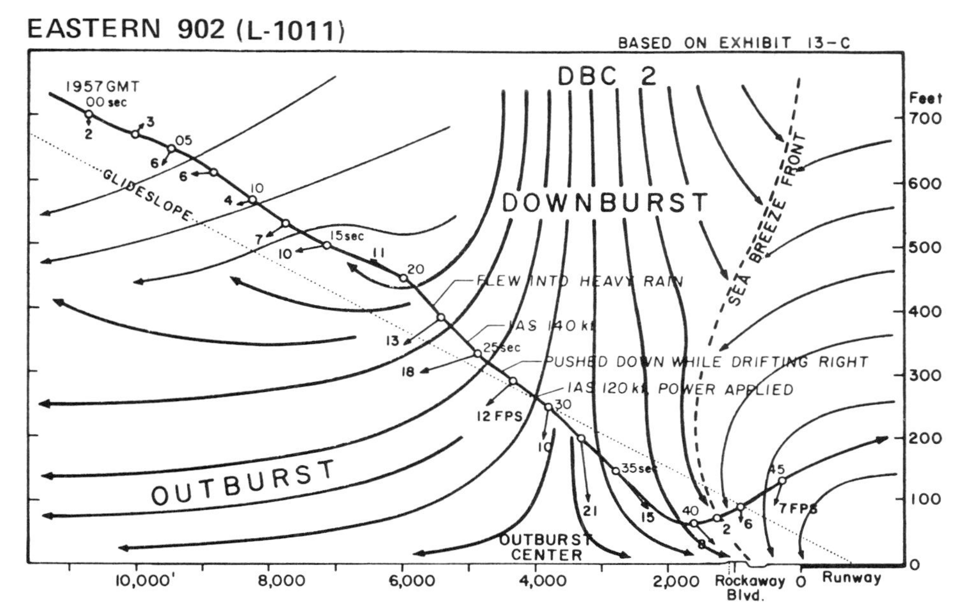

Chart from SMRP Research Paper 137 (March 1976) "Spearhead Echo . . ." T.T. Fujita. See following article for discussion of report contents.

bly less severe than those encountered by Eastern 66. All of these flights managed to negotiate the conditions without mishap, but not without difficulty. The captain of Flying Tiger 161 stated that after he recognized the shear he needed near-maximum thrust to keep his aircraft from losing altitude. At that point, he was not sure of his aircraft's missed-approach capability and he had to continue to a landing.

The pilot of Eastern 902 had no forward visibility when he penetrated the area of the most severe wind changes. Therefore, he was flying his aircraft solely by reference to flight instruments. It is obvious from the DFDR traces that he immediately recognized the downward acceleration of his aircraft and responded with the addition of thrust and noseup pitch changes. Nevertheless, the aircraft descended about 120 feet below the glideslope and within about 70 feet of the elevation of the approach lights.

The pilot of Finnair 105 anticipated the adverse wind conditions and added 20 to 25 kn to his normal approach reference airspeed. Although he too experienced an increase in the rate of descent as a result of the downdraft and horizontal wind changes, the total effect and control corrections required to decrease the rate of descent were probably lessened by the higher airspeed. The pilot apparently detected the effect of the wind and responded rapidly to maintain flightpath control.

Likewise, the pilot of N240V, a Beechcraft Baron, was able to limit the altitude loss caused by the wind conditions with less difficulty because of the different flight characteristics of his smaller aircraft and because he was flying it at a higher-than-normal approach speed.

The flightcrew of Eastern 66 was made aware of the adverse wind conditions by Eastern 902's report on wind shear, and they, too, added 10 to 15 kn to their normal approach reference speed. Both theory and simulator test results indicate that increasing final approach airspeed is advantageous when an aircraft is flying through dynamic wind conditions. However, too much airspeed can lead to a potentially hazardous situation for landing, particularly when the runway is wet. Since the captain of Eastern 66 inquired about the braking conditions, he was concerned about stopping the aircraft after landing. Therefore, after considering all of the approach conditions, the Safety Board believes that the addition of a 10- to 15-kn airspeed margin was reasonable. Simulator tests showed that even with this airspeed margin, the pilot must recognize immediately the aircraft's descent below the glideslope. He then must make rapid and pronounced pitch attitude and thrust changes to stop the aircraft's descent and prevent impact short of the runway.

There were no voice comments or sounds, until shortly before impact, which indicated that the flightcrew was either aware of or concerned about the increased rate of descent. Throughout the time period, the captain probably was looking outside, because about 6 seconds before the rate of descent began to increase he called "I have approach lights" and about 7 seconds after the rate began to increase he called "runway in sight." At the time of the latter call, the airplane was descending rapidly through 150 feet and was about 80 feet below the glideslope—twice the distance that would have produced a full-scale "fly up" indication on the related flight instruments if the glideslope signal was reliable. The Safety Board believes that the first officer's immediate response, "I got it," to the captain's identification of the runway indicates that the first officer also had probably been looking outside or was alternating his scan between the flight instruments and the approach lights. Although the aircraft was in heavy rain, the absence of significant turbulence might have caused him to underestimate the severity of the wind's effects.

Even though the first officer might have detected some of the glideslope, airspeed, and rate of descent excursions, simulator tests suggested that he probably reacted with insufficient thrust and pitch corrections to alter the excursions before he switched to visual references. These tests showed that large pitch and thrust changes were needed to stop the descent, and that the pilots often applied less sufficient changes than were needed because of the control forces involved and their reluctance to alter their instrument scan to verify the thrust settings.

Because of the low visibility, the flightcrew probably realized too late how rapidly they were descending and the magnitude of the corrections which were needed to stop the descent. By the time the first officer called for takeoff thrust, impact was inevitable.

The Safety Board recognizes the tendency of the pilot who is flying the aircraft to transfer at the earliest opportunity from instruments to visual references. In fact, this tendency is probably greater on approaches to runways like runway 22L at the Kennedy Airport because the ILS glideslope is designated as unusable below 200 feet. However, the Safety Board continues to believe that the visual references available to a pilot under conditions of rain and reduced visibility are often inadequate to provide timely recognition of flightpath deviations, such as those which can occur when traversing adverse wind conditions. This accident and others like it emphasize the need for air carriers to educate their flightcrews on the effect of a wind shear encounter, and to review instrument approach procedures which are related to flightcrew duties. The Safety Board believes that these procedures should stress that at least one pilot must scan the instruments until sufficient exterior references are visible to provide vertical guidance. Also, the Safety Board believes that research must be continued to develop a better method to transition from instrument flight to visual flight. High intensity VASI's on all runways served by instrument approaches, the "heads-up" displays, and the monitoring of flight instruments until touchdown as practiced by some air carriers are three concepts that appear promising.

Even with these landing aids, an approach which places an airplane in or near a thunderstorm at low altitude is hazardous. The wind conditions which might exist can place the airplane in a position from which recovery is impossible–even if both the pilot and the airplane perform perfectly. The number of recent approach and landing accidents which have been caused by the airplane's passage through or near localized thunderstorm cells indicates that many pilots and air traffic controllers do not have the proper appreciation for the hazards involved.

Approach Operations to Runway 22L

Since the thunderstorm astride the localizer course to runway 22L was obvious and since there was a relatively clear approach path to at least one of the northwest runways (31L), the Safety Board sought to determine why approach operations to runway 22L were continued, particularly after both pilots and controllers had been warned that severe wind shear conditions existed along the final approach to the runway.

According to the Kennedy tower local controller, he did not consider a runway change, either before or after he received the recommendation from Flying Tiger 161, because the surface winds were most nearly aligned with runway 22L. He further stated that he was too busy to pass the recommendation to the assistant tower chief who was responsible for initiating runway changes. Although the runway-use program did not require that runway selection be based on alignment with the wind, the criteria did require that, if conditions permitted, another set of runways be used for noise abatement because runways 31L/R had been in use for more than 6 hours. Therefore, because noise abatement favored the use of runways 22L/R, which were most nearly aligned with the wind, the control tower personnel apparently believed that they were operating with the best runway configuration.

However, the Safety Board concludes that had the thunderstorm activity been evaluated properly, it should have been apparent that the approach to runway 22L was unsafe and that approaches to that runway should have been discontinued. The Safety Board believes that ATC did not consider a runway change either before or after the Flying Tiger captain's recommendation because a change of runways would have further increased traffic delays and would have increased the already heavy workload.

When operating at capacity, the air traffic system in a high density terminal area tends to resist changes that disrupt or further delay the orderly flow of traffic. Delays have a compounding effect unless they can be absorbed at departure terminals or within the en route system. Consequently, controllers and pilots tend to keep the traffic moving, particularly the arrival traffic because delays involve the consumption of fuel and tardy or missed connections with other flights, which could lead to further complications. As weather conditions worsen, the system becomes even less flexible.

Although ATC has major responsibilities in the safe conduct of air operations, under current regulations and procedures, the pilot-in-command is the final authority on whether he will pursue a certain course of action, including whether he will conduct an instrument approach through a thunderstorm or other adverse conditions.

In view of the above, the Safety Board sought to determine why the captain of Eastern 66 continued his approach to runway 22L. The captain had received only one report of adverse conditions—the report from Eastern 902. This report apparently disturbed the captain ("... this is asinine"), but it also apparently was quickly rationalized to some degree ("I wonder if they're covering for themselves"). Had the captain known that two flights had reported adverse conditions, rationalization probably would have been more difficult. However, had he decided to make his approach to a different runway, he probably would have been delayed up to an additional 30 minutes because simultaneous instrument approach operations could not be conducted to two different runways. A 30-minute delay would have reduced substantially his fuel reserve of about 1 hour. Considering the thunderstorm activity affecting the New York City area, including his alternate airport, LaGuardia, his fuel reserve would have been minimal.

It is uncertain when the captain of Eastern 66 made his final decision to continue the approach. He apparently had not made a final determination when the flight was 5 miles from the OM and was cleared for the approach because he told the final vector controller, "...we'll let you know about conditions." Also, about a minute later, he explained to the first officer, "I have the radar on standby in case I need it...", which suggests he was thinking about the possibility of either not making the approach or having to abandon it. However, because pilots commonly rely on the degree of successes achieved by pilots of preceding flights when they are confronted with common hazards, it is likely that he continued the approach pending receipt of information on the progress of the two flights which were immediately ahead of him. By the time the second of these two flights had landed without reported difficulty, the captain of Eastern 66 was apparently committed to the approach, which discloses the hazards of a reliance on the success of pilots of preceding flights when dynamic and severe weather conditions exist. Within minutes, flight conditions can change drastically in or near mature thunderstorms. Moreover, pilot and controller workloads, and communication frequency congestion, can lead to omissions and assumptions, and confusion about who is aware of what.

In summary, the accident involving Eastern 66 and the near-accidents involving Flying Tiger 161 and Eastern 902 were the results of an underestimation of the significance of relatively severe and dynamic weather conditions in a high density terminal area by all parties involved in the movement of air traffic in the airspace system. The Safety Board, therefore, believes that no useful purpose would be served by dwelling critically on individual actions or judgments within the system, but that the actions and judgments required to correct and improve the system should be reviewed. All parts of the system must recognize the serious hazards that are associated with thunderstorms in terminal areas.

DOWNBURST CELL LINKED TO 727 CRASH AT KENNEDY AIRPORT

by YEARBOOK Staff

Research by Dr. Theodore Fujita, Professor of Meteorology at the University of Chicago, has pinpointed a hitherto unidentified form of thunderstorm downdraft cell as one of the major causes of the Eastern 727 crash at Kennedy Airport on June 24, 1975.

Initial speculation that tornadic winds might have been involved in the accident led Dr. Fujita to carefully examine pilot reports, films of radar returns, and other information about weather conditions at JFK that day.

Such examination of the meteorological conditions at the time of the accident led to the following conclusions:

–The thunderstorm off the approach end of Runway 22-L was at the peak of its growth at the time of the accident.

–Ground radar returns showed a fast-moving, sharply-pointed "spearhead echo" as the leading edge of the storm.

–This spearhead contained "downburst cells," or areas of intense downdraft and outflow of air. Similar downbursts have been measured at 92 mph at 40,000 feet.

–Aircraft flying between downburst cells off the approach end of Runway 22-L had no trouble landing.

–Aircraft flying through downburst cells moving across the approach path had considerable difficulty. One aircraft encountered a 60+ knot crosswind; an L-1011 executed a missed approach after being blown to the right of the approach path and sinking to 60 feet above the ground despite near-maximum thrust and an abnormally nose-high attitude; the 727 crashed.

FORMATION OF DOWNBURST CELLS

Downburst cells track rapidly across the ground, while maintaining very strong downward air currents near the surface. This ground speed and intense downward flow is directly related to how they are formed, according to Dr. Fujita.

The "normal" downdraft in a thunderstorm is formed in the mid-troposphere, with air injected from the side of the cloud. A downburst cell is formed when a thunderstorm develops an "overshooting top" above the anvil. Such tops may reach 45,000 to 70,000 feet, well into the lowermost stratosphere. The top then collapses, undershooting into the anvil, and creating a downburst cell fed by fast-moving jet stream or stratospheric air.

The high-speed stratospheric air feeding the downburst cell generates intense downflow (exceeding a minimum of 12 feet per second), strong surface winds or outflow, and a rapid ground track.

NOTE: The above information is based on SMRP Research Paper 137 (March 1976). A copy of the complete report may be obtained from Dr. T.T. Fujita, Department of Geophysical Sciences, University of Chicago, Chicago, Illinois 60637 or from the Flight Safety Department, MIACK, Eastern Airlines, Miami International Airport, Miami, Florida 33148.

WIND SHEAR CAUSES STAPLETON CRASH

Released by NTSB
June 24, 1976

The Continental Air Lines Boeing 727 takeoff crash at Denver last August 7 was caused by a severe wind shear from a nearby thunderstorm which halted the jetliner's climb and forced it back to the ground, the National Transportation Safety Board reported today.

The three-engine aircraft was flying at or near its maximum aerodynamic lift capability, but at about 100 feet in its takeoff climb had insufficient altitude or airspeed to overcome the effects of the wind shear, the Board held.

The Board concluded that the wind shear—in this case a rapid change from a headwind to a tailwind—caused a 41-knot reduction in the aircraft's airspeed in just five seconds. Impact came less than 12 seconds after the wind shear encounter began.

None of the 127 passengers and seven crewmembers was killed as the 727 pancaked onto the ground near the departure end of Stapleton International Airport's Runway 35 Left. Ten passengers and five crewmembers were injured. There was no fire, and an emergency evacuation was completed in three to four minutes without serious injury.

The Safety Board's accident investigation included a 727 performance analysis. From it the Board concluded that the accident "was unavoidable because the aircraft was performing near its maximum capability when it encountered the wind shear." Whether different takeoff procedures might have enabled penetration of the shear could not be determined.

Last June 9 the Safety Board recommended a joint government-industry study of existing airline takeoff procedures and a search for any which might enable pilots to better cope with low-altitude wind shear.

HIGH-BASED CUMULUS CLOUDS MAY BE SIGN OF TURBULENCE OR DOWNFLOW

by YEARBOOK Staff

High-based cumulus clouds may be a sign of a potentially dangerous downdraft condition, an airline Flight Safety program has discovered.

On two occasions, airliners were making landing approaches in seemingly good weather. There were high-base (9–10,000 ft.) scattered cumulus clouds in the area; some with *virga* (wisps or streaks of water or ice falling from the clouds but evaporating before reaching the ground). The surface temperatures were typically warm to hot for summer (80° to 89° Fahrenheit); the humidity was low (40° to 50° dew point spread); and the surface winds were light. One aircraft encountered a strong vertical wind shear; the other encountered moderate to severe turbulence. Each aircraft successfully completed the landing, although not without difficulty.

A comparison of the two incidents has resulted in the following explanation. Rain falling from the high-based clouds chills the air beneath the clouds and causes a downflow of air. The extremely dry air into which the rain falls causes additional chilling of the air through evaporation. These two actions (rain chilling and evaporative cooling) generate a cascade of falling air. This cascade will either be dissipated by the winds, or will reach the ground causing gusty surface winds. Light rain showers will be caused by unevaporated rain.

Pilots should be aware therefore of the probability of wind shear or turbulence whenever these four conditions occur together:

–High-based cumulus clouds with *virga*.

–Very dry surface air (dew point spread of at least 35°F).

–Weak (less than 15 knots) winds from the ground to the cloud base. (Stronger winds would cause the mixing and dissipation of the cascade.)

–Surface temperature greater than 75°F.

The National Transportation Safety Board has issued its report on the near-collision of an American Airlines DC-10 and a Trans World Airlines Lockheed 1011 at 35,000 feet near Carleton, Michigan on November 26, 1975. Both aircraft were operating in instrument meteorological conditions, within positive control airspace, and while under the control of the Federal Aviation Administration's Cleveland air route traffic control center. As a result of the evasive maneuver that had to be executed by the captain of the DC-10, three aircraft occupants were injured seriously, 21 were injured slightly, and the cabin interior was damaged extensively. None of the occupants of the L-1011 was injured.

The Safety Board determined that the probable cause of this near-collision "was the failure of the radar controller to apply prescribed separation criteria when he first became aware of a potential traffic conflict which necessitated an abrupt collision avoidance maneuver. He also allowed secondary duties to interfere with the timely detection of the impending traffic conflict when it was displayed clearly on his radarscope." Contributing to the accident the Board said "was an incomplete sector briefing during the change of controller personnel—about 1 minute before the accident."

American Airlines Flight 182, a Douglas DC-10-10, departed Chicago at 1839 e.s.t. for Newark, New Jersey, with 13 crewmembers and 179 passengers aboard. Trans World Airlines Flight 37, a Lockheed 1011, departed Philadelphia for Los Angeles at 1815 with 11 crewmembers and 103 passengers aboard.

At the time of the near-collision American 182 and TWA 37 were operating under the jurisdiction of the Wayne sector of the Cleveland Center, which is responsible for aircraft flying at or above 35,000 feet. The radar beacon signals from both aircraft were being received by the national airspace system Stage A Digitized (narrow band) Radar System and processed by the radar data equipment at the Cleveland Center which generates the information displayed on the radar controller's Plan View Display scope. This display, for each aircraft, consisted of a symbol for the aircraft's position and an alphanumeric data block that includes the aircraft's identification and assigned altitude. In the case of American 182, which was climbing, the data also included reported actual altitude, updated every 12 seconds.

Information retrieved from the radar log at the Cleveland Center showed that the target for TWA 37 was first processed for the Wayne sector Plan View Display at 1903:44 and showed the target at its assigned altitude of 35,000 feet, tracking approximately 290 degrees true at a ground speed of 408 knots. The target was about 105 nautical miles southeast of Carleton on J-34 airway. The target representing American 182 was initially processed for the Wayne sector PVD at 1914:24 and showed the aircraft to be about 100 nautical miles west of Carleton, climbing through 26,600 feet—at about 1,000 feet per minute—to its assigned altitude of 37,000 feet. At 1921:19 American 182, still climbing, was about 40 nautical miles west of TWA 37 and reporting at 33,000 feet. The two aircraft were on reciprocal courses and closing at a speed of about 850 knots.

The circumstances which led to the near-collision developed while the Wayne sector was being manned by two controllers; a "radar controller" responsible for radar control of traffic within his sector and a "manual controller" who maintains current flight data on progress strips, issues departure clearances, and coordinates with other air traffic facilities. A third controller, assigned to the "handoff controller" position, and who assists the radar controller, was at lunch and his duties were being carried out by the radar controller.

The radar controller stated that when he accepted the handoff of American 182 from Chicago he realized that there might be a traffic conflict between that flight and TWA 37. However, he said that his previous experiences that day had shown that several flights climbing eastbound out of Chicago to 37,000 feet had been leveling off a "considerable distance west" of where the near-collision incident later occurred. He thought that by "keeping an eye" on the situation he could turn the aircraft if the required separation could not be met.

When asked about any operational factors that might have distracted him, the radar controller recalled that about the time American 182 reported at 28,000 feet the Chicago Center called with a manual handoff of a Learjet aircraft. After he accepted the handoff he became occupied for about five minutes as he attempted to insert a route change for the Learjet into the computer. Normally, the manual controller would have handled this computer function but the radar controller took over the task as he thought the manual controller was busy and his own workload was moderate at that time. Thus, the radar controller "became preoccupied with secondary duties," the Board said, "and failed to see the impending traffic conflict displayed on his radarscope."

Then, at 1922, or about one minute before the near-collision, the radar controller was relieved by the third controller who had returned from lunch. Both controllers stated that during the briefing associated with the transfer of duties TWA 37, the Learjet, and several other aircraft were mentioned but American 182 was not. In a transfer situation, the Board said, the controller being relieved is responsible for the completeness and accuracy of the briefing.

At the time of the transfer of duties both controllers failed to notice the "unresolved conflict" of the TWA and American traffic. However, about 50 seconds after taking over the position and while scanning the radarscope, the relieving controller detected the conflict and called American 182 to verify its altitude. American 182 replied "Passing through 34.7 at this time. We can see stars above but we're still in the area of clouds." As soon as this 7-second transmission was completed the relieving controller instantly directed American 182 to

"descend immediately" to 33,000. Two seconds later American 182 responded "descending to 330 at this time." When the relieving controller issued the clearance the aircraft were about a mile apart and he saw the radar targets "merge and then separate" as the two aircraft came within 100 feet of each other.

The Board noted that since the relieving controller had no reason to expect that the responsibility he accepted included an acute problem it is fortunate that he noticed the problem within 50 seconds after taking over the position. However, this timely discovery does not exonerate both controllers from their failure to notice the conflict during the transfer of duties. The briefing was "incomplete," the Board concluded, because neither controller reviewed the actual situation as depicted on the radarscope.

The Board said that an understanding of the circumstances that led to this near-collision accident should give a controller a better insight into his critical role in air safety. Consequently, the Board recommended today that the Federal Aviation Administration "distribute the Safety Board's report on this near-collision accident to all FAA Air Traffic Control personnel and discuss it in their training programs in order to alert them to the catastrophic potential of distraction."

VARIG 707 HAD TOILET FIRE

Reprinted by Permission FLIGHT INTERNATIONAL (April 17, 1976)

A fire in the rear starboard toilet caused the crash of a Varig 707 on July 11, 1973, as the aircraft was descending towards Orly at the end of a scheduled flight from Rio do Janeiro, according to the report of the French commission of inquiry. The aircraft was in contact with Orly Approach and flying at Flight Level 80 when a radio transmission announced that there was a fire on board and requested immediate descent clearance. Orly cleared the 707 to descend to 3,000ft for a visual landing on runway 07; at that time the aircraft was 22 n.m. out and in line with the runway. As the captain was transmitting more details of the fire he was informed by the chief steward that the cabin was filling with fumes and that passengers were being asphyxiated. Almost immediately fumes were detected on the flight deck. The aircraft was recleared to 2,000ft and acknowledgement of this call was the last transmission the crew were able to make. The transponder response was visible on the ground radar for about one minute after this call.

The flight crew put on oxygen masks and goggles but the cockpit filled with enough black smoke to obscure the instruments. The captain decided that even with the direct-vision windows open he could not continue the approach safely and elected to make a forced landing. This was achieved with little damage to the aircraft, which came to rest some five minutes after the beginning of the emergency.

Of the ten crew members who survived the accident, the two pilots were injured by trees which smashed the windscreen. The remainder suffered from fumes. Four of the crew escaped through each of the flight-deck side windows and one through each of the two forward cabin doors.

When the aircraft came to rest only a small amount of smoke was visible externally at the rear of the fuselage, but within the cabin smoke and fumes were so dense that neither the surviving crew nor nearby farm workers who ran to help were able to do anything to help the passengers. When firemen arrived six minutes after impact the fire had broken through the fuselage skin and there was no sign of life in the cabin. The firemen removed four passengers, one of whom subsequently recovered. Seven crew members (the aircraft was carrying a relief crew) and 116 passengers died.

There was so little damage to the aircraft as a result of the forced landing that the underfin was intact and the door of the forward cargo hold was still in working order; from this standpoint the accident was survivable. The extent of carbon-monoxide poisoning found in 93 bodies was sufficient to cause death; it was high enough in 11 more cases for death to have been probable from this cause alone. Some doubt surrounds the

cause of another 16 deaths, but carbon-monoxide poisoning could not be ruled out in the opinion of the commission. Analysis of the fumes given off by the cabin furnishing showed a high proportion of carbon-monoxide.

In its analysis the commission says that the fire undoubtedly originated in the rear-toilet area. Although, it says, American airworthiness regulations stipulate standards to prevent fire developing, they do not demand the containment of fumes. Toilet-furnishing materials are not the worst fire risk; this stems from the quantity of waste paper (towels, etc.) likely to accumulate during a long flight. Although, says the commission, the possibility of an electrical fire must be considered, no circuits actually pass near the toilet waste bins. A cigarette end in the waste bin is the alternative explanation of the fire.

The commission recommends that passengers should be warned of the dangers of smoking in toilets. It calls for the wider use of non-flammable materials, preferably metal, in the construction of aircraft toilets, and for the provision of adequate fire extinguishers and oxygen masks in toilet areas.

CUTTING THE ODDS OF SURVIVAL

Reprinted by Permission
AIR LINE PILOT (April 1976)

by Captain B.V. Hewes, (DAL) Chairman, ALPA Rescue and Fire Committee

To judge US airport fire and rescue facilities today, we must go back to the end of World War Two, when commercial aviation was given one of its greatest pushes forward by the availability of surplus military equipment. Besides aircraft, the government was turning over many military airfields for civilian use. Along with an airfield, surplus fire and rescue equipment was often included, usually for $1.00 per unit.

Most were old 500-gallon water trucks with a few CO_2 bottles and not much else on board. They were quite good for extinguishing grass fires but that is about all. In the meantime, the Air Force was reequipping its fire and rescue units with the latest crash trucks capable of discharging a new extinguishing agent—protein foam.

There were, however, very few fire trucks being built for civilian use. The Walter Motor Truck Company was producing a few units for airports such as those owned by the Port of New York Authority and for FAA's airport at Washington National. A few Walters and LaFrance trucks were also sold to some of the larger commercial airports throughout the US.

Even in the 1950's, when Chicago's Midway Airport was the world's busiest, its equipment consisted of two structural pumpers: an old Cardox truck and, of all things, a hook-and-ladder truck that stayed around until the late 1960's when Midway gave up its position to the new O'Hare airport.

Things were so bleak in the 1950's that ALPA advised its members that when they had inflight problems that could result in a landing accident to head to the nearest military airfield, where they at least stood a sporting chance of survival in the event of a crash fire.

The 1960's brought further military advances and, late in the decade, the large P-2 fire truck appeared at Strategic Air Command bases. The US Navy also began reequipping and several of its old surplus

MB-5 trucks made their way to a few of the more than 500 US airports used by the scheduled airlines. But like most surplus military equipment, these fire trucks were obsolete and almost completely worn out. They were expensive to operate and very difficult to maintain.

At the same time, there were no airport fire training schools in the US except those run by the military. A few individual commercial airports conducted training programs, but many of the instructors were structural firemen who had little or no experience in the specialized field of fighting aircraft fires.

Only one word—pathetic—can describe the airport crash fire fighting situation in the 1960's in the US.

ALPA was well aware of this sad state of affairs. As early as 1950, the Association formed a Rescue and Fire Committee. Its mandate: See what can be done to improve the airport fire-fighting situation. ALPA became an active member of the National Fire Protection Association (NFPA) and other industry organizations.

Progress was so slow, however, that more drastic action was needed. So ALPA went to Congress to seek enactment of a law that would require FAA to upgrade airport safety requirements.

One thing that politicians react to is statistics. The facts had to be determined before ALPA could expect any legislative action. So in 1968, the Association conducted its first Airport Survey with follow-up studies over the next three years. Existing airport fire and rescue resources were rated by NFPA recommendations (See Table A).

The results helped push the Congress toward enactment of corrective legislation. Despite industry and FAA opposition, the Airport and Airways Development Act of 1970 contained the first requirement for airport certification.

Of particular interest to ALPA's Rescue and Fire Committee, the Congress inserted a paragraph specifically ordering FAA to require adequate crash equipment capable of rapid access to all parts of an airport.

FAA, however, was totally unprepared—it took two years for it to establish the Federal Aviation Regulation Part 139 required by the new law. The

TABLE A

Year	Total Surveyed	A	B	C	D
1968	562	138	114	94	216
1969	544	106	143	99	196
1970	523	128	154	95	146
1971	488	123	152	84	129

A—CFR equipment meeting the recommendations of the NFPA 403.
B—Good equipment available but less than NFPA recommendations for the type of aircraft using the airport.
C—Only token fire equipment available.
D—No crash rescue equipment on the airport.

TABLE B

PROTEIN FOAM/DRY CHEMICAL

GALLONS / POUNDS

U.S. AIR CARRIER AIRCRAFT	NFPA	ICAO	FAA 5210-6B	FAA PART 139
CV 580 M-404	1,500 / 300	1,900 / 300	1,830 / 500	* / 500
DC-9 Electra B-737	3,000 / 500	3,200 / 500	3,180 / 750	1,500 / 500
DC-8, B-707 B-727	5,000 / 500	4,800 / 500	4,820 / 1,000	3,000 / 500
DC-10, DC-8-61 L-1011	7,500 / 1,000	7,200 / 1,000	7,290 / 1,500	4,000 / 500
B-747	10,000 / 1,000	9,600 / 1,000	9,770 / 1,500	6,000 / 500

*Part 139 Index "A" Minimum for Turbine Operation 500 gallons/300 pounds.

regulation includes standards for security, fencing, disaster plans, as well as airport fire and rescue equipment. However, pressure from airport management resulted in an inadequate FAR 139 which does not fulfill the Congressional mandate.

Extinguishing Agents

Where do we stand today regarding the airport crash-fire-rescue situation? The International Civil Aviation Organization (ICAO) has made new recommendations regarding fire and rescue capability, including extinguishing agent quantities. NFPA has its own recommendations which are constantly being upgraded in keeping with the state of the art. And FAA has two sets of standards—FAR 139 and Advisory Circular 5210-6B. Agent quantity requirements in the FAA Advisory Circular are very similar to those of ICAO and NFPA. As published, however, FAR 139 requires agent quantities that are considerably less than those deemed necessary by the expertise of the world's fire and rescue organizations (See Table B). The Advisory Circular, however, is only a recommendation. FAR 139.49 is the regulation for US airport authorities and it is woefully inadequate.

The Movement Factor

Another disturbing factor is the widely held belief that the number of accidents at a given airport is related to the number of aircraft movements. In reality, there are as many accidents at smaller airports as at larger facilities.

The reasons are operational: Smaller airports have shorter runways, fewer navigational facilities, inadequate or nonexistent air traffic control facilities and poor weather reporting. To compensate for these deficiencies, smaller airports should actually have better fire and rescue equipment.

ICAO uses the factor of 700 movements in the busiest three months of a year to establish the standard for fire and rescue equipment. If there are less than 700 movements of the largest aircraft in the busiest three-month period, then fire protection for the next largest aircraft may be provided. (A movement is a landing *or* a takeoff.)

ALPA does not agree with this theory but FAA has topped it by using an average of five scheduled *departures* a day of air carrier aircraft, estimated on a yearly basis. If there are less than five departures a day of the largest aircraft, capability can be reduced to the requirement for the next largest aircraft. (Nine-hundred movements in three months.)

Here, however, is the real problem: If there are a total of less than five air carrier departures a day, the airport must only provide 500 pounds of dry chemical. However, a small combination vehicle containing 500 gallons of water and 300 pounds of dry chemical must be provided if any of the aircraft are turbojets. For example, if an airport is served by only four B-747 departures a day, 500 gallons of water and 300 pounds of dry chemical are the only requirements. This situation, of course, is not likely to exist in this country. But many airports with less than five daily departures are served by Boeing 727's, which should normally require 5,000 gallons of water and 500 pounds of dry chemical, instead of the FAA standard of 500 gallons and 300 pounds.

Somebody is off the track when the capability for extinguishing a fire is related to the number of airplanes operating into an airport instead of the size of the potential fire.

Response Time

Fire and rescue response time is another subject in which ALPA is vitally interested. Survival time in an aircraft fire can vary from 10 seconds to 10 minutes depending on fuselage integrity and fuel spillage. For practical purposes, however, ALPA estimates that the average survival time is from two to four minutes if the fuselage is reasonably intact.

In airport fires, response time is all important. The finest training, the best extinguishing agent and the greatest crash trucks in the world are of no use if this equipment can't get to the scene of the fire in minimum time. Reducing response time costs money and has therefore been fought bitterly by many in the industry.

Life-saving response time is dependent upon: (1) an adequate alarm and location system, (2) fast, efficient fire trucks, (3) strategic locations of fire houses, (4) adequate access roads to all parts of the operational area and (5) training.

Chart C shows approximate locations of airport accidents between 1964 and 1975. About 75% of all air carrier accidents occur at airports, and of these 99% are on or adjacent to the active runway. Statistics show that many aircraft finish beyond the runway threshold at distances up to 2,000 feet. Few accidents beyond this point are survivable due to high speed impact and terrain conditions. Therefore, the ideal location for a fire station is within 2,000-3,000 feet of the ends of each major operational runway. At airports with multiple runways, this could involve more firehouses than fire trucks, which is not practicable. ALPA's answer is to locate firehouses at the midpoint of each main instrument runway—not in some obscure corner of the airport.

IFALPA and ALPA's response time requirement is for a maximum of two minutes. ICAO's Fire and Rescue Panel

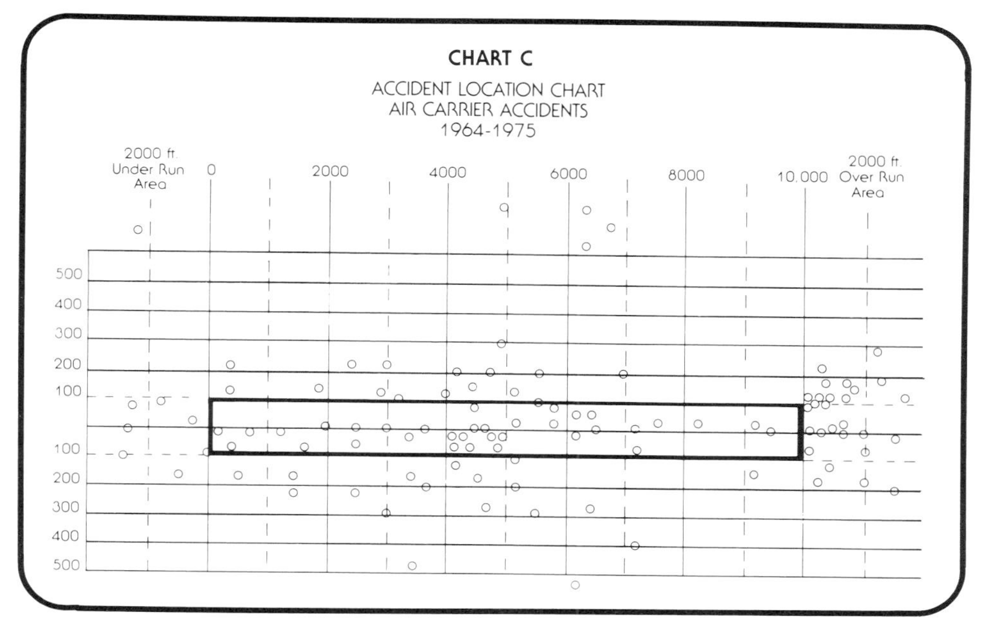

recommends a response time of three minutes, preferably two minutes to any operational area point. FAR 139 requires three minutes for the first vehicle, three-and-a-half minutes for the second and four minutes for all others—but only to the *midpoint* of the furthest runway. The result is that under FAR 139, it can take up to six minutes to reach the ends of principal instrument runways. A rather foolish requirement since so many accidents occur in the runway overrun areas.

Training

Another vital element in adequate response time is training. Still, only two civilian training schools exist in the US and they are short courses conducted by two chapters of the American Association of Airport Executives and are considered recurrent training. There is no basic airport fire school such as the British CAA school at Stanstead in England.

Over a year ago, the Congress enacted the Fire Prevention and Control Act of 1974. ALPA was instrumental in obtaining two requirements relating to aircraft operations: a training curriculum and a program of aircraft fire research. It will probably be five or six years before anything useful will come out of the Fire Prevention and Control Administration. Meanwhile, some research is being conducted by the military manufacturers and at FAA's National Aviation Facilities Experimental Center in Atlantic City, but still no national fire training school exists.

What about the future of aircraft fire protection? Airline pilots would like nothing better than to see all airport fire fighters unemployed by eliminating or actually reducing crash fire potential. But that is not to be for many years to come.

ALPA is, however, actively trying to incorporate new fire-resistant materials in cabin interiors that do not release toxic gas when exposed to heat.

Other things pilots would like to see include: (1) strategically located fire barriers in the ceilings and wing roots of all widebodied aircraft; (2) onboard Halon extinguishing systems for certain cabin and galley areas, as well as the lavatories where several recent fires have occurred; (3) advances in the field of fuel containment and/or anti-misting fuel additives.

Locator transmitters

A way must still be found to locate a crashed aircraft when there is minimum visibility. Both NFPA and ICAO state that two or three minute response times should be in optimum weather and visibility conditions. Airline pilots do not buy this statement. The minimum response time must be made in all weather and visibility conditions experienced when aircraft are operating. ALPA suggests that all commercial aircraft be equipped with emergency locator transmitters (ELT's). All emergency vehicles should then be equipped with an automatic direction-finding system that would lead them to the accident site.

Radar could be a big help in these circumstances. In the US, however, few airports have ground radar—only 15 out of 500 commercial airports at the latest estimate. Any system used must be reasonably inexpensive and available at every airport, not just a selected few. In the meantime, landing and takeoff minimums have been reduced to zero ceiling and 700-foot visibility. Emergency protection must be available during these conditions.

Federal Aid

Fire protection for the scheduled airlines is at long last becoming a reality. When FAR 139 became law in 1972, over 250 temporary waivers had to be issued because existing fire equipment was even below FAA's inadequate standards.

The Airport and Airways Development Act includes federal financial aid to provide airport fire equipment—82% of all costs for fire vehicles and safety equipment (saws, lights, axes, protective clothing, etc.) are covered. The federal government also provides 50% of the funds for building fire stations. However, salaries to man the equipment must be paid out of local operating funds.

US airports, therefore, have a peculiar assortment of firemen. At Port of New York and New Jersey Authority airports, firemen are also trained as policemen and double up in airport security duties. At many airports, firemen are members of the city structural fire department and have little or no training in the specialized field of aircraft fire fighting. In other cases, firemen double as emergency medical technicians and do laudable work in responding to medical emergencies within the airport complex.

Most smaller airports can't afford professional firemen, so they use maintenance personnel, grass cutters, snow removers, electricians, etc., give them minimum training and call them certified fire fighters. Most of these people are paid an absolute minimum salary, which in turn gives a minimum quality of fire protection.

Just as aircraft cannot fly without a skilled pilot at the controls, fire equipment needs skilled fire fighters to efficiently conduct a rescue operation. Professional, dedicated personnel are sorely needed. Hopefully, someday ALPA's objective will be reached. Until that time, airline pilots will continue to exert all available pressure to correct situations that are cutting the odds of survival.

B. V. (Vic) Hewes has been an active ALPA member since 1949. This article is an adaptation of a presentation he gave as guest lecturer at the British CAA Fire Service Seminar held at the Training School in Stanstead, England, October 20-24, 1975.

REPORT TO THE FEDERAL AVIATION ADMINISTRATION

Reprinted
by Permission
AIR LINE PILOT
(April 1976)

by Captains Art Ashworth (FAL), Charles E. Bassett (PAA), Robert N. Buck (TWA), Don McBain (UAL), Walter P. Moran (AAL) and Paul Soderlind (NWA).

(Editor's Note: In March 1975 the six retired pilots named above, five of whom are ALPA members, were invited by the FAA to form the Special Air Safety Advisory Group (SASAG) "to determine actions that can be taken to improve flight safety with special emphasis placed on approach and landing problems." The creation of the group was occasioned by the severe criticism leveled at the FAA in the aftermath of the crash of a TWA 727 into a Virginia mountainside the previous December. The group's report was completed on July 30, 1975. Since it contained information critical of FAA, release of the report was refused until late January 1976 in order to give the agency time to prepare a rebuttal. The Group's report is a classic in air safety literature. It [is] published with FAA's anonymous comments.)

INTRODUCTION

The Special Air Safety Advisory Group effort has been unique in safety investigation and analysis. During a period of some four months, it viewed the real world of airline flying through extensive exposure to the performance of pilot duties in the cockpit, the operation of aircraft along their scheduled routes, the procedures used during widely varying flying conditions, and the actions of support people in the routine operation of the Air Traffic Control.

Of special value has been the Group's ability to understand and analyze what they saw and heard because of their personal backgrounds and experience. The six SASAG Captains have logged nearly 160,000 hours in commercial aircraft in both domestic and international operations. The Group's members have recently retired from active airline service and are beholden to no one. Their only aim is to make flying safer. Feeling responsible to an industry they have devoted their lives to, they have attempted to be objective and unbiased.

After observation of some 600 flights on 27 different US carriers, the Group finds that basically the scheduled airlines are providing the public with safe, reliable transportation. The accident records show the airlines of the United States are one of the safest means of travel in the world.

However, air travel is not as safe as it could be. The potential for a catastrophic accident is always present and is often avoided by slim and, at times, nervous margins. Failure of either man or equipment operating in a highly complex and interdependent air transportation system can cause disaster. Misjudgment can be fatal.

The standardization of flight crew procedures and the increase in aircraft reliability, along with the building of an extensive ground support system including the latest navigation and terminal aids, have made air transportation increasingly safe. But in the process of building this system, new hazards have developed which are a product of the very support systems designed to make flying safer. For the various parts of the air transportation system to work together to reenforce safety demands a high degree of discipline and near faultless performance on the part of pilots, controllers, and support personnel. People are the paramount factor in safety: each person must know his job and do it.

The task of protecting against human error has not always taken the correct form. Attempts to compensate for natural deficiencies sometimes have brought on additional unrealistic rules or procedures rather than technological improvements or necessary restraints on the system. At the same time, equipment sometimes fails at a critical time. In almost all cases, it is the human who must compensate for malfunction of the equipment or the system. Because of too many demands upon the human, he becomes seriously overloaded. Reduction of the demands upon the human so that he can do his best and perform safely is the goal of the Special Air Safety Advisory Group's recommendations.

The pressure of economics and expediency often work to the detriment of simplifying the human workload. Partial solutions or temporary imposing of additional procedures or mechanical stop-gap measures seriously complicate the smooth, safe functioning of the air transportation system. This trend toward partial solutions, which are expedient, must be reversed. If a reversal of this trend is not accomplished, the safety record of the past will constantly deteriorate.

FAA Comment:

The FAA has long recognized the need for improving the airline safety record, with respect to accidents occurring during the approach and landing phase of flight. FAA, of course, has placed great emphasis in this area. The overall aviation safety record has been excellent, and there was a marked improvement during 1975. Scheduled and supplemental air carriers had only two fatal accidents during the year with a total of 122 fatalities, the best record since 1957. Even though this represents a significant reduction in fatal accidents and fatalities, the FAA is not satisfied. Aviation safety has always been the agency's first priority and will remain so. For instance, following are some of the major FAA actions taken, or underway, designed to improve system safety:

1. In 1971, FAA developed a new criteria designed to improve the safety of nonprecision approaches. The criteria calls for additional facilities, and funds were programmed for this purpose during FY 1976.
2. A flight procedure improvement program was implemented to identify and resolve potential problem areas in the development of standard instrument approach procedures, the publication of associated charts, and interpretation of charted information by pilots and traffic controllers.
3. Last June, FAA established a procedure requiring traffic controllers to issue an alert to a radar-identified aircraft if the controller notes an automatic altitude report on radar showing the aircraft to be at an altitude that might place it in unsafe proximity to terrain or obstructions.
4. A minimum safe altitude warning system that could be added to existing ARTS III systems is being developed. This system would provide a method of alerting traffic controllers and, in turn, pilots operating aircraft being tracked by ARTS III who have penetrated or are predicted to penetrate below a predetermined safe altitude.
5. To minimize personnel injury and aircraft damage whenever approach light

supports are struck by aircraft, improved frangible supports have been developed to replace the present wood or steel structures. FAA is providing frangible supports for new systems in FY 1976 and plans to retrofit existing systems in future years.

6. Recent approach and landing accidents and incidents appear to involve the wind shear phenomenon. Accordingly, all segments of the aviation industry have become aware of the need for a concentrated effort to discover acceptable means of reducing the hazards of wind shear. The FAA has established, along with other government agencies, a program to find near-term solutions to aid pilots in coping with low altitude wind shear.

7. On Dec. 18, 1974, FAA adopted a rule requiring each airline turbine-powered airplane to be equipped with a ground proximity warning system to alert pilots to impending unsafe proximity to terrain. A companion rule was recently issued requiring these airplanes to be also equipped with a glideslope monitoring system to alert pilots when they have descended too far below the ILS glideslope. Both systems must be operative no later than Sept. 2, 1976.

8. FAA, last May, initiated a "no-fault" aviation reporting system to stimulate the free and unrestricted flow of information concerning deficiencies in the aviation system. These reports are evaluated and system improvements, when warranted, will be undertaken.

9. Issued rule December 1974 requiring pilot altitude management when cleared for approach.

The report states that the goal of SASAG is aimed toward reduction of the demands upon the human so that he can do his best and perform safely. We believe that the intent of the report was to do just that and it does point out many areas from a pilot's point-of-view that need attention. We also believe that the fact-finding visits to ATC facilities highlighted only cursory impressions and do not reflect a complete understanding of the ATC system and requirements. This viewpoint is reinforced several times in the report as it relates to the workings or complexities of the ATC ground system and controllers not understanding cockpit demands. In the final analysis, we believe the report is a genuine attempt to enhance the safety of the air transportation system. With this thought in mind, the Air Traffic Service will take a positive approach toward each recommendation as it relates to the ATC system.

SASAG observed nonstabilized approaches caused by the efforts of ATC to "keep-em-high" which resulted in descent rates as much as 2,200 feet per minute to less than 100 feet above the runway. SASAG also observed touchdown speeds as high as 170 knots in a 727 and stated that, "The entire keep-em-high exercise is an invitation to disaster."

The primary purposes of the "keep-em-high" program are to enhance safety and to provide noise relief for airport neighbors. This is done by minimizing the amount of time high-performance aircraft are allowed to operate at low altitudes within the ATC system. This is where the greatest midair potential exists. The policy of "keep-em-high" is to have the aircraft enter the terminal area at or above 10,000 feet AGL and begin their descent from 10,000 feet AGL at a distance of not more than 40 flight miles from the airport and not closer than 30 flight miles from the airport. Descent below 5,000 feet AGL is normally limited to prescribed descent area where final descent and glideslope intercept can be made without exceeding terminal instrument approach procedures (TERP's) criteria.

Like all air traffic procedures, the "keep-em-high" program has undergone routine review to determine adherence and applicability to the changing environment. Evaluations of the "keep-em-high" program have disclosed nothing unsafe about the procedures. A review of all air carrier accidents for the past two years clearly indicates that it has not been a contributing cause of any accident. However, since we have had reports from pilots that they did not receive their descent clearances as early as desired, the facilities have been instructed to closely monitor its application. In view of our experience to date, it appears the observations reported by the SASAG were isolated situations and not indicative of the system's normal performance.

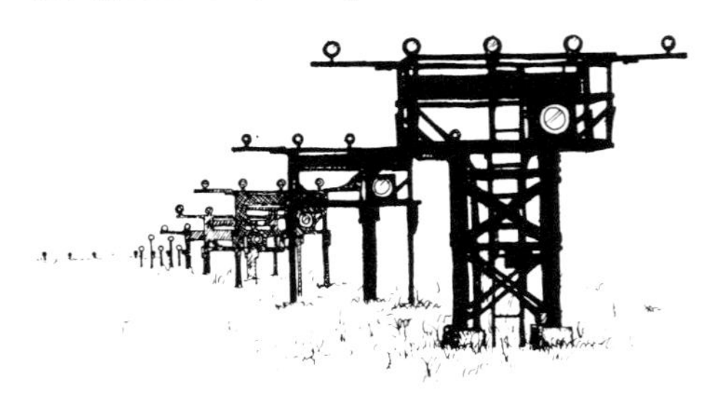

APPROACH AND LANDING

The primary objective of this Committee, as charged in Mr. Skully's letter of March 3, 1975, is "to determine actions that can be taken to improve flight safety with special emphasis placed on approach and landing problems."

Accordingly, SASAG has explored this with the managements of air carriers, with nearly 1,000 pilots, with several training centers, with the three major aircraft manufacturers, with ATC, approach and tower facilities, and with several FAA Field Offices.

Vertical guidance—primary problem

It was the unanimous opinion of every airline pilot interrogated that the lack of vertical flight path guidance is the number one shortcoming in the approach and landing phase of operations. Reporting on the Iberian Airlines accident at Boston on Dec. 17, 1973, the NTSB stated, "The poor visual cues available because of the low ceiling and visibility made the visual detection of the aircraft's pitch attitude and rate of descent difficult; runway 33L was not equipped with a visual approach slope indicator." NTSB also recommended VASI on all ILS runways.

SASAG cannot emphasize strongly enough the danger in lack of vertical guidance on all types of approaches and landings. The Group recommends VASI or similar equipment such as the Douglas "Pulse Light" for all runways on an urgent basis. Such installations will bring a marked improvement to the safety of approaches and landings.

In addition, SASAG is convinced that DME should also be provided on every ILS.

Cockpit distractions during critical flight regimes

In a continuing effort to find the distractions that lead to approach and landing problems, we were impressed by a bulletin recently issued by TWA. It was entitled, "Do Not Disturb, Plus Three, Minus Eight." This means that, "beginning with the takeoff roll and for the three minutes that follow and during the approach for the final eight minutes of the flight, the cockpit door will not be opened and the flight attendants are requested not to knock, enter or call during this critical phase of flight. Additionally, during this time, conversation in the cockpit should be kept to a minimum consistent with good cockpit management and operating procedures."

SASAG is in agreement with this principle, but from the many observations in flight, believes the time periods spelled out by TWA are too short. Longer times for complete cockpit sterilization during both takeoff and landing might better be adjusted to some altitude . . . such as the period the aircraft is under 10,000 feet. It is especially important that no ATC communications be required from the outer marker, or an equivalent distance, to the ground.

Additional crew members, check pilots and FAA inspectors should not talk to any other crew member during the sterile period. Conversation between jump seat riders and the flight engineer are especially distracting to the pilots.

ATC distractions

The greatest intrusion into the cockpit is undoubtedly approach control and the tower, since any comments made from these sources are outside the operating environment of the cockpit, and distract the attention of the flight crew from its primary obligation, flying the airplane precisely to the touchdown point on the runway. Too often, ATC interferes (unintentionally, of course) with the safety of aircraft rather than provides a service.

Copilot duties suffer during excess communication

On many takeoffs and landings, it was observed that the copilot was taken completely out of the operating loop because he was relegated to the function of a communicator. The numerous communications demands of ATC caused a serious deterioration of the safety of the operation. To some extent, this could be decreased by eliminating the need for the copilot to have a microphone in his hand during much of the low level operation, both on takeoff and landing. Various types of boom microphones are available and the Group urges their adoption on all carriers. Pan Am and United are successfully using them, as are many foreign carriers.

Improper seating gives improper visibility

Members of the [Group] noted a surprising number of pilots who sat too low. Each of the manufacturers, as well as the airline training schools, has made a point of the fact that airplanes are designed around a predetermined eye position, and that only if a pilot locates his eyes in this position can he see everything necessary to safely operate the airplane. This includes traffic while in climb, approach lights when on approach, the runway, and all of the instruments required to conduct an instrument approach. The February 1975 "DC-10 Flight Crew Newsletter" published by McDonnell Douglas shows that a seat position three inches lower than the designed position results in reducing the down visibility by 11 degrees, a very serious degradation.

Sitting too low tends to make pilots "duck under" the glideslope as they become contact and stretch up and forward to see the ground. The "duck under," of course, is highly hazardous.

The stabilized approach

The industry and the FAA must join together in insisting upon a return to the stabilized approach. Certainly everyone wishes to do everything possible to reduce noise for our airport neighbors, but when it comes to seriously reducing the level of safety, the compromise is unacceptable.

Speed control should be left to the discretion of the pilot under all circumstances for at least the final eight miles of the approach. All approaches should be stabilized on glideslope and on localizer by the outer marker, or equivalent distance from the runway on non-precision approaches. The stabilized condition, including landing flaps and stabilized speed and thrust, should be accomplished not later than 1,000 feet above the runway.

Some members of SASAG have had the opportunity to fly the two-segment approach as well as to interview other pilots who have flown it. We conclude that it is hazardous, and recommend against its adoption.

SASAG observed non-stabilized approaches caused by the efforts of ATC to "keep 'em high," which resulted in descent rates of as much as 2,200 feet per minute to less than 100 feet above the runway! SASAG also observed touchdown speeds as high as 170 knots in a 727. These can only be considered as aerobatic maneuvers. The entire "keep 'em high" exercise is an invitation to disaster.

Approach call-outs

The whole approach call-out philosophy has gotten out of hand, and SASAG believes approach call-outs may be overdone. A practical approach call-out procedure is desirable, but the industry should develop a uniform call-out system with full consideration for the contributions that radar altimeter warnings and ground proximity warning systems will provide. The call-outs of air speed, altitude, deviation from localizer and glideslopes, etc., have become a potential distraction at a time when concentration is mandatory.

Copilot landings in marginal weather

Four out of five of the most recent accidents have involved the first officer flying the airplane. We do not interpret this to mean that the first officers are incompetent. We are more inclined to believe that this indicates that captains are sometimes prone to be less helpful copilots than are first officers, and that copilots, when flying, take over functions that disregard procedures and remove the captain from the loop.

The captain cannot abandon his responsibility for the safety of the airplane just because he allows the first officer to make an approach and landing under what turns out to be marginal weather conditions. Therefore, he has to observe the actual flare and touchdown at any time the first officer is flying. This means no one is watching the ILS, altitude, and other necessary parameters during at least some portion of the final 100 to 200 feet of the approach. The complete procedure and discipline of the approach suffers. It must be recognized that under many marginal weather conditions the captain should fly the aircraft so that flight crew coordination and disciplines are maintained to the utmost.

The virtually universal practice of giving the copilot every other landing, regardless of weather or experience level, is wrong. Captains should not volunteer it; copilots should not expect it!

Missed approach

In many cases an accident or near accident occurs because the pilot elected to continue an approach when everything was not "just right," obviously hoping he could catch up. We are convinced that missed approaches are not being executed when, in fact, weather/runway conditions dictate that they should be. In the case of the Eastern accident at New York, the pilot of another airplane elected to execute a missed approach within eight minutes prior to the accident. No one was injured on that flight. Without in any way presuming to determine the probable cause of the accident, it seems safe to assume that the airplane would not have crashed if a missed approach had been executed at some earlier stage of the approach.

The points made in this section of SASAG's report are the most urgent. However, the entire report reflects on these areas and should be read with that in mind. It is difficult to point out any single area in aircraft operation without considering the entire complex, interwoven pattern. An event which occurs minutes, or even hours, before landing may finally make that landing unsafe.

SASAG recommendations:

1. VASI, the Douglas "Pulse Light" or similar vertical guidance aids should be installed on all runways. Funds should be diverted from other areas as necessary to accomplish this as soon as possible.

FAA Comment: *We concur with the first part of this recommendation that vertical guidance should be provided for all (air carrier) runways. However, funding priorities must be established. FAA developed, a number of years ago, criteria for improving the safety of nonprecision instrument approach procedures. This criteria is in the final stages of coordination with the TERP's Advisory Committee and will provide for VASI to give pilots visual vertical guidance to the runway.*

2. DME should be installed on all airports served by nonprecision approaches. A fix on these approaches should be located such that the final descent path to the runway does not exceed three degrees.

FAA Comment: *The criteria referred to in the FAA comment [on] Recommendation 1 will require a visual descent fix (VDF) provided by either ground based distance measuring equipment (DME) or marker beacons to give the pilot a point in space to begin a normal rate of descent to the runway from his minimum descent altitude. As indicated under Recommendation 1, the criteria also calls for visual approach slope indicators (VASI's) so pilots will have visual vertical guidance to the runway. We reprogrammed funds for FY 1976 to establish these facilities for runways served by airline jet aircraft. Additional funding will also be sought for lower priority runways served by airline prop aircraft and general aviation.*

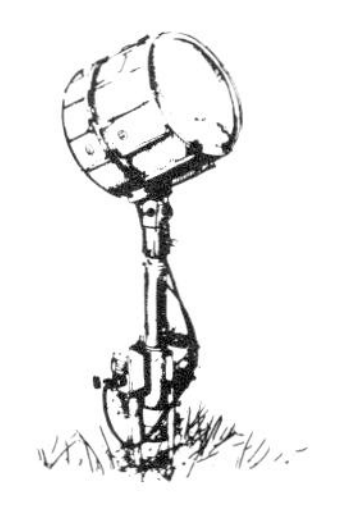

3. ATC communications during the final stages of the approach should be held to an absolute minimum. Except in an emergency, no ATC communications should occur between the outer marker and the end of the landing roll.

FAA Comment: *While we agree there is merit to keeping communication to a minimum during the final stages of the approach, it must be recognized that the exchange of certain information is essential. For example, certain fluctuations in wind direction or velocity, runway visual range, traffic situations which were not anticipated, course deviations on parallel approaches, and low altitude alerts must be issued in a timely manner to the pilot regardless of his position. We have made improvements over the years in reducing unnecessary communication through educational programs and discussions with both pilots and controllers. Our efforts will continue.*

4. Training on the hazards of sitting too low must be reviewed and emphasized.

FAA Comment: *We concur with this recommendation and will reemphasize the hazards of pilots sitting too low to the airlines and FAA air carrier inspectors.*

5. Immediately establish ATC procedures that will provide a stabilized approach from the outer marker, or equivalent distance, to the runway. Choice of speed must be left to the pilot for at least the final eight miles of the approach.

FAA Comment: *Our existing procedures provide for stabilized approaches. Aircraft must be vectored to intercept the final approach course before reaching the approach gate and, in the case of precision approaches, before intercepting the glideslope.*

Speed control is not applied to aircraft: A. FL 290 or above without consent of the pilot; B. conducting a published high altitude approach; C. cleared for an approach, except that when action is necessary to maintain or achieve safe spacing until the aircraft reaches: (1) for ILS approaches—the outer marker; (2) for visual approaches—the turn to base leg, a geographical point, or a navaid at least five miles from the runway; (3) for other than ILS or visual approaches—the final approach fix or a point five miles from the runway, whichever is farther from the runway.

We believe situations involving unstabilized approaches are a result of inadvertent actions on the part of a controller/pilot rather than a deficiency in procedures. However, our procedural review will continue along with our educational efforts.

6. Educate controllers on the fact that the characteristics of current jet transports make it difficult to slow down and descend at the same time.

FAA Comment: *We are aware that certain descent rates preclude any simultaneous substantial reduction in speed, and our controllers have been so informed. It is reasonable to conclude that during periods of heavy activity a controller may occasionally inadvertently give less consideration to this factor than desired. Therefore, we have and will continue to stimulate continuous awareness by periodic issuance of articles or reminders.*

7. Efforts to promote the two-segment approach should be abandoned.

FAA Comment: *Although FAA was considering withdrawal of its support of the two-segment procedure, EPA requested FAA to issue an NPRM proposing the two-segment approach in addition to two other noise abatement procedures. This NPRM was issued on September 25, and a public hearing was held on November 5. The comments received during the hearing are being analyzed.*

8. Review immediately the conditions under which copilots are allowed to make the approach and landing.

FAA Comment: *We concur with this recommendation, as too many recent accidents and incidents have occurred when the copilot was flying. We will request the airlines to review conditions under which copilots conduct approaches and landings and ask that they issue appropriate guidelines to flightcrews. We will also request our inspectors to monitor copilot performance during en route inspections.*

9. Immediate action should be taken by air carriers to establish practical and uniform callout procedures.

FAA Comment: *We concur with this recommendation and have established a project to revise the FAA recommended callout procedures. We are presently evaluating the procedures of one major* *airline, which are considered among the best in the industry.*

10. Captains must be reminded that a go-around, where necessary, is a mark of good judgment. ATC procedures must be revised such that an airplane that makes a missed approach does not lose its rightful place in the approach sequence.

FAA Comment: *We concur that the airlines must reemphasize to their pilots that whenever the outcome of an approach is in doubt a missed approach should be immediately executed. We will issue instructions to our air carrier inspectors to see to it that the air carriers emphasize this during initial and recurrent pilot training.*

ATC has no specific procedures to provide for resequencing of missed approaches. The basic concept of the ATC system is first-come-first-served. Aircraft that execute a missed approach and desire another approach, are placed in the approach sequence as soon as traffic conditions permit.

Normally, controllers do everything they can to reestablish these aircraft in the sequence as soon as possible. Additionally, if a pilot of a missed-approach aircraft advises of a problem requiring expeditious handling, controllers provide maximum assistance to the aircraft and expedite handling.

FAA Comment: "Internal cockpit distractions should be eliminated." *Although SASAG did not make a formal recommendation, we concur with the discussion in the report. Accordingly, we will ask the airlines to include in their manuals appropriate instructions prohibiting unnecessary conversations with flightcrews by flight attendants and jumpseat riders. We will also furnish similar instructions to our air carrier inspectors.*

FAA Comment: "Provide flightcrews with boom microphones to reduce cockpit workload." *Again, the discussion was not followed by a formal recommendation. Nevertheless, we concur and will encourage the airlines to provide flightcrews with boom microphones.*

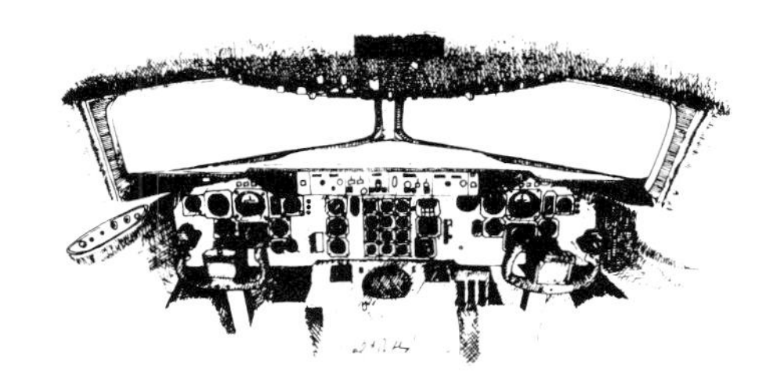

VERTICAL GUIDANCE

Most approach and landing accidents happen in weather above the lowest minimums allowed. The hazardous approach is one with just enough visibility so the pilot sees the ground or lights well in advance of the runway. Problems develop when the pilot begins to use discontinuous visual information outside the aircraft, succumbs to a sensory illusion, or both.

Approach lights help the pilot align himself with the runway, but there is no external aid helping him find and remain on the proper flight path in the vertical without aids such as VASI or the Douglas "pulse light." Without such aids the pilot attempts to orient himself in relation to the proper flight path in the vertical by looking back into the cockpit for glideslope information, then, after a quick look, lifting his eyes to look out. This is a difficult and dangerous way of guiding the aircraft since reorientation is needed each time the pilot looks in and out, and the time available is only seconds or parts of seconds. Many ILS glideslopes, also, are unusable below certain attitudes in the final stage of the approach.

At least one carrier requires that one pair of eyes monitor the ILS throughout the final approach and touchdown. When the captain is flying not knowing whether he will break out at minimums or have to go around, the first officer remains on instruments. Several fatal accidents in marginal weather in recent years suggest that both pilots were looking for the runway and both became subject to sensory illusions.

Approach lights, while necessary for runway alignment, give no vertical guidance, but actually give false indications of height during low visibility operations; the impression usually is that the aircraft is higher than it actually is. This is a sensory illusion that reaches beyond physiological capabilities, therefore other help must be provided. The pilot must be given vertical guidance in this critical area where the aircraft is close to the ground and its path capable of being changed from safe to unsafe in seconds.

The emphasis to date has been on the electronic glideslope. But it only serves within the cockpit, and even then is often unusable at the most critical stage of the approach.

Though VASI is an important and needed aid immediately available, it is not the complete answer. Other aids also should be provided. One is the heads-up-display in which independent and precise vertical flight path reference is constantly available as the pilot looks through the windshield.

Heads-up-displays have been developed and tested to a very high degree. They are being operationally used in other countries. They have not been used operationally in the United States due to differences of opinion and economic constraints. SASAG feels a decision on certification and use of heads-up-display should be made promptly. This has been far too long in coming.

In non-precision approaches, where an airport must be found from some distance, the possibility of sensory illusions becomes stronger. A single light or group of lights will often give a false indication of bank. A few lights in the night ahead actually present false height indication. The ground, during daylight, with fog, snow or similar conditions, reduces contrast and makes everything appear in the same plane making judgment of height difficult.

Many approaches were observed where under night or low visibility conditions, sensory illusions could become almost overpowering. These involved unusual, but not uncommon, combinations of sloping runways, city lights at different elevation than the airport, and airports in areas with either no surrounding ground lights, or few such lights. In some cases the runways involved had no VASI's or vertical flight path guidance of any kind. Installation of ILS, REIL's, VASI's and/or other path guidance aids should be made at such airports on a priority basis.

DME at airports will give the pilot substantial help in establishing more precise descent paths on non-precision approaches. REIL's or their equivalent are needed to help the pilot find the runway. VASI, the Douglas "pulse light" or similar aids must then be provided to give the necessary vertical flight path guidance. Perfection and installation of independent, on-board flight path angle instrumentation (including heads-up-displays) should be accelerated. Such instrumentation will provide precise flight path guidance, even on runways with non-precision aids. Availability of such aids, with training in their proper use and limitations, and on the sensory illusion problem is essential to reduce approach and landing accidents.

Cases were also observed where approach strobe lights were not turned on during day approaches even though low visibility conditions would have made their availability distinctly advantageous. In several cases, such as on runway 22 at LGA and 13L at JFK, the VASI's were left off because their glideslope did not agree with that of the ILS. Although such installations should be corrected to provide the desired glideslope agreement, the VASI should be on even when its descent path does not agree with the ILS glideslope, as long as it provides adequate obstruction clearance. The Boston DC-10 accident may well have been prevented if VASI had been available.

FAA Comment: Vertical guidance. *In our opinion, the discussion in the SASAG report concerning the need for vertical guidance has been overstated. A lay reader will surely come away with the impression that an instrument approach conducted without vertical visual guidance external to the cockpit is a dangerous undertaking. VASI's do serve to enhance safety, and FAA's installation program is one of our continuing system improvements.*

The report also makes a very strong pitch for heads-up displays and states that they have not been used operationally in the US "due to differences of opinions and economic constraints." The report also leads one to believe that heads-up-displays have been fully developed and are available for operational use by the airlines. The former statement does not consider the technical opinions of other experts within the aviation industry, many of whom believe that heads-down displays must be used for complex all weather operations. For instance, we have been advised that, "Many airline experts believe that heads-up-displays, while superficially attractive for general use—and potentially useful for some special uses under VFR conditions—will find little serious application for complex approach operations, especially under poor visual conditions. It is widely felt that adequate system integrity is not likely to be achieved and inherent optical and eye adaptation problems of such displays relative to the outside scene are likely to be limiting . . ." With respect to the latter statement in the report, the information we have is that one domestic foreign airline in November 1974 installed a heads-up-display in eight aircraft. Category III minima approved for this air carrier are based only on the fail passive autoland system installed in its aircraft. The heads-up-display (HUD) is not required by the civil aviation authorities. To date, very little operating experience has been obtained by this air carrier using a HUD because of the unusually good weather conditions experienced last winter. In view of the foregoing, we cannot agree that heads-up-displays "have been developed and tested to a very high degree" as stated in the report. However, as indicated in our comment [on Recommendation 2 following], FAA intends to evaluate the operational use of HUD.

SASAG Recommendations:

1. Accelerate the installation of VASI's on all air carrier runways, as well as REIL's on non-ILS runways.

FAA Comment: *We have previously discussed the recommendation concerning VASI's. Although the recommendation regarding REIL's has merit, current FAA criteria establishes priorities for installing REIL's on non-ILS runways. Priorities are established based on safety considerations and are installed to the extent possible consistent with budgetary limitations. As a matter of fact, many non-ILS runways are equipped with REIL's.*

2. The industry and FAA should accelerate perfection and installation of heads-

up-displays, including independent onboard flight path angle instrumentation.

FAA Comment: *The FAA, in its efforts to evaluate airborne wind shear detection equipment, among other things, is considering the feasibility of the use of heads-up-displays and flight path angle instrumentation. Our Systems Research and Development Service has contracted with the Stanford Research Institute to assist in the evaluation of wind shear detection systems, which may be heads-up or heads-down-displays. In our view, we are proceeding as rapidly as is prudent to determine the role of heads-up-displays in the conduct of instrument approach procedures.*

APPROACH PROCEDURES AND CHARTS

During studies of approach procedures and related matters, a letter from a line pilot to his airline's safety department came spontaneously and unsolicited to SASAG's attention. Although the writer's criticisms may be thought a bit harsh (and in one area they are misdirected as explained below), he has stated the basic problem so well, pertinent portions of his letter are quoted here.

" 'En route and local Terminal Charts will be readily available for use.' I appreciate the intent of this but essentially it is nonsense. The fact is the whole Jeppesen Chart system is not relevant to the radar environment. Terminal approaches on a 360° azimuth completely upended the viability of the charts now in use, a fact which seems unrecognized in spite of the body count involved. We can wallpaper the cockpit with area charts and it won't resolve the gut fact that vectoring is a sophisticated game of blind man's bluff, and the only real protection the crew has is common sense. Jeppesen, with the apparent blessing of the FAA and ATA, has met the jet and radar age head on by cramming the maximum amount of information into the minimum amount of space. The result is a monument to the printer's art about as useful to the crews as the Lord's Prayer on the head of a pin.

"Most of this info is not relevant to the realities of the radar environment and actually produces negative results in obscuring what is really important. The recent Jeppesen workbook was very helpful, but it also illustrated the above point of view. I went through it in the comfort of my living room chair, well rested, and in a good light and still had trouble finding some of the goodies involved such as those lethal little notes that are playfully distributed in a random way designed to keep a man on his toes. As a legal document these charts are superb. As a work tool, they are a disaster.

"Radar has irrevocably abridged cockpit autonomy. In my view the entire philosophy of approach and en route charts needs total revision to bring them into compatability with this fact. No one person has the answer to the problem, but it is certainly available in the collective sense. The first step is to recognize that the problem exists. So far all I see is the usual paste on approach, don't rock the boat, etc."

The writer erred, certainly unwittingly, in laying the lion's share of the blame on Jeppesen. It does not belong there.

Jeppesen is but one of two primary chart publishing agencies (the other being NOS) who must publish procedure information supplied by FAA. Although the charts can be improved, a function the publishing agencies can control, the basic problem lies in the FAA procedures and criteria on which they are based.

The crews flown with were of the unanimous opinion—as was every member of SASAG—that many procedures are complicated beyond reason. Specifically: a. charts are cluttered with unnecessary information; b. information the pilot needs (e.g., MVA's) does not appear; c. chart clutter results in vital information (e.g., minimum altitudes) being shown in print too small to read under night or turbulent conditions; d. MDA and DH data are so complicated it is often impossible in the time available to determine what the legal minimums are; e. tabular information requires interpretation that could be eliminated by graphic presentations; f. chart approach and departure control frequencies are often not the ones actually in use; g. approach procedures, SID's, STAR's, and avigation and area charts are unduly complicated and impractical in many cases and h. holding pattern entry procedures require a degree of attention far out of proportion to the results achieved. They interfere with proper management of the airplane at a critical time when such distractions must be kept to an absolute minimum.

The fact [that] one carrier found it necessary to publish a 61 page training workbook on the use and meaning of approach charts—charts that should and can be largely self-explanatory—is testimony to the inadequacies of the present system.

From studies of the procedures and observation of their use by hundreds of crews, it is clear the fundamental problem is lack of practical pilot input in criteria and procedure development. The New York La Guardia Runway 4 ILS missed approach procedure illustrates the point: "PULL UP: Climb to 5,000' outbnd LGA R-046 to intercept and proceed inbnd on CMK VOR R-200 to CMK VOR & hold NORTH, LEFT turns, 196° inbnd. Cross SCARSDALE INT (LGA R-046/14.0 DME & DPK R-315) at 4,000'; cross STAMFORD INT (LGA R-046/21.0 DME & DPK R-332) at 5,000'."

Among other things, the procedure: a. requires setting LGA VOR frequencies on NAV receiver(s); b. requires selection of new VOR courses; c. implies that the first level-off altitude is 5,000 feet when it is 4,000 feet; d. requires DME monitoring for Scarsdale and Stamford Intersections; e. if DME is inoperative or not available, the procedure requires tuning a NAV receiver to DPK and selection of two different intersection radials; f. requires selecting CMK frequency on two NAV receivers; g. requires selection of the inbound-to-CMK course on both course indicators and h. requires selection of the holding course on both course indicators.

The procedure is hopelessly complicated, impractical, and contradictory. It is clear that traffic control requirements were primary, and consideration of the pilot's problem secondary. And while it may seem unusual, given the infrequent use of missed approach procedures, to use one in this context, it perfectly illustrates the thread of impracticality that runs through a vast number of existing procedures.

Specific examples

The flight is cleared for a VOR DME-B approach at Missoula, Mont. Overhead the VOR, the copilot, who is flying, turns to the 152 radial outbound and starts to descend below the last airway MEA. The captain orders the descent stopped since the profile suggests that the MEA should be maintained. The copilot, who intended to descend to 8,600 feet was right since he had found the "hidden" note under the VOR identification/frequency box that permitted descent to 8,600 feet. The flight was on solid instruments throughout this discussion, and everyone—including the SASAG observer—had missed the hidden note. The confusion occurred on an approach to a station that is surrounded by high mountain peaks.

A flight is arriving Pocatello, Ida., (PIH) via V269 and is at the MEA of 7,000 feet. When about 15 WSW of the VOR, the flight is cleared for a VOR DME Runway 21 approach via the 15 DME arc. The flight transitions to the DME arc at 7,000 feet via the 358 radial. When asked what minimum altitude applied on radial 358, the captain said: "I guess the altitude I'm at." The chart gives no minimum altitude along the 358 radial, and the MSA (which must be visualized instead of being graphically displayed) is 6,500 feet. Since this is below the DME arc minimum altitude—but doubtless above the MVA, on which the pilot has no information—confusion reigns.

A flight is arriving Seattle, Wash., and is cleared for approach when at 6,000 feet 20 NE of the airport on a vector heading of 240°. The captain, who is busy flying, asks the copilot and second officer what minimum altitude he can descend to, and nobody can give him a definite answer.

In about one out of every five cases SASAG observed, either the approach or departure control frequency shown on the chart was not the one actually being used. In three recently received Jeppesen revisions, the only reason for the revision of some 85 pages was that these particular frequencies had been changed. In several of the cases, immediately following the revision, the frequencies actually used were still not those on the charts. Jeppesen reports that 29% of its revision pages are made solely because of frequency changes. In the first place, the information is unnecessary on the chart, and in the second place, a vast amount of money and pilot effort is spent on useless changes.

Numerous ILS approach procedures call for a descent of as little as 100 feet after procedure turn completion. This is not only ridiculous, it is dangerous. Several cases have been observed where such a descent has not been stopped at the minimum altitude because of distractions that occurred. What possible reason can there be for an unnecessary step that exposes a flight to additional hazard?

A flight approaching Los Angeles is diverted to Ontario when the approach is missed. The crew is so busy they have no time to determine which minimum actually applies in the limited time available. Chart minimums are complicated beyond reason.

The navigation and area charts show many intersections for which there seems little need. Referring to the San Francisco area chart, for example, is it really necessary to have Mt. Day and Mt. Hamilton intersections when they are only five miles apart on V107? And why is it necessary to designate the Santa Rosa radial 200 to mark Freestone intersection when the existing radial 204 forms an intersection within one mile of the indicated position?

A cursory examination of the SFO area chart indicates that at least 54 intersection radial lines, radial numbers, three-letter identifiers, and intersection radial frequency number groups could be eliminated. By careful screening and with practical criteria, it should be possible to eliminate hundreds if not thousands of such chart clutter items in the over-all airway charting system.

While many more examples of impractical procedures and cluttered charts could be offered, the above should be enough to make the point. There is much that can be done to improve these areas.

SASAG Recommendations:

1. Revise criteria for approach pro-

cedures, SID's, STAR's, and related matters with practical pilot input from air carriers.

FAA Comment: *We believe this recommendation pertains principally to the specifications for depicting these procedures on the instrument approach and terminal area charts rather than the criteria for developing these procedures. The FAA has undertaken a flight procedure improvement program to identify and resolve potential problem areas in the development of and charting of these procedures.*

A number of potential problem areas have been identified and just recently we completed a review of all NOS approach charts to identify potential problems of pilot interpretation.

A review of the entire SID/STAR (departure/arrival) program is underway and emphasis is being placed on charting techniques. Simplification of selected SID's/STAR's is expected to be achieved as one of the objectives of the review.

The Terminal Instrument Approach Procedures (TERP's) Advisory Committee is also reviewing instrument approach procedures and developing departure procedures to update the TERP's Handbook.

We believe that substantial improvements can be made in the area of charting. It should be noted that pilot/user organizations (such as ALPA, NBAA, ATA, AOPA, etc.) are represented on the Flight Information Advisory Committee (FIAC) which has long been established to provide input to the Interagency Coordination Committee which is the body responsible for developing charting criteria.

2. Provide that missed approach procedures be shown graphically.

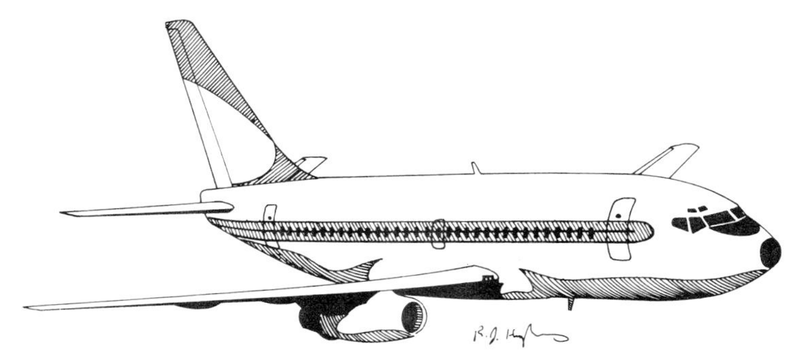

FAA Comment: *We will investigate the feasibility of graphically portraying the missed approach and will be included in our procedures improvement program.*

3. Eliminate approach minimum categories A, B, C, and D, and provide a single minimums value instead. Improvements in airplane handling characteristics and instrumentation make the existing minimums differences and complications unnecessary.

FAA Comment: *This recommendation has merit, as it pertains to straight-in ILS instrument approach procedures. We do not concur that these categories can be eliminated with respect to circling approaches because of the different turning radii of the several aircraft categories. We will review the TERP's criteria with respect to the categorization of aircraft as they now apply to straight-in instrument approach procedures. This, of course,* *may mean in many cases that weather minima would be lowered for category C and D aircraft.*

4. Revise and simplify criteria for minimums with approach components inoperative.

FAA Comment: *Again, we believe this recommendation pertains to the description or depiction of these minimums on the Jeppesen approach charts. Jeppesen present depiction of minimums for each type of operator, plus consideration of the inoperative component table has introduced an element of complexity. Jeppesen is planning to issue a new format by the first part of 1976. Under existing operating procedures, the air carriers and airline pilots can work directly with Jeppesen on charting matters. The FAA will continue to work toward simplification in instrument procedure design and charting.*

5. Provide Minimum Vectoring Altitude (MVA) information on approach charts (in graphic, not text fashion) along with the means for the pilot to know his position within the bounds of the various MVA areas.

FAA Comment: *Previous efforts to provide an effective means for displaying MVA information on approach charts have not been successful. The FAA, in conjunction with NOS, is presently developing experimental charts to display terrain information; and the USAF, in an independent effort is planning a military evaluation of approach charts with MVA information. These various techniques will be evaluated to determine the effectiveness of this type of supplementary information.*

6. Eliminate approach and departure control frequencies from approach pages along with other superfluous and unnecessary data.

FAA Comment: *Such [a] concept requires careful study before implementing. Arbitrary removal of frequencies can result in an increased workload for air traffic facilities, particularly where there is a high ratio of VFR traffic.*

A recent letter was sent to the field to encourage a reduction of the number of frequencies on IAP charts; e.g., removal of multiple sector frequencies. The objective was to list one VHF and one UHF frequency for use as a backup frequency *or for use of VFR pilots. A recent letter was also sent to the field to elimiate superfluous notes and unnecessary information from SID's and STAR's. Both actions will take place on a routine basis to minimize user costs.*

Also, removal of all frequencies would require pilots to cross reference a separate document when an initial contact frequency was incorrect, inoperative, or misunderstood. Some regions have expressed concern and a need for most or all sector frequencies to continue to be published. Flight publications should be designed to reduce "head-in-the-cockpit" time, not increase it.

7. Eliminate existing holding pattern entry procedures. Require only that arrival at the holding fix be at or below maximum holding speed and that entry be accomplished by turning the shortest way to enter the pattern. Revise holding pattern airspace criteria if necessary to accommodate this recommendation.

FAA Comment: *We believe that the benefits of this recommendation are not sufficient to justify the enormous costs that would be required. Enlarging holding patterns so as to be more convenient rather than involving a safety issue, would substantially hamper efficient movement* *of large volumes of traffic. Enlarging holding patterns would require the redesignation, relocation, and recharting of numerous intersections and holding fixes throughout the national airspace system. These changes would also require substantial reprogramming of all ARTCC computers and most stored flight plans.*

We are not aware of any pilots experiencing difficulty in this regard. We believe significant progress has been made to reduce holding situations through various management programs designed not only to minimize holding but to conserve fuel.

COMMAND DETERIORATION

In some airline crews SASAG found a deterioration in command that adversely affects safety. Various factors have created this condition.

As the airline industry and the aviation governing agencies grew, more procedures and precise methods of operation have been developed. Air Traffic Control is a large part of this; pilots are strongly directed by ATC, with the need for precise performance of ATC's requests. Company procedures have resulted in computer flight planning that gives the pilot little time, opportunity or motivation for making the final flight plan selection. The movement toward all weather operation generates a subconscious, but real, feeling in pilots that flights are no longer canceled because of weather. Weather information is often inadequate and presented to the pilot in a form not conducive to a full judgment of conditions. Cabin attendants do not relate to the flight crews, often are part of marketing and not part of the safety team and do not understand they are under command of the captain.

With all this has come a new type pilot, sometimes from the military, who is accustomed to following orders and flight plans prescribed by military necessity. All members of the flight crew are often of a similar age and, in some cases, copilots are older than the captain.

These factors have developed pilots in command who sometimes are confused as to where their command begins and ends. They tend to do what the system tells them, responding precisely to ATC requests with little question even though a flight path may be through hazardous weather. They follow flight routes the computer calls for even though better routes are available for weather avoidance or economy. They tend to leave the problem of terrain clearance to the ATC controller. They are not commander of the cabin team, and conditions have occurred in which the cabin team, on its own, initiated emergency action that was not needed and created dangers.

The cockpit often develops too much of a "buddy" atmosphere with excessive kidding, and with the copilot attempting to overrule and domineer the captain, and doing it successfully.

While SASAG does not take any position in the two man-three man crew controversy, there is no question that the *policies* carriers apply to the third man often result in significant cockpit distractions.

Management, not wanting the third man, tries to get rid of him by the ineffective practice of giving him no duties. This generates non-standardization and distractions in the cockpit since the third man must respond to whatever individual captains want, or, when the captain is silent on the matter, devises his own procedures. This does not contribute to safety, but detracts from it.

It is almost automatic that captains and copilots alternate flight legs and in many cases, because pilots are equal in age and experience, the captain is reticent and slow to correct any errors of the copilot and seemingly embarrassed to take over a flight leg when the weather is bad. SASAG witnessed extremes such as a copilot making a landing during winds of 50 knots, doing a poor job of flying, and touching down at reference speed plus 40 knots, with the captain making no move to correct him.

This is not a criticism of the pilots, but rather airline managements, the FAA and the pilots' unions for allowing the situation to develop.

When pilots are upgraded to captain, little is done to include training in the command function, its limitations and responsibilities within the system, either by the airlines or the FAA. The unions avoid their responsibility in promoting command responsibility by allowing economic negotiations to be the primary factor. They could, by realizing their responsibility, promote the stature of the captain and his command responsibility.

Aviation has not reached a stage where it can be automated with inflexible procedure and method. The science of weather forecasting, and reporting, has not reached an exact completely reliable state, and an aircraft commander is faced, almost daily, with unexpected weather changes and conditions that require judgments only he is in a position to make. Aircraft, though highly reliable, still have malfunctions with emergencies that must be overcome by the aircraft commander.

In summation, it must be realized that aircraft operation still requires command by an individual. Once the aircraft is in motion, it must be that individual's responsibility for the safety of his passengers and crew. The industry, airline, government, pilots, and their unions, have allowed this command to deteriorate and become indefinite. It is their responsibility to see that it is fully restored.

SASAG Recommendations:

1. Carriers provide training in crew coordination and command problem-solving to upgrading captains and periodically during recurrent training.

FAA Comment: *We concur and our current regulatory project that proposes better air carrier line check procedures includes a provision for training in the command function. We will request our inspectors to be sure that air carrier training programs are adequate in this regard. We will also ask our inspectors during en route inspections to be alert to any deterioration in the command function.*

2. Unions direct communications to members concerning command and the captain-copilot relationship.

FAA Comment: *This recommendation is not directed to FAA.*

3. Where there is a third crewmember, he should be given specific duties.

FAA Comment: *It would be more appropriate for the airlines to consider this recommendation since the aircraft discussed in the report are those certificated by FAA for a crew of two pilots. Thus, FAA does not consider it necessary for a third pilot crewmember to perform any duties or functions on these aircraft. We have serious reservations about a third crewmember performing any duties other than reading checklists and handling communications.*

AIR TRAFFIC SYSTEM

The air traffic control system is a major force in aircraft safety. In accomplishing its task of traffic separation and flow, it has created hazards, slowed traffic, restricted productive flight by all segments of the aviation industry, and used energy in frightening amounts.

SASAG's particular scrutiny has been the approach and landing phase of flight. While searching this area we find that the air traffic control system has loaded the cockpit with extra work to the extent that, during critical phases of flight, the air traffic function uses most of one crew member's time removing him as a useful navigational and system's management assistant. This, in turn, decreases the safeguards of redundant monitoring, callouts and relief of the workload from the operating pilot. All this creates dangers in the approach and landing portion of flight as well as other areas.

Heavy communication load between ground and aircraft is not only distracting but by its nature is susceptible to errors . . . radio contacts in terminal areas often average more than one every five seconds for busy flight periods. Many occur with repeats, misunderstandings and improper response to requests. SASAG observed an IFR room where a mistake in altitude caused a flight to be at 7,000 feet when it should have been at 6,000 feet. The reason was a communications error. How many of these go unnoticed?

The atmosphere of an IFR room is a jumble of confusion with demands on the human of the highest order.

The paper slips, which record individual aircraft and establish priority, are handled by people. The slips are printed out by computer in many cases, but regardless of the priority, stacking is done by hand.

Radar scopes demand the complete attention of the operator. Hand-offs from one scope operator to another are oral, with "eye balling" the major method of traffic separation and flow.

It is indeed amazing that more conflicts have not occurred. There have been accidents and unless dramatic remedial action is taken, there will be more. It is frightening to visualize the ever present potential catastrophe of two jumbo jets in collision.

The pressure of such a system, which demands that both the traffic controller and the pilot have the highest order of concentration, not only creates distraction but conflict as well. On many occasions tempers become short, improper communications are used and arguments develop.

The controllers often show disdain for pilots and do not always understand the pilot's problem and the fact that flying an aircraft is more than just air traffic control. It must be navigated, its complex systems kept in proper order and communications accomplished. The controller has one task, although often difficult and of high pressure; traffic separation and movement. Talks with controllers frequently bring out their lack of understanding of the pilot's job.

It is the nature of the controller's job to make him somewhat cocky. He must have strong confidence in himself to "eye ball" traffic, move it around as on a chess board, and feel confident of his moves. This develops an atmosphere in which traffic controllers attempt to usurp pilot command authority.

The cockpit resents this intrusion upon pilot's command. The pilot often is not being consulted concerning decisions for which he ultimately will be responsible and which only he is competent to make.

These attitudes create conflict between ground and air that promote emotional response in an environment where it cannot be tolerated.

This is not meant to be a finger pointing criticism of controllers or pilots, but rather of the system. The conditions do not exist everywhere and often the operation is relaxed and friendly. But as traffic becomes heavy, this friendliness disappears.

The system is a jumble of people, radar scopes, communications lines and stacks of paper strips, people communicating by voice, by radio to impersonal aircraft out in space, while in the control room people are milling about, talking and creating distractions.

It is obvious that the system is reaching a point of "critical mass." Its methods, confusion, and piling up of radar and people as traffic grows, will not take care of the future. It detracts from efficient aircraft operation and creates hazards which lead to accidents other than collision.

The direction our government is taking toward future air traffic control is not commensurate with available technology. A program of top priority by an independent body is required to create a system that will reduce human error to a minimum.

A new system cannot be developed quickly. It is urgent that the existing system be improved even though such changes are only crutches for the ailing and ancient system we now have.

SASAG, with a deep sense of urgency, recommends these immediate actions, but these should not hinder development and installation of a new system. It is desperately needed for safety.

The pseudo authority ATC maintains over flight operations causes command deterioration. The matter of who keeps aircraft clear of terrain has been falsely relegated to ATC in the minds of pilots. This has been proven by various accidents. It simply cannot be allowed and the primary responsibility for terrain clearance must remain in the cockpit.

However, the cockpit crew does not always have the necessary information during radar vectors to know exactly where the aircraft is relative to terrain. DME information on the airport will help considerably; but it is evident that some form of electronic pictorial navigational information for the pilot is also needed.

In a similar vein, ATC demands that pilots fly weather that is hazardous. Vectors are often given into thunderstorms even when the pilot requests other routing. One case is the pilot who told ATC their vector would take him into a thunderstorm and that he did not want to fly it. The reply from ATC was, ". . . either turn left [into the storm] or you will have a violation." Such action simply cannot be allowed! Regulations should be adopted prohibiting ATC from denying a pilot clearance to circumnavigate hazardous weather upon his request.

ATC often refuses requests pilots make to avoid turbulence. An example was a flight in the Denver area in which ATC cleared the flight to descend over mountains into an area of turbulence. The captain asked to delay the descent, but his request was refused. In the descent, he encountered moderate turbulence.

ATC also decides routes it feels best for weather avoidance and often a pilot is required to fly a route he did not file because of it. This judgment does not belong within ATC. It is an operational function, not a traffic separation matter, and the pilots, airline and meteorological services are better judges of routing. There are case histories to prove ATC's poor operational judgment. An occurrence was witnessed during evening departures from Kennedy when ATC denied the routes filed in favor of a "better weather" route. The ATC route proved, when flights became airborne, to be full of thunderstorms and violent weather while the routes pilots requested were in the clear. ATC's job is to provide for traffic separation, but it is not ATC's job to make flying judgments. They create hazards in doing so. ATC does not have all the information available a pilot has, nor do they have the knowledge and ability to do the job . . . in short, they are not pilots flying that particular flight.

ATC appears not to understand that there is more to flying than traffic separation. They seem to expect the pilot to accommodate every wish . . . and at times these demands are not possible.

Speed-up and slow-down requests are typified by a DC-10 approach and landing in which the DC-10 was told to slow to 200 knots 40 miles from the airport, and then again to 160 knots 20 miles out. The pilot requested from ATC the reason for this slow-down and was told a light plane was making a practice ILS approach. Certainly a better solution could be found.

SASAG members continually observed landing clearances from too high, too close to the airport that resulted in high descent rates too close to the ground: ". . . flight cleared for approach when 1 mile from outer marker at 5,500 feet . . . chart altitude called for 2,115 feet." Another: "Flight cleared for approach when 4 miles from outer marker at 7,000 feet and 250 knots. Chart glideslope for outer marker was 2,200 feet." A 747 landing was witnessed at SFO during a non-precision omni-approach to the West with a 900 foot ceiling. Approach Control did not clear the flight for approach until it was 7.5 miles from the field and still at 4,000 feet. This required a high descent rate to low altitude and then recovery, which is difficult and dangerous. In addition, the cockpit was so busy attempting to obtain the clearance from Approach Control, and responding to their demands, that the copilot did not have time to aid the captain in back-up radio navigation. During much of the last part of the flight, the copilot was too occupied as a communicator to call out altitudes, speeds, etc., until far into the approach, and the captain was flying the 747 solo!

ATC demands flight maneuvers that are unreasonable such as the witnessed case in which the landing runway was changed three times when the aircraft was less than 5 miles from the airport.

Clearances are changed during critical stages of flight and one case revealed a new SID being given a flight during takeoff roll. This means a restudy by the crew of SID departure procedures, changing radios and readjusting to a complete new mental orientation, when attention is more importantly needed in flight and aircraft management.

The number of transponder code and radio frequency changes issued demonstrates ATC's lack of knowledge of the cockpit workload.

There does not seem to be an understanding that right after takeoff the pilots are busy flying the aircraft through a critical phase. Departure communication procedures must be revised to eliminate these cockpit distractions during takeoff and initial climb regime. The airlines should equip all their aircraft with dual selector VHF tuners so that changing frequency after takeoff would be simply a matter of flipping a switch instead of one pilot looking down to select a new frequency.

ATC vectors often require an aircraft to maneuver long distances and for protracted periods at low altitudes. Flight at low altitude puts aircraft in the path of general aviation and other low altitude aircraft. Paradoxically, aircraft being vectored at low altitudes (under 10,000 feet) will finally be held at some intermediate altitude so close to the runway that a high descent rate is required to get in the airport. ATC seems to require more and more space for vectoring and this is a wasteful and dangerous use of air space.

ATC does not appear to adequately police itself. Radio procedures deteriorate, confusing non-procedure language is frequent. Background noise, that occasionally includes laughter, makes it difficult for pilots to understand transmissions from ATC. While ATC has its checking methods, they do not do the job. The Administrator should provide for the effective policing of these vital functions.

SID's and STAR's appear to be designed by ATC without sufficient operational input. This should be corrected.

The list of air traffic system problems is long and detailed, but basically they reduce to these; a system too dependent on the human element; a system that has grown from old concepts with complex fixes applied to it in an attempt to accommodate its inadequacies. This, in turn, has created a monster of procedures, rules, methods and confusing interplay between people who are separated by distance, technical knowledge, and understanding of each other's problems.

FAA Comment: *SASAG stated . . . that, "It is indeed amazing that more conflicts have not occurred. There have been accidents, and unless dramatic remedial action is taken, there will be more."*

Statements that the IFR room is a jumble of confusion and a breeding ground for incidents and accidents are inaccurate and misleading. Such description undoubtedly results from a lack of knowledge of the methodology used in these facilities.

We recognize that any environment requiring high demands on the human element, and one which is largely dependent upon human judgments and reactions, is also one that is susceptible to human failings. FAA has a keen awareness of these demands and continuously strives to identify and eliminate factors that cause or contribute to human failures.

We also note that SASAG states . . . "The number of transponder code and radio frequency changes issued demonstrates ATC's lack of knowledge of cockpit workload."

At the present time, there is not too much that can be done about the number of frequency changes. As an aircraft flies from one area of responsibility to another, a frequency change is necessary.

We believe we have reduced the number of transponder code changes with the advent of automation. Beginning early in 1976, most pilots flying in the 48 contiguous states will be able to keep one individual radar identification code from takeoff to landing.

We have programs to familiarize the controller with cockpit workload. It is part of the controller's training that certain ATC procedures are based on cockpit workload; the SF-160 familiarization program which permits controllers to fly in the "jump seat" and observe ATC from the air.

SASAG Recommendations:

1. To assure realistic traffic management, with priority going always to flight safety, SASAG strongly urges FAA to provide adequate policing of ATC operations.

FAA Comment: *We agree with this philosophy. Safety is our foremost concern, and we employ several means to aid us in identifying weaknesses. Our National Evaluation Staff is charged with this responsibility at headquarters level. Our regional Air Traffic Divisions also maintain evaluation staffs and work even closer with facilities to accomplish their goal. There are local facility programs geared to eliminate errors and improve situations identified as questionable as they relate to safety. Though we believe these programs and methods to be effective, there is always room for improvement. We are taking steps to refine and improve these programs.*

2. Establish a reliable system of ATC performance accountability when traffic control is substandard.

FAA Comment: *We believe our current system of accountability for substandard performance is adequate. The action we intend to take to improve the evaluation of the air traffic system will result in better enforcement of present policies. This recommendation is closely related to Recommendation 1 [in this section] and as related in the response to that recommendation, we are taking steps to refine and improve our existing evaluation programs.*

3. Conduct a study to determine whether the air traffic system would be operated more efficiently with advanced technology as an independent public company.

FAA Comment: *There is nothing in the SASAG report indicating that a "public company" could do the job any better. The existing NAS systems, specifically the NAS Stage A en route system, the ARTS terminal systems, and the flight service station systems, are quantum leaps ahead of any ATC system in the world. Yet, the US system, with unparalleled flexibility to accommodate all users, enjoys an enviable and unmatched safety record. Certainly, the FAA would like to take advantage of all of the latest technological advances, but we must operate within budgetary and fiscal constraints.*

4. Develop pictorial area navigation instrumentation using modern technology.

FAA Comment: *We believe that our program for improving charting will obviate any need for regulations that would require the airlines to install pictorial area navigation equipment.*

5. Abolish high, too close to the runway, ATC clearances that result in high descent rates.

FAA Comment: *We believe that this problem may be in reference to a combination of keep-em-high and visual approach procedures. Operations of this nature, which could compromise a pilot's handling of his aircraft, are to be avoided. Our facilities have an awareness of these situations and are taking steps to preclude high descent rates which could compromise safety.*

6. Abolish communication by ATC when the aircraft is inside the outer marker or equivalent distance and during the critical period after takeoff.

FAA Comment: *This is a rewording of Recommendation 3 [following "Ap-*

proach and Landing" section] with the injection of departing aircraft. The same rationale applies. We agree that communications with departing aircraft should be kept to the absolute minimum.

7. Require all airline aircraft be equipped with dual tuning VHF communications heads.

FAA Comment: *Although this recommendation has merit, we do not believe that regulatory action is needed at this time. Nevertheless, we will encourage airlines to install this equipment.*

8. Reduce excessive vectors, especially at low altitudes, to a minimum.

FAA Comment: *Air traffic procedures specify that vectors should be used when it is necessary for separation purposes, required for noise abatement considerations, or there is an operational advantage to the pilot, or controller, or when requested by the pilot. We agree that excessive radar vectoring should be avoided, especially at low altitudes. We have identified some locations at which lengthy low altitude, slow speed, vectors are used frequently.*

We are working on methods to reduce excessive vectors by modifying procedural applications at some locations. An example is the flow control application for the Dallas/Fort Worth Regional Airport. This procedure is designed to reduce delays, provide improved traffic flow, conserve fuel, and improve the operational interface between the terminal and en route options. By design, it minimizes excessive vectoring. Because of the complexities of airspace structure at some locations, this method cannot be applied. When this becomes the case, alternative methods must be explored.

9. In order to better define the responsibility and authority of FAA controllers, change the term "controller" to "coordinator."

FAA Comment: *An air traffic controller controls air traffic. Any other connotation would only serve to confuse the issue. We do not believe adoption of this recommendation would improve the safety and efficiency of the ATC system.*

10. Adopt regulations prohibiting ATC from denying a pilot clearance to circumnavigate hazardous weather upon his request.

FAA Comment: *We believe that consideration and understanding of weather hazards must be thoroughly understood by controllers. We will continue to stress this issue. Part of the air traffic controller's training is centered around weather and its effect on that ATC system. The only reason an aircraft would ever be denied approval to circumnavigate weather is other traffic. The controller must provide separation between aircraft, regardless of weather. The adoption of this recommendation could create other undesirable situations.*

NON-STANDARD COMMUNICATIONS

SASAG noticed many cases of nonstandard communications by ATC controllers and flight crew members. This leads to misunderstanding of clearances by the flight crew or the controller and jeopardizes safety. Some controllers in high density areas run their words together and speak too fast, thereby causing crew members to ask for clearances to be repeated. This hinders communication. In high density areas some controllers give "shot gun" clearances to three or more aircraft without stopping long enough for individual flights to respond, yet expect the pilot to respond accurately and quickly in sequence after clearances to many aircraft are given. This also "blocks out" the chance for a pilot to make inquiries or ask for corrections which might be of immediate importance.

Many cases of non-standard phraseology were observed: Phrases such as "Forget five thousand," or, "I want you out of seven," are common. Statements like, "I want you at one four thousand by Hampton," or such non-standard rapid usage creates the feeling in the cockpit that the controller is trying to fly the airplane.

Some controllers are dictatorial in demanding that air crews respond to their commands. Often flight crews hear background interference from ATC radios with irrelevant conversations, including laughter, that make primary communications difficult.

At times controllers become sloppy in aircraft vectoring. There are cases where the controller simply "forgot" the flight and extended it with excessive vectors. On other occasions, cases were observed where the aircraft was vectored through the localizer course without the pilot being forewarned.

Departure control frequency changes are often demanded too early following takeoff, causing pilots to have their heads in the cockpit changing frequencies at a critically low altitude.

These irregularities are simply examples of serious departures from proper procedures and lead SASAG to feel that proper checking and proficiency maintenance of the controllers and communications is not being accomplished. Correction is needed.

SASAG Recommendations:

1. Require controllers to issue clearances in concise words and phrases.

FAA Comment: *We agree improvement in this area is possible, and we will continue to pursue it. We have concise words and phrases (phraseology). The Air Traffic Service Director has made it clear he expects adherence to them. In the August 1975 edition of the ATS Bulletin, we stressed the importance of correct phraseology. We also made a video tape on correct phraseology and disseminated it to our field facilities.*

2. Require that pilots positively acknowledge a clearance immediately after issuance and that controllers issue and get response to one clearance at a time.

FAA Comment: *For the most part, this is done. No doubt there have been instances where it should have been done and wasn't. There are situations that must be treated separately; e.g., the final phase of a GCA (PAR) approach, or situations requiring immediate action to prevent impending traffic conflicts, etc.*

3. Take positive steps to eliminate all background noises from ATC facilities, including unnecessary chatter in control cabs, laughter, etc., and "bleed over" from one frequency to another.

FAA Comment: *Background noise in our facilities has been a problem for a number of years, especially since the introduction of the lightweight headset. We have contracted to test a quiet headset at Cleveland ARTCC in the near future. We have also improved the acoustical qualities of our facilities. We agree improvement is possible and will re-enforce our efforts to achieve it.*

4. Require air carriers to provide initial and recurrent training for their pilots on the necessity for standard phraseology.

FAA Comment: *Although such training is presently required by FAR 121, we will ask our inspectors to have their assigned carriers re-emphasize such training during initial and recurrent pilot training.*

AUTOMATIC TERMINAL INFORMATION SERVICE (ATIS)

ATIS is being cluttered with NOTAMS and non-essential information. Some ATIS broadcasts are so lengthy crew members often have to listen to two or more transmissions to receive the entire information as their concentration on ATIS, during such a lengthy time, is often interrupted by routine traffic control communications or aircraft management duties.

Approach Control is not very sympathetic to this and in some cases, when the flight crew requested what type approach was in use, was told, "You should have gotten it from ATIS. Get it and call me back."

Many times ATIS information is not current. A typical case was an ATIS that reported the ILS operating on test. When the crew contacted Approach Control, they were cleared for an ILS approach. The crew questioned this because of the ATIS information and was told, "That's old information, it's okay now." ATIS altimeter settings often disagree with other QNH information.

ATIS was originally designed to give the pilot notice of the type approach in use so he could prepare for it in advance. But it has grown as various ideas, apart from its original purpose, were added. The present handbook states, "ATIS information should provide advance information to arriving and departing pilots concerning operational, meteorological, and Notice-To-Airman data." The handbook lists "message content" as: a. airport identification and phonetic alphabet code by the ATIS message; b. weather sequence and time; c. departure runway; d. instrument approach in use, runway if different from instrument approach runway; e. pertinent NOTAMS and airman advisories; f. if the airport is above 2,000 MSL and temperature is above 85°F, the statement, "check density altitude"; g. other operational information as local conditions dictate in coordination with ATC. This may include such items as VFR arrival frequencies, mu-meter readings, temporary airport conditions, etc; h. instructions for pilots to acknowledge receipt of the ATIS message on initial contact with appropriate control and state the phonetic alphabet code.

This is far too much information and is almost ludicrous as it infringes on the boundaries of primary instruction and the NOTAM and weather system. All these added items make ATIS lose its effectiveness and add to cockpit distraction instead of reducing it.

Arrival information should be separated from departure information and both include only "need to know" information.

The "beep" at the end of the ATIS is annoying and tends to distract the pilot. Item 8 "instructions for pilots to acknowledge receipt of ATIS, etc." is superfluous and adds length to the ATIS. Any pilot aware of ATIS knows what to do with it.

SASAG Recommendation:

Arrival information should contain only: a. airport identification and phonetic code of ATIS information; b. ceiling and visibility where below certain values; c. wind direction and velocity; d. type of approach in use with runway when needed.

FAA Comment: *Controllers were required to provide pilots with certain information before the advent of ATIS. This required information can be used by the pilot in the departure as well as arrival phases of flight. The purpose of ATIS is to reduce frequency congestion and con-*

troller workload by automating certain repetitive noncontrol information. The ATIS also permits the pilot to receive this information at their leisure times when cockpit duties are at a minimum. The contents of the present ATIS broadcast have evolved from all user input and contain all of the above items plus the altimeter setting. We believe it to be the best product possible to satisfy the requirements of the majority of users. We continually emphasize improvement of the ATIS program. This is done by recent procedural changes in updating information on the ATIS and periodic articles in our ATS bulletin.

HAZARDOUS WEATHER

Although thunderstorms, "clear air turbulence," and wind shear have caused many accidents, serious deficiencies were observed in information available to flight crews on hazardous weather.

Except for hourly sequence reports, the only thunderstorm information available is usually the NWS Radar Summary. This information is: a. nearly one hour old when first available off the printer; b. available only on the ground; c. not available for the full 24 hour period and d. not specific enough to provide adequate correlation with the intended route, nor with potential storm avoidance routes.

Information on low level wind shear was virtually nonexistent, and pilot training on practical handling of low level shear appears minimal.

Information on "clear air turbulence" was frequently inadequate and often totally lacking. Although knowledge of tropopause height is extremely important, few carriers provided adequate tropopause height information in the flight papers. Most that did provided it on an "area" rather than the more desirable "point" basis.

The record is clear that these phenomena need far greater attention than they are now getting. In 259 incidents in the five year period from 1969 through 1973, turbulence was the prevalent first cause. It occurred more than three times as often as the next most frequent cause. Of the 107 accidents occurring during the in-flight phase, 75 were caused by turbulence.

In 1974, of the 38 non-fatal carrier accidents, 17 were caused by turbulence in which passengers and crew members suffered broken legs, broken necks, and other serious injuries. As of April 28, 1975, two of the eight non-fatal accidents were associated with turbulence. It appears that the recent fatal 727 accident at JFK was related to shear associated with thunderstorm.

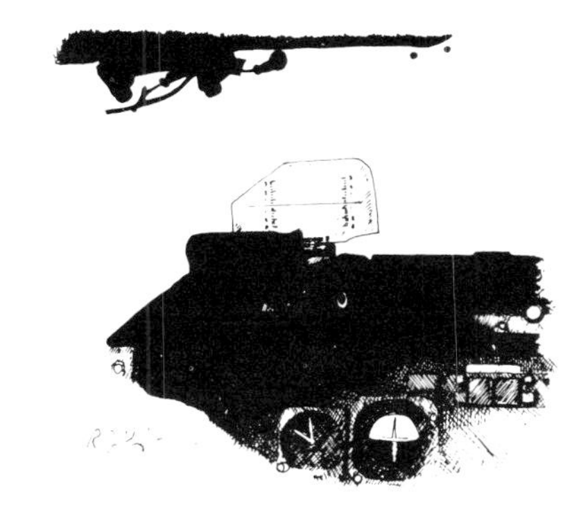

Specific cases

Case 1—On July 23, 1973, thirty-eight persons lost their lives in an air carrier accident at St. Louis. Results of the NTSB investigation indicated that: a. the flight continued an ILS approach into a thunderstorm; b. there was a lack of timely issuance of severe weather information and c. the terminal weather was improperly assessed.

Case 2—On Dec. 17, 1973, a DC-10 with 153 passengers and 14 crew members crashed at Boston. Fifteen passengers and one flight attendant were injured, seriously in three cases. The airplane suffered major damage.

The airplane struck approach light piers when an increased descent rate developed due to low level wind shear. The captain's ability to detect and arrest the increased descent rate was adversely affected by lack of information on the existing wind shear and the marginal visual cues available.

Case 3–On Jan. 4, 1971, between 2152 and 0023Z, nine air carrier flights missed approaches at JFK with ceilings and visibilities well above minimums. Two flights diverted to alternates, one having missed the approach twice. During the same period, a DC-3 crashed on approach to LGA. In all cases the cause was low level wind shear.

Case 4–On May 13, 1974, twelve people were injured in a turbulence incident in which the airplane involved was in moderate to severe turbulence for some 240 miles. The flight was, for this distance, continuously in an area of heavy thunderstorms of increasing intensity, flew through both the SW and SE ends of an intense line of thunderstorms imbedded in the area, and flew through the cell with the highest reported tops in the area.

Case 5–On Dec. 8, 1968, an Aero Commander was destroyed in flight in extreme turbulence near Wolf Creek, Mont., with fatal injuries to all occupants. The only relevant weather forecast available to the pilot indicated: "MTNS AND PASSES NRN IDA AND MONT W CONT DVD FQTLY OBSCD BY CLDS OR SNW. LCLY MDT TURB BLO 120 WRN HALF MONT AND NRN IDA WITH WINDS GSTG LCLY TO 40 KTS." The flight's chosen cruise altitude was 12,000 feet.

Only one carrier–Northwest Airlines–has a practical and proven hazardous weather information and avoidance system. It is known as "Turbulence Plot" and provides up-to-the-minute, graphic information on thunderstorm activity, "clear air turbulence," and low level wind shear.

In every one of the above cases, the Northwest Turbulence Plot (TP) system provided specific, timely, and graphic information on the hazardous weather that caused the accidents.

For example: At the time of the DC-10 crash on Dec. 17, 1973, the Northwest TP message gave specific information on the existing shear condition, its crews verified the shear's existence and reported same to the tower. In the Jan. 4, 1971 missed approach series at JFK (and the DC-3 accident at LGA), the only flight to land successfully on the first attempt was Northwest 747 Flight 232. That crew had specific information on the existing shear and was prepared to cope with it without difficulty.

In Case 4, the Northwest TP system graphically showed the area of heavy thunderstorms, the imbedded line, the individual cell referred to, and that the complete area could have been avoided by minor detour. In the fatal accident of Case 5, the Northwest TP message advised avoidance of the specific area and altitude where the accident occurred.

Many other documented cases exist in which Northwest flights, by virtue of the TP system and its attendant training, were able to avoid difficulties due to turbulence and wind shear while other flights in the same area did not. In view of these facts, it is difficult to conceive of a reason for other carriers not adopting a similar system. Nor is there any practical reason why a similar system cannot be made available to general aviation.

SASAG Recommendations:

1. Require all scheduled carriers to adopt a system similar to the Northwest Turbulence Plot system.
FAA Comment: *The development of a regulatory project along the lines of this recommendation is nearing completion.*

2. Provide a means for distribution of this service to general aviation.
FAA Comment: *We will pursue this recommendation further with the National Weather Service.*

3. Label aviation chart segments where wave turbulence is common.
FAA Comment: *Although this recommendation has merit, it needs to be thoroughly explored by FAA, the airlines, the National Weather Service, and the charting agencies. We will explore the feasibility of this recommendation further to see if wave turbulence information can be provided on the charts without further clutter.*

4. Revise jet routes to coincide with established wave bypass routes. This will give a large measure of "automatic" wave turbulence protection, even to flights without TP information.
FAA Comment: *At an FAA/meteorological meeting Aug. 26-27, 1975, with FAA, airlines, NWS, ATA, and others, it was agreed "Northwest Airlines and United Airlines would submit recommended mountain wave deviation routes to ATA. ATA will work with FAA to establish these high altitude jet routes."*

5. Cancel the proposed digital radar code plan that would grid the ground weather radar scope into a series of 22 mile grid boxes. This would coarsen the basic source data to an extent that would effectively "kill" the thunderstorm portion of the Turbulence Plot program.
FAA Comment: *The NWS tested the new digital radar code plan at Charleston, S.C.; Grand Island, Neb,; and Waycross, Ga. for one month. The NWS is currently evaluating the results of the test. The NWS is also reviewing user comments and will advise the FAA of their conclusions before implementation of the new product.*

TRAINING

FAR training requirements for airline pilots need to be rewritten. Pilots who are trained in accordance with FAR 61, 121 and Appendices A, E, and F, and who have passed their flight checks, have not been trained for the tasks they will be obliged to perform. Because regulations, as they are now written, emphasize proficiency requirements on individual exercises, each one separate and distinct from others, blending of them as seen during airline flying is overlooked.

The system now used for training, by isolating each exercise into a separate study increment, seems designed to serve the examiner rather than the candidate. Training is synthesized in a way that provides for a "grade" for each exercise and overlooks the important need for the "harmonizing" of flight maneuvering . . . the essence of an airline pilot's daily work.

Some improvements have been made as in the case of substitution of simulator in lieu of actual aircraft flight training. But the wide gap that now exists between "training" as presently practiced, and the realities of airline flying is of great concern to operators and pilots. A totally inadequate accommodation to this need is the regulatory requirement for the captain's "25 hour" line check after transition training. Administration of this requirement has been left to the airlines. Its effectiveness is subject to the whims of airline operations' managements and in most cases pilots receive little or no benefit from it other than the experience they gain under the surveillance of a "safety-pilot." No provision exists for training in "line" functions. Problems involving weather, communication, cockpit discipline, in-flight emergencies (with their attendant practical problem solving), route planning, dispatch, equipment deficiencies and crew management are ignored. No training is required in the application of things newly learned to experiences previously acquired.

Airlines today hurry to obey the regulations now in effect in order to be "legal." Staffs usually are not inclined to perform research needed to improve training programs.

FAA in the past has been progressive in its encouragement of aircraft simulation development and flight training program's revisions associated with them. Practical progress, however, has limited itself to accommodating the old flight training rituals permitting most required exercises to be performed in the aircraft flight simulator. A new departure is needed to more closely align the simulator aircraft training program with the actual line requirements.

Accidents happening today, as in the past, have not been caused by crewmen not having been trained. In cases of "pilot error" the cause has often been traced to his not having put into action training already received.

A strong FAR line reinforcement "coaching" requirement is needed.

SASAG Recommendations:

1. Rewrite FAR 61, 121 and Appendices A, E, and F to consolidate the simulator and aircraft training exercises into a "line flying phase of flight" requirement.

FAA Comment: *While we basically agree with this recommendation, the report does not cite any training deficiencies with respect to pilots observed by SASAG. We have a proposal in the Biennial Operations Review that closely parallels this recommendation, and it will be considered for possible rule-making action.*

2. Carefully study each training exercise and maneuver to validate its usefulness and write off those not needed.

FAA Comment: *Same as 1 above.*

3. Require training programs to always emphasize the line proficiency tasks; require that they be programmed accordingly.

FAA Comment: *Same as 1 above.*

4. Require a line "coaching" program be associated with all pilot training involving transition, line checks or requalification by the air carrier.

FAA Comment: *Our present regulations require that pilots transitioning to new type aircraft obtain line operating experience under the supervision of an airline check pilot. This regulation does not prohibit coaching of the transitioning pilot by the check pilot. In fact, the FAA encourages this. With respect to the periodic line checks required by the regulations, the pilots must satisfactorily complete the check without coaching by the check pilot.*

FLIGHT AND ON-DUTY TIME LIMITATIONS

Cases were observed during the final landings of multi-segment flights in which crews used check lists lackadaisically, if at all, responded by rote rather than with action, and otherwise conducted flight with reduced safety margins, all as a clear and direct result of fatigue. Carriers presently schedule many crew patterns to the limit the regulations allow, with no consideration for the fatiguing aspects involved. Present limitations are antiquated and relegated to contract negotiations. The resulting crew schedules often are designed for economy and personal time-off rather than for safety.

Previous research into the physiology of fatigue has done little to relieve the problem in the cockpit. FAA has failed to grasp the significance of fatigue because those who administer the regulations, and those who ride in airline cockpits during line checks (the ACI's) do not experience the cumulative results of protracted crew schedule mismanagement. FAA is necessarily influenced by these people. The deficiency within the rule making system as it applies to this issue, is the inability, until now, of the FAA to obtain knowledgeable and authoritative information.

As a result of past and present scheduling practices, all members of SASAG have personally operated under conditions of excessive fatigue during their airline careers. This has been of considerable aid in recognizing the deficiencies.

SASAG Recommendation:

- Research and act upon all facets of the fatigue problem, including: a. on-duty and schedule originating time; b. number of landings; c. diurnal period; d. time zone changes; e. geographical areas . . . compression of route and terminal traffic density.

FAA Comment: *One of the proposals in the Biennial Operations Review pertained to flight time limitations. The FAA intends to review the entire subject of pilot flight and duty time limitations.*

AIRPORTS

Several airports were observed to have either procedures or physical characteristics that were hazardous. Specific examples exist at Los Angeles International, Houston, Santa Barbara, Boston, and others. These hazards include, among other things: a. ditches or holes close to the runway ends; b. displaced thresholds that deny pilots use of the full runway length under adverse conditions and c. requirement to land downwind into opposite direction traffic taking off on parallel runways.

The existence of such obstacles as ditches near the end of runway can turn an under- or over-shoot into a disaster. The filling of such ditches should not only be simple, it would have the positive effect of turning a potential catastrophe into an incident.

An immediate study of all airports should be made and the highest priority given to the elimination of these hazards. In cases such as the displaced threshold on Boston's runway 4R, which has been a dangerous situation for many years, the minimums should simply be raised until such time as the basic problem can be corrected.

It seems reasonable for FAA to prohibit air carrier operations into airports and on runways that can turn a simple incident into a horrible disaster, until the hazards are removed.

SASAG Recommendations:

1. The FAA initiate an immediate and comprehensive survey by competent people to identify and correct airport hazards.

FAA Comment: *In our view, the current FAA guidelines and procedures concerning airport hazards are adequate, nevertheless, additional emphasis may be necessary.*

2. Until these hazards are eliminated, runways should be closed, minimums raised, and other interim fixes applied as appropriate.

FAA Comment: *The kinds of actions suggested in this recommendation have been taken on many occasions in the past; nevertheless, Flight Standards, in conjunction with Airports Service, will look further into this matter.*

THREE POINTER ALTIMETER

There are basically three types of barometric altimeter presentations in common use: the three-pointer (3P), the drum-pointer (DP), and the counter-pointer (CP) versions. Although experience clearly shows the 3P altimeter can be easily misread, a relatively large number remain in use in large, turbine-powered airplanes.

Naval Research Laboratory tests established that in 1,080 trials pilots misread the 3P altimeter 80 times. Considering the thousands of operations conducted with 3P altimeters each month, it is obvious exposure to dangerous reading errors is high.

For each reported 3P misreading incident there is little doubt there are many unreported cases. The following cases have been fully documented.

Specific cases:

Case 1–On Oct. 20, 1967, a DC-8 inbound to San Francisco on a clear, moonless night, began descent over the ocean to its clearance altitude of 6,000 feet. The first officer was flying and the captain was performing the normal first officer duties.

The captain made the required call-outs at 30,000 and 20,000 feet, but as 10,000 feet was penetrated, he became distracted in an attempt to locate reported traffic on his radar. This distraction caused him to miss the 10,000 feet call-out. Simultaneously, the first officer was distracted by trouble obtaining a "lock on" to the VOR he had just tuned to. The second officer was occupied with normal panel monitoring duties and the navigator was stowing his equipment.

As the flight was approaching what was thought to be 10,000 feet, the first officer began reducing speed to 250 knots. Simultaneously, the captain made a remark to the first officer about the need to reduce speed.

Suddenly, and apparently simultaneously, the captain, first officer, and second officer all realized they were approaching sea level rather than 10,000 feet. An immediate pullup was made and the flight continued and landed at San Francisco. The captain voluntarily reported the incident to his company.

Subsequent readout of the flight recorder revealed that: the pullup generated a load factor of +1.4g and the minimum height above the water's surface was 125 feet.

The captain involved had years of DC-8 experience and was regarded as a very competent and professional pilot. The particular version of the DC-8 involved had been in service only five months. If it had crashed, it is virtually certain the cause would never have been known, and this "new, unproven airplane" could well have been grounded. The airplane was equipped with 3P altimeters which had been misread approaching 10,000 feet.

Case 2–On Dec. 18, 1970, a 707 began its descent into the Chicago area under instrument conditions. Ceilings reported in the area were 100 to 200 feet AGL. As the crew approached what they believed to be 11,000 feet, the ground became dimly visible and the first officer noted they were "down among the trees." An immediate pullup was made and the flight continued and landed at Chicago O'Hare. Subsequent readout of the flight recorder revealed that: a. the minimum MSL height reached was 910 feet; b. at that point the flight was in the immediate vicinity of terrain above 1,000 feet MSL; c. the pullup generated a load factor of +2.46g and d. the flight had actually cruised at 19,000 feet instead of its assigned 29,000 feet.

The airplane was equipped with 3P altimeters which had been misread by 10,000 feet initially at 19,000 feet, and again by the same amount approaching 11,000 feet. The carrier involved has since replaced all its 3P altimeters with the latest type counter-pointer versions.

In addition to the above, there is good reason to believe that a 727 crash in Lake Michigan, with fatalities to all aboard, was caused by a similar 10,000 feet, 3P altimeter reading error.

SASAG Recommendation:

- Forbid, after a reasonable but minimal time period, use of the three-pointer altimeter in all large, turbine-powered airplanes.

FAA Comment: *We concur, and a rules project will be established which would propose to eliminate this altimeter from all large turbine-powered airplanes.*

CABIN SAFETY

While not directly related to the task of this Group, many important discrepancies of a safety nature were seen in the passenger cabin. These demand attention and correction.

Although the paramount image of cabin attendants is one of serving the passengers, their primary job is really directed to safety. Their role of aiding passengers in the event of an accident, plus seeing certain safety precautions are taken on every flight, is very important.

It has been found that airlines often reverse these roles. The emphasis given the cabin attendant position is the service role and not safety. In many cases, the cabin attendant is under the supervision of the marketing department which causes a split in authority to the extent that certain cases have come to light in which cabin attendants have told captains that they have no authority over them in flight. This is an untenable situation because there must be a complete chain of command through the entire airplane and crew. From the viewpoint of safety the cabin teams are part of operations, and should be under its supervision.

Most cabin attendants receive a safety review yearly. This is generally a one-day affair and many include all types of aircraft of an airline even though the attendant may not fly some of them except on rare occasions. A cabin atten-

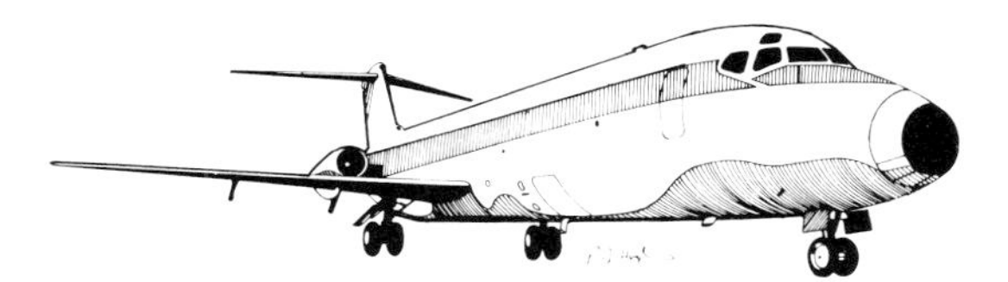

dant may not have flown an aircraft for a number of years and then have one trip with only the yearly review to fall back on for knowledge of the aircraft and its equipment. Since the review is a classroom session, real familiarity with the aircraft is missing. There needs to be a strict standard in keeping cabin teams current. Cabin attendants should be supervised by operations departments and not by marketing.

These various discrepancies were observed during the Group's flights: The constant complaint of cabin attendants, and one easily observed, is excessive hand luggage and packages. The ground personnel allow passengers with arms full of packages and suitcases to go on the aircraft. It then becomes the cabin attendant's task to stow these packages under seats. Because of size and bulk of the packages, this is often impossible. The cabin becomes crowded with small suitcases, packages, brief cases and clothes carriers. One aircraft carries a life raft in the coat compartment, the only place for excess carry-on luggage. This excess was put on top of the life raft. During quick evacuation, it would be impossible to remove the raft. Airlines must assume this excessive carry-on luggage responsibility at the gate, and not pass it on to the cabin attendants.

Cabin attendant extra seats on certain 727's and 707's fold out into the aisle. In the event of an accident and injury to the cabin attendant, this seat could easily block the aisle and prevent evacuation of the aircraft. Often seats are near the galley where the galley equipment, which is certain to come loose in a severe accident (it sometimes does even on a maximum stop after normal landing), would incapacitate the cabin attendant.

Overhead racks often have heavy things stowed in them, including ten or more magazines with leather covers equipped with metal fittings.

Life rafts in overhead racks, which resemble baggage racks, do not have adequate means to prevent passengers from inadvertently opening them. The rafts are heavy and could do considerable damage if they fell.

Cabin attendants have been observed serving trays with wine bottles that are not secure on the tray. The bottles roll around and, in even mildly choppy air, may fall on a passenger. This has occurred!

There is not sufficient room for garbage on many aircraft and it eventually is stuffed into bags which are stowed in the galley before landing, and in front of a door which would be blocked in the event of an emergency evacuation.

Overhead baggage compartments are placarded for 76 pounds on certain revised 727's. Some cabin attendants are told to open them while taxiing so passengers can remove luggage as soon as the plane stops. The 76 pounds of luggage could easily fall during the taxiing, sudden stop, swerve, etc. The compartments should be kept closed until the aircraft is parked at the terminal.

On some DC-10's bar carts cannot be fastened down adequately and only have a simple door-stop to keep them from sliding around. On one carrier's DC-10 there were wall fittings for bar cart fastening, but no fittings on the carts. These are heavy service items, filled with bottles, etc., and would be unguided missiles in an accident.

While we all delight in good service, it must be subservient to safety. In the zest for competitive advantage, safety is often forgotten as is the fact that an airplane is a means of transportation and not a restaurant/bar combination, nor the cabin attendants waiters and barpersons.

SASAG Recommendations:

1. Cabin crews be placed under the same flight operations administration as the other flight crew members and subjected to the same training standards and operational disciplines.

FAA Comment: *We believe this recommendation has merit and therefore will recommend to the airlines that they consider placing flight attendants under the jurisdiction of the airline flight operations department. We will also ask the airlines to make sure that flight attendants understand that during flight time they are under the supervision of the pilot-in-command.*

2. Training and currency of cabin attendants needs to be fortified and

raised to the same level as all other flight crewmembers.

FAA Comment: *See comment following Recommendation No. 3.*

3. Cabin crew seats, galley components and other cabin equipment be designed, certificated so as not to constitute a hazard for both the seat's occupants and the potential for blocking the emergency exit routes of passengers.

FAA Comment: *With respect to Recommendations 2 and 3, FAA has taken the following action and we look for improvement in these areas: a. Notices of Proposed Rule Making proposing to require improved flight attendant seating installations, restraint systems, and surrounding working environment; b. Proposed rule changes are currently under consideration to increase the amount and type of flight attendant training. In addition, a project has been established to evaluate current flight attendant training programs with particular emphasis on emergency equipment and procedures; c. Proposed rulemaking currently under consideration with respect to: lower deck service compartments and improved safety standards for flight attendants and retention of items of mass in passenger/crew compartments, galleys and compartment interiors.*

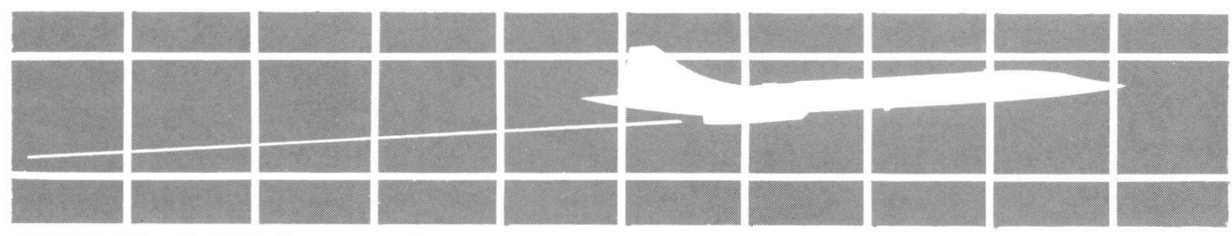

THE YEAR OF THE CONCORDE

CONCORDE SERVICE RECORD

Reprinted by Permission FLIGHT INTERNATIONAL (July 31, 1976)

Scheduled supersonic services were inaugurated, amid equal measures of doubt and ballyhoo, by British Airways and Air France on January 21. Now, six months on, supersonic routes link Europe with three continents and the mood has changed to one of guarded optimism. Already Concorde travel is becoming routine and the talk is all of "when" rather than "if" the New York gateway will open. JOHN BELSON in London and JACK GEE in Paris discuss Concorde operations with the national flag carriers.

In the space of six months British Airways has inaugurated supersonic routes to Bahrain and Washington. At the turn of the year the first production Concordes were being delivered to Air France and British Airways. The two airlines, lacking approval for services to the US, decided respectively on routes to Rio de Janeiro via Dakar, and Bahrain. While the French route is straightforward and mainly supersonic, the British sector is in many ways the more operationally demanding, requiring careful planning and accurate tracking while flying supersonic over Middle Eastern countries.

Bahrain was seen from the outset as only the first leg of an eventual 13-½ hr route via Singapore to Australia, but even so British Airways is carrying double the amount of first-class city-pair traffic predicted for the route; load factor for the six months was 48.5 per cent. Bahrain, with its excellent runway and the offer of a supersonic service to London, is becoming an increasingly important hub for the area. Both British Airways and pool partner Gulf Air serve the area well, and many other carriers are keen to pick up interline traffic.

British Airways originally estimated 25 passengers per flight, or about 4,000 first-class passengers in the year. That figure has now more than doubled and no significant reduction in subsonic traffic is expected.

In an effort to spot trends, the British flag carrier is collecting as much information as possible from its supersonic travellers. Many Concorde passengers are experienced air travellers who take a close interest in the whole operation. Some have already made repeat journeys, and one American passenger has clocked up three return trips to Bahrain. Good despatch reliability has doubtless helped win business passengers, and Mr Gordon Davidson, British Airways' Concorde director, tells *Flight* that he receives far more laudatory letters than complaints. "Too much food" and "no female cabin staff" were too early comments (although hostesses have since qualified for Concorde service), and about the only serious problem has been the lack of leg room beneath tables. Says Davidson: "People are thrilled to fly in Concorde, and anyway they don't have time to nit-pick—on a 747 there is plenty of time."

Conceived from the outset for North

Atlantic operations, Concorde finally arrived at Washington Dulles on May 24 to inaugurate scheduled supersonic services from Europe. Complaints from local residents on the day were few, and eight thousand people were at Dulles to welcome the SST at the start of a 16-month trial period. Since then British Airways has recorded a 93.3 per cent load factor on the service and forward bookings for August and September passed 70 per cent some time ago. "Washington loads are most satisfying, but don't kid yourself we are making money," says Davidson. "Concorde is making a very satisfactory contribution to revenue, with Washington loads giving revenue way ahead of out-of-pocket expenses."

Unexpectedly, businessmen do not predominate on British Airways' Concorde passenger lists. One flight to Washington carried an unaccompanied lady of 93, and several families with children have flown Concorde. Although the early flights undoubtedly benefited from their novelty, regular operations are showing that people from all walks of life are using the supersonic services.

Utilisation is very poor at this stage and will continue at a low level until a New York service is opened. Already committed to buying five supersonic transports, British Airways must use its Concordes intensively in an endeavour to spread the amortisation costs as widely as possible. Davidson says that, given a conservative utilisation of 2,750hr per year and a ten-year depreciation, break-even load factor is 65 per cent; this falls by five points if the depreciation period is stretched to 14 years. "We will never make lots of money," says Davidson, "but who can put a value on national pride? Concorde has done better than we

dared hope. It is doing what the makers said it would, and is doing much for Britian and France."

In keeping with its guarantee of Mach 2 speed, British Airways offers the Concorde passenger terminal facilities which are geared to a short and painless check-in, as well as first-class-plus service. Operationally, the French press sees British Airways as being "extraordinarily prudent" in its calculation of fuel reserves. Although Air France is prepared to make a technical stop (it has not yet had to), British Airways does not currently entertain this possibility. The British carrier nevertheless feels that fuel reserves *may* be too high, claiming that at supersonic speeds airfield weather forecasts are often tantamount to actual reports.

British Prime Minister James Callaghan chartered a Concorde to fly to Puerto Rico on June 26. But the 4,182 n.m. leg has been bettered in Air France service twice when flying Caracas to Paris direct, a distance of 4,330 n.m.

With little Concorde operating experience built up as yet, the learning curve remains fairly flat. But available loads are growing daily and Davidson is confident that various small operational changes will restore the 100-seat payload for nearly every sector. Already, refinement of the baggage-weight calculation has increased the number of seats on offer. Normal procedure is to offer a minimum number of seats for advance booking. Current Washington-London offers are 72 in July, 78 during August, 81 in September, and 100 for the rest of the year. A flight plan is compiled one day before departure, but only half of the additional seat total thus calculated is sold. Then, on the day of the flight, seats up to the maximum calculated number are made available.

British Airways first started "shipside sales" at Washington, using the public-address system to persuade would-be subsonic first-class passengers to fill available seats. British Airways does not overbook its Concorde services, but the no-show rate for supersonic passengers is apparently similar to that of their subsonic counterparts. During the first month (ten flights) on London-Washington there were 56 no-show passengers. Surprisingly, only ten booked passengers failed to travel on Washington-London flights. On one occasion 13 Washington-bound passengers failed to arrive at Heathrow.

Noise levels recorded during the first few operations into and out of Washington soon established that the SST was louder than most Boeing 707s on take-off, but somewhat quieter on landing. The Federal Aviation Administration is issuing monthly reports throughout the 16-month trial period. So far Concorde has been flown as it was during the Environmental Impact Statement assessments. The data produced have verified the performance figures offered by the manufacturers, and British Airways is now keen to make slight changes aimed at reducing community noise nuisance. An FAA noise survey of a sample 2,000 people, carried out before the beginning of Concorde operations, showed 45.7 per cent in favour of the 16-month trial period, with 17.6 per cent strongly in favour: 36.9 per cent objected, of which 22.3 per cent strongly disapproved. The same 2,000 people will be approached again now that Concorde is in service.

The US report for May shows that 47 complaints were made about Concorde, compared with 11 favourable comments. The 45 Concorde movements in June generated 85 complaints (including eight made on days when Concorde did not operate) and two favourable comments. Typically, 60 complaints concerning subsonic aircraft were made in June.

The choice of a mid-day Washington departure stems from the need to attract interline traffic. Nearly two-thirds of the United States' population now lives within the Concorde catchment area for a same-day flight to Europe, and Davidson is confident that by this time next year British Airways will not only be offering 100 seats, but will probably be carrying that many passengers on every flight out of the US.

Plans to fly to Dallas-Fort Worth remain fluid, but Davidson agrees that his earlier demand for a 40-passenger guarantee can be moderated. There is ample time for Concorde to fly on to Dallas the same day, while the 0900hr morning departure from Dallas permits the established Washington Dulles departure time to be maintained.

A revised centre-of-gravity schedule, increased contingency power (up from four to seven per cent) and higher tyre rolling-speed limits offer substantial payload improvements for little cost. Where Concorde take-off performance is limited by tyre speed for the abort (at Singapore, for instance) each additional knot gained on allowable tyre speed can yield 700kg of additional payload. Profile changes to the nose cone, dorsal fin and intake lips should each contribute worthwhile payload improvements, and removal of the airframe de-icing equipment, found to be unnecessary so far, has also been suggested. All these refinements, though cheap to produce, will require flight-testing and cannot therefore be introduced overnight.

British Airways had handled the question of spares extremely well. By stressing its needs to many key but disillusioned component manufacturers, British Airways has so far not gone short. Contracts placed years ago, when hundreds of Concordes were expected to be sold, have long since been cut back, and many suppliers are not interested in small batches. But the airline has to have the spares and is in some cases only too thankful to foot the bill.

Despatch reliability has been good, says British Airways, largely because of the leisurely way in which it has been able to operate during the first six months. Most line engineers think of Concorde as something different, but the airline is stressing that Concorde must be maintained to the same standards as the rest of the fleet. Wear limits and system handling will naturally improve with increasing familiarity, but British Airways tells *Flight* that—with the exception of the carbon brakes, which are showing higher-than-predicted wear—Concorde systems are generally giving less trouble than expected.

The key decision

A New York decision can be expected towards the end of the year. Assuming that it is favourable, British Airways will go for a daily service to New York starting in January 1977. The supply of trained flight crews will be the deciding factor. To complete his training a captain must fly eight three-crew sectors plus six supervised legs. With so few routes available, this schedule can take a long time to complete.

Only if New York slams the door in Concorde's face will British Airways look eastwards first. Australia, having seen and heard the aircraft, has approved an overland supersonic route into Melbourne. No bilateral problems remain there or at Singapore and Bahrain, but there may be overflight problems with Indonesia.

Supersonic-corridor negotiations with India will be held in September. A technical stop might be needed at Colombo if, as seems likely, India remains firm in its resolve not to grant overflying rights. There are other options, such as a stop at Sharjah or Bombay, but generally British Airways would rather lose payload than make a regular stop.

Prospects of a service to Tokyo are, in Davidson's opinion, firmly dependent on the ability of Russia's Tu-144 to fly a similar route. Even if that proviso is met, there still remain the environmental objections of the Japanese. These may only be overcome finally by a favourable New York decision and continued operations beyond the 16-month trial period.

Whatever happens, New York is the key to Concorde's future. Without this city the SST is simply not viable. But if the British and French are permitted to serve New York permanently, the 195,000 business journeys into and out of Kennedy which are forecast for this year (larger than any other market and far larger than Washington's 35,000) will allow Concorde to show its paces with a good service to Europe and beyond.

Air France: six months' satisfaction

Satisfaction with current operations to Washington, Rio and Caracas and restrained optimism about the prospect of opening the New York gateway are the keynotes of Air France's thinking six months after putting Concorde into commercial service.

M Jean-Claude Martin, Air France Concorde co-ordinator, tells *Flight*: "The first half year does better than confirm our original view that Concorde is a technical success and ready for regular service. We now know that there is a worldwide clientele to fly in Concorde. Within 15 or 20 years a far from negligible part of civil air transport will be aboard supersonic aircraft, and Concorde will be making a major contribution."

Air France opened its fare-paying passenger service to Rio de Janeiro from Paris Charles de Gaulle on January 21 with the simultaneous departure of British Airways' flight from London Heathrow to Bahrain. Capt Pierre Chanoine released the brakes of Concorde No 5 (F-BVFA) at zero count with his British colleague Capt Brian Calvert. The aircraft landed at Dakar 2hr 43min later. But airport festivities, a speech by Senegal President Leopold Senghor and a late returning passenger extended the stopover by 27min. Later, a malfunctioning air inlet on the No 2 engine delayed take-off, and the aircraft reached Rio in 7hr 42min block time instead of the scheduled 7hr.

On the return flight to Paris the next day Air France again lost time in Dakar when the fuel meter on a tanker lorry failed to function.

But these were merely teething troubles. Martin says: "If all sectors flown by Air France are taken into account, Concorde's regularity is very good. Using 15min within schedule as the yardstick, 80 per cent of flights between Paris-Dakar-Rio, 75 per cent on Paris-Azores-Caracas and 88 per cent on Paris-Washington arrive on time. The Caracas figures are distorted because two flights were made non-stop to Paris and arrived 90min early.

Air France's Concorde flights to Rio leave Paris twice a week, on Wednesday and Sunday at 1300hr local time, and return at 0730hr the following day. The schedule has been chosen to enable travellers from other European centres to reach Paris in good time and, on their return, to get back to their own countries by mid-morning.

Streamlined check-in and baggage procedures have worked well at Charles de Gaulle from the outset, and initial problems at Rio have since been ironed out.

First-class handling facilities

The smoothness of handling interline passengers soon became a hallmark of Air France's Concorde service. A passenger from Germany for the first scheduled Concorde flight to Rio arrived at the gate of the Concorde terminal at Charles de Gaulle 15min before departure time. He was aboard with his baggage seven minutes later—just as the doors were about to be closed. Passenger boarding has been efficient and fast: it takes 10min with passengers boarding in three groups, beginning with the rear cabin.

Fears that Concorde's baggage hold might not be large enough have not materialised, but average weight per passenger has been 53lb compared with 37lb for first-class passengers on the airline's North Atlantic routes.

Air France Concordes have also been carrying 1,500lb of spares and support equipment so that such items need not be stocked down the route. Morale was boosted in February by US Transport Secretary William T. Coleman's decision to authorise Concorde's services into New York and Washington for a 16-month trial period.

While awaiting approval for Washington operations, Air France opened its second Concorde route on April 9 with a weekly service between Paris and Caracas, Venezuela. The inaugural flight took off 1min ahead of schedule at 1900hr local from Charles de Gaulle and arrived at Caracas 6hr 12min later. A 1hr 7min refuelling stop was made at Santa Maria in the Portuguese Azores.

The aircraft left Caracas the next morning and arrived back in Paris 6hr 17min later after suffering a 22min air-traffic delay at Caracas.

Inauguration of the Paris-Caracas route coincided with delivery of Air France's second production aircraft (207) on April 8. The airline is due to receive its third aircraft (209) before the end of this year and its fourth and final Concorde (211) in spring 1977.

Air France will then return to Aérospatiale the back-up aircraft (203), which was used for the route-proving flights to Rio and Caracas last summer. Deliveries of the last two production aircraft were originally scheduled for this summer and December. But a postponement was decided because of the slow start in services to the United States and Aérospatiale's desire to stretch Concorde production and provide work for its labour force.

Air France's Paris-Caracas Concorde ticket costs Fr5,245 and applies the same 20 per cent surcharge on first-class fares as do the fares for Rio (Fr6,425) and Washington (Fr4,205).

Martin says: "Caracas was not in our original plan. While waiting for the United States gateways to open we had to keep our Concordes flying. South America is a traditional market for Air France. Flying Concorde there could also help to promote the aircraft and, in the longer term, encourage sales to South American airlines."

Caracas is certainly not a money-maker for Concorde. The schedule (departures from Paris at 1900hr on Friday and from Caracas at 0900hr on Saturday) is hardly a bait for French businessmen. Martin adds: "We shall certainly continue to fly to Venezuela as long as we have Concordes available, even if and when New York opens up. We hope to enrich our South American network by offering Concorde services beyond Rio and Caracas to Panama, Peru and Bogota. But if demand for Concorde elsewhere in the world expands, we shall have to make a choice between the most profitable destinations."

Air France expected load factors of about 60 per cent to Rio before opening the service, and is very satisfied with the average of 69.44 per cent on the first 50 round trips.

Martin says: "Quality-wise we are even more satisfied with the results. We expected 40 per cent of the traffic to originate from South America, with France and the rest of Europe sharing the remainder. Instead, under one passenger in three is a South American, the French account for only 33 per cent and other Europeans for 37."

Air France remarks that the introduction of Concorde has actually increased first-class bookings on its own 747 flights from Rio and on Varig's DC-10s. Many passengers fly Concorde one way and widebody the other.

Because Concorde is so relaxing compared with subsonic aircraft, businessmen who used to concentrate their trips to South America into a weekly stay once a year are now making more frequent, briefer, visits. The airline reports that 36 per cent of its Concorde passengers to Rio have never flown Air France before. Many have already made two and even three journeys during the past six months.

These findings bear out market studies made last year by both Air France and the Port Authority of New York and New Jersey which indicated that 20 per cent of first-class passengers and those paying full economy fares would travel more often when Concorde entered service, thus generating a new market.

The opening of the Paris-Washington route on May 24 was the second major event in Air France's Concorde programme. As for the inauguration of their first commercial service, the French and British airlines made it a joint occasion, this time with a simultaneous arrival.

Air France's Concorde, F-BVFA–with Capt Pierre Dudal in the left-hand seat, Richard Puyperaux, first officer, and Martial Destiennek, flight engineer–covered the 6,000km from Paris in 4hr 07min block to block, with 3hr 50min flying time, of which 2hr 55min were supersonic cruise.

The service is operating three times a week, with departures from Paris Charles de Gaulle at 2000hr local and arrivals at Washington Dulles at 1755hr local on Wednesday, Friday and Sunday. Flights from Washington depart on Monday, Tuesday and Saturday at 1245hr local and arrive in Paris at 2250hr local.

The first 20 flights–excluding the inaugural and the following flight, which were principally organised for VIPs (there were only 49 paying passengers out of 80 on the first flight)–showed an average load factor up to July 12 of 88.91 per cent. This is reckoned on the basis of 80 passengers–a figure at which a refuelling stop is unnecessary.

Martin says: "These results are very much better than I hoped. Of course, beforehand we were obsessed by the idea that without New York we could find no salvation in North America. But now we know Washington can be a major gateway to the rest of the United States. Even if New York were to remain closed we could still serve the rest of the USA by improving our Washington service.

"We never planned to fly only to Washington. We had reckoned that, with flights going into New York as well, Washington would produce a load factor of between 50 and 55 per cent on three weekly flights. Now we are getting almost 90 per cent."

After the seasonal lull in business traffic which is expected during the summer, Air France will review the situation and decide whether to put on a fourth weekly flight between Paris and Washington.

Air France is relatively satisfied with the FAA's latest report on Dulles noise monitoring, and is amused by the news that Concorde's low-frequency vibrations have less effect on private houses than those from a vacuum cleaner. "Well, there has never been any question of banning vacuum cleaners from American villas."

Martin notes that Concorde pilots taking off from Dulles can choose between making more noise near the runway, where the FAA microphone is located, or over neighbouring villages. He claims that the first method, coupled with Concorde's admirable turning abilities, can reduce noise levels by 15 PNdB once off the runway. Over Stirling Park noise is from 80-85 PNdB—quieter than the noise level on an urban motorway.

Martin comments: "British Airways is more systematic and less individualistic. Its Concordes make more noise on arrival and less on departure, thus conforming with FAR 36. It nurses departures and is therefore more successful than us with the Americans.

"Our British friends arrive in hotter temperatures, so they are bound to make more noise."

Martin recalls that before flights began to Washington a poll in the United States showed that residents of cities which received the most subsonic visitors were the most worried by Concorde. Home owners were also more concerned than tenants because of the fear of deterioration and depreciation of their property.

In June, when Concorde made four times more visits to Washington than in May, there were 85 calls, of which two were favourable. Martin considers that local resistance to the aircraft is now negligible; he describes the complaints about phantom flights which did not occur as "worth their weight in mustard" in vindicating Concorde. Except from French environmentalist bodies, complaints at Charles de Gaulle have practically ceased.

With a load factor close to 90 per cent, Air France is refusing passengers on most of its Washington flights. Recently, when the meteorological reports were particularly favourable, Concorde left with 82 passengers instead of the usual 80, which "is not inflexible."

Longest route so far

Air France intends to take a close look at the met predictions for the winter before deciding whether to increase the number of Washington seats. Already, the chances of Concorde's having to refuel at Gander are five times less than those for a Boeing 747.

Air France has demonstrated Concorde's range capabilities by flying Caracas-Charles de Gaulle twice non-stop during the past few weeks. On May 29, Capt Duchange decided not to make an intermediate stop after receiving an adverse met report from Santa Maria. He brought his 42 passengers to Charles de Gaulle in 4hr 34min block time. Flying time was 4hr 19 min, of which 3hr 37min were supersonic. This was bettered on July 10 by Capt Dudal, who flew the route in 4hr 24min with 55 passengers.

Air France Concorde crews were enormously encouraged by this proof of what the aircraft can do. Morale on the flight deck is excellent, says Martin. "They feel motivated and turn up in large numbers at debriefings in order to discuss procedures. They are keen to promote Concorde. One pilot became a big TV success in the United States when he went on a week's tour with one of our public-relations team." (sic)

Cabin staff are also happy with their work aboard Concorde in spite of the difficulty in manipulating trolleys on the sloping, narrow aisle during climb.

Air France awaits the American court decisions which will allow the two airlines to take Concorde to New York. If all goes well, Air France will be ready to open its New York service in late December or early January. The airline will send its Concorde pilots with an empty aircraft to New York to try out all the procedures beforehand.

Technical problems have been minimal, with only three minor incidents, one of which involved a cracked windshield. In each case the trouble occurred on departure from Roissy (once on a Rio-bound aircraft, twice Washington-bound), and the back-up Concorde was immediately available as a replacement.

Air France has had no problems involving holding over airports during its first six months of Concorde operations, since Rio, Caracas and Washington are not busy.

Martin concludes: "We are not being given any privileges at Dulles. Concorde is subject to the same severe constraints as other aircraft. We encounter no hostility from American traffic control. They behave like disciplined US military people with no bad faith, but no favours or bouquets either."

THE NEW METAL ANGEL

The arrow of twentieth century man's progress has aviation as its point. Manned, powered, controlled flight began with the Wrights at Kittihawk, USA, in 1903. No lifetime in history has seen such aston-

CONCORDE NOISE COMPARISONS

		Take-off	Landing
Measured at Heathrow (PNdB):			
Concorde		107–119	102–127
Peak recordings for	**Trident**	112	124
"	**Boeing 707**	118	126
"	**Boeing 747**	115	115
Measured by US authorities (EPNdB):			
EIS assessment	**Concorde**	119·5	116·5
	Boeing 707-300	113	118
FAA May report	**Concorde**	111·2–125·2	109·6–120·6
FAA June report	**Concorde**	105–130	115–130

CONCORDE OPERATIONAL PERFORMANCE

	Destination		
British Airways*:	**Bahrain**		**Washington**
Number of trips	44		16
Passengers carried	3,578		2,410
Load factor (per cent)	48·5		93·3
	Destination		
Air France:**	**Rio de Janeiro**	**Caracas**	**Washington**
Number of trips	88	22	26
Passengers carried	6,182	1,210	1,825
Load factor (per cent)	70·25	55·0	87·8

*To July 20. **To June 24.

EXTERNAL DIMENSIONS

Wing span:	83ft 10in	25·56m
Wing-root chord:	90ft 9in	27·66m
Wing aspect-ratio:	1·7	
Wing area:	3,856sq ft	358·25m^2
Length:	202ft 3·6in	61·66m
Height:	37ft 1in	11·30m
Wheel track:	25ft 4in	7·72m
Wheelbase:	59ft 8in	18·19m

PERFORMANCE (envelope explored)

Altitude:	68,000ft,	20,725m
Airspeed:	1,260kt,	2,340km/hr
Mach No:	2·23	
Minimum airborne speed:	119kt,	221km/hr
Incidence:	23·7°	
Take-off weight:	400,580lb,	181,700kg
Maximum cross-wind:	25kt,	46·5km/hr
Longest subsonic flight:	6hr 15min	

INTERNAL DIMENSIONS

Cabin length:	129ft	39·32m
Aisle:	1ft 5in	0·43m
Cabin width:	8ft 7in	2·63m
Cabin volume:	8,440cu ft	238·5m^3
Window size:	6in × 4in	13cm × 8·5cm
Ceiling height:	6ft 5in	1·96m
Baggage compartment		
underfloor:	237cu ft	6·71m^3
rear fuselage:	470cu ft	13·31m^3

NORMAL PERFORMANCE

Cruise: Mach 2·05 or 530kt corrected airspeed CAS (whichever is less) at 51,300ft
530kt CAS is equivalent to true airspeed TAS of 1,176kt, 1,354 m.p.h.

Range with 25,000lb payload: 3,800n.m.
with 20,000lb payload: 4,000n.m.

Both include fuel reserves for 230n.m. diversion plus 30min hold

Runway length required for max range:
Take-off: 10,500ft
Landing: 7,700ft

ishing accomplishments, or such majesty of mind over matter. Man went supersonic in 1947, not knowing what would happen beyond the speed of sound. Aviation put a man on the moon in 1969. Now supersonic flight is on the market.

The devil as well as the angels have used the aeroplane. Warplanes have killed a quarter of a million people. But airliners—the "metal angels"—bring redemption. They draw people closer together, stop at no national frontiers, and are the enemy of misunderstanding and prejudice. Now a new metal angel brightens the future of civilised man.

SUPERSONIC WAYPOINTS

Reprinted by Permission
FLIGHT INTERNATIONAL (January 17, 1976)
Edited for space

A special supplement to FLIGHT INTERNATIONAL to mark the inauguration by Britain and France of the world's first supersonic passenger services. A plain man's guide to Europe's greatest technical and transport achievement.

Supersonic Man's Twelve-hour World

From the evolution of homo sapiens until about 1840 the distance a man could travel in his working day—say 12 hours—was about 75 miles. By the early nineteenth century man's twelve-hour world became 500 miles. Then technology produced the aeroplane. By 1958 a man could travel 7,000 miles in his working hours. In 1976, thanks to Concorde, man's twelve-hour world will be the world itself.

Editor J.M. Ramsden
Assistant editor Stephen Broadbent
Design Peter Slocombe

Nearly 30 years have elapsed from the first level flight faster than sound to the first supersonic passenger service. This gap is longer than was predicted when US Air Force test pilot Major Charles Yeager took his rocket-propelled *Bell XS-1* to

SEATING

Maximum:	140 at 32in pitch
Standard:	128 at 34in pitch
Mixed:	108 (72 at 34in pitch plus 36 at 42in pitch)
Superior class:	108 at 38in pitch

WEIGHTS

Maximum taxiing:	404,000lb	182,500kg
Maximum take-off:	400,000lb	181,435kg
Maximum payload:	28,000lb	12,680kg
Maximum landing:	245,000lb	111,000kg
Fuel capacity:	211,100lb	95,500kg

ENGINES

Type: 4 × Rolls-Royce/Snecma Olympus 593 Mk 610 axial-flow, two-spool turbo-jets with reheat.
Low-pressure (LP) compressor: 7-stage (titanium alloy)
High-pressure (HP) compressor: 7-stage (titanium and nickel alloys)
Combustion chamber: annular, nickel-alloy material
Pressure ratio: 11·6
HP turbine: single-stage nickel alloys
LP turbine: single-stage nickel alloys
Take-off thrust, sea-level static, reheat on: 38,050lb
Cruise thrust, Mach 2, 53,000ft: 10,030lb
Cruise fuel consumption (total; four engines at cruise thrust) 47,750lb/hr (approx 4·25gal/mile)
Weight: dry 6,780lb

BOEING SST AND McDONNELL DOUGLAS AST

	SST	AST
Cruise Mach	2·7	2·2
Skin temperature (°F)	360-420	225-240
Structure	All Ti	70% Ti/30% Al
Max T/O weight	750,000lb	750,000lb
Length overall (ft)	280	310
Wing span (ft)	141	135·5
Height overall (ft)	50	54·8
Wing area (sq ft)	7,700	10,000
Leading-edge sweep, inboard	74°	71°
outboard	50° 30′	57°
Trailing-edge sweep	zero	17° 30′
Controls	Automatic	Mechanical reversion
PERFORMANCE*		
Noise (EPNdB) at FAR 36 points		
Approach	109 (108) †	107
Sideline	119·5 (112) †	105
Take-off (noise-abatement)	108 (108) †	108
Cruise lift/drag ratio	7·95	9·6
Range with full passengers (n.m.)	3,650	4,400
Max passenger capacity	261	273
Fuel burn per aircraft-mile (lb)	86	73

* With General Electric "mini-bypass" cycle (BPR—0·1 : 1). † Goal for 1975-80.

COMPARATIVE JOURNEY TIMES

	Concorde	Subsonic
London-New York	3hr 30min	7hr 05min
Paris-Tokyo	6hr 50min	14hr 40min
London-Sydney	13hr 05min	24hr

COSTS

Price, fully equipped: about $60 million

Total Anglo-French "launch investment", including Research & Development and non-recurring production-tooling costs, 1962-1976 £1,200 million

PRIME CONTRACTORS

Aérospatiale, France, and British Aircraft Corporation, England.

just over Mach 1 in October 1947. Within 10 years, American aircraft designers were forecasting that a supersonic airliner would be in passenger service by 1965. One manufacturer was even convinced that the correct supersonic speed was Mach 3 and that this would be achieved in airline service before the end of the 'sixties.

In fact, the perfection of passenger supersonics will have taken longer than from the biplane fighters of World War I to the Bell XS-1.

At Göttingen in 1933, the German scientist Buseman had examined in a Mach 1.5 wind tunnel the compressibility drag rise which occurs as the speed of sound is approached. For some years it had been widely thought that this would present a permanent barrier to practical flight faster than sound. German scientists showed how the drag rise could be reduced or delayed by using thin aerofoils and sweptback wings.

The problem of how to manufacture structurally sound thin sections, and how to make them generate lift at subsonic speed, seemed daunting. So did the attainment of the great engine powers required. Coincidentally, the jet engine was being independently developed in Britain and in Germany during the 'thirties. The jet engine was, of course, to prove crucial to the achievement of practical supersonics, although a rocket engine would first propel man through the "sound barrier," as the compressibility drag rise became popularly known.

The competition of war in 1939-45 brought home to both sides the speed limit of the traditional piston-driven fighter. The squadron pilot discovered in combat, often by accident, the effects of compressibility and wave drag control.

By the end of the war, Europe was well ahead with high-subsonic developments. The first swept-wing aircraft to enter combat was the rocket-powered Messerschmitt *Me 163*. It was a sharply swept, tailless fighter designed to intercept the high-flying US Eighth Air Force B-17G Flying Fortresses.

The Me 163's swept wing was not troubled by compressibility until well over Mach 0.8, when the pilot felt the lack of a tail to trim out the rearward CP shift and resultant pitch-down. In fact, the tail-less swept wing was not the best way into supersonics. In Britain, the de Havilland company tried this formula, and the *D.H. 108*—basically a Goblin-engined Vampire nacelle with a sharply swept wing—was flying as early as 1945.

The Swallow, as it was popularly known, was intended to be a scale flying model of the D.H. 106 (later the Comet). It was also a research vehicle for advanced new military types.

The D.H. 108 broke up and killed its test pilot in September 1946 during practice for an attempt on the world air speed record. There is a plaque at Hatfield to the effect that the test pilot, Geoffrey de Havilland, "died while flying at a speed greater than had previously been attained by man."

Nobody was yet talking about supersonic passenger flying. The race to develop military supersonics was now on. The United States was first through the speed of sound with the Bell XS-1. Six years later, in 1953, the *Douglas Skyrocket* became the first manned aircraft to exceed Mach 2—twice the speed of sound—in level flight.

The American industry was drawing ahead. The French, having lost their aircraft industry in the war, were at least as determined as the British to prove their technology. With hindsight we can now see how the Anglo-French rivalry of the 'fifties laid the foundations of Concorde.

One of the first French experimental supersonic aircraft was the *SO 9050 Trident II*, which had a very thin wing of rectangular plan-form and was powered by wingtip-mounted jet and a rocket motor installed in the rear fuselage. The Trident set several speed records, reaching Mach 1.8 in December 1956 and Mach 1.9 in the following January.

Two delta-wing Mach 2 research aircraft, one British and one French, were at the front of the European stage in the mid-fifties. One was the *Fairey FD.2*; the other was the *Griffon* turbo-ramjet by Nord, an antecedent of Aérospatiale.

The supersonic hours were building up. The Americans were the first to gain operational supersonic experience, with the 1953 *F-100 Super Sabre*–to which goes the distinction of being the world's first level-supersonic operational fighter.

In the Soviet Union the MiGs 17 and 19 followed the MiG 15 into supersonic service. Though they were from a different design bureau, these types contributed to the formidable Russian aerospace capability which produced a rival for Concorde, the *Tupolev Tu-144*.

The Fairey FD.2 had a major influence on Concorde. In 1956–in the days before space flight made such records meaningless–it won for Britain the world air speed record at over 1,000 m.p.h. (1,600 km/hr). The FD.2 was one of the first supersonic aircraft to explore and develop the delta formula, and to come up with the "droop snoot" answer to the pilot-visibility problem at high angles of attack. Together with the *H.P. 115* low-speed narrow delta, it explored the whole range of stability and control from below 100 m.p.h. (160 km/hr) up to the kinetic-heating regime of over Mach 2.

The FD.2 contributed even more directly to Concorde. The little angular delta was remodelled with the ogive delta which had been chosen for Concorde. Known in its new guise as the *BAC-221*, it verified the plan-form which has been one of the foundations of Concorde's aerodynamic success.

The choice was not then as obviously right as it may seem now. The very thin straight wing of the *Lockheed F-104*, which was adopted and manufactured by leading European members of Nato as their standard supersonic fighter, was very different from the deltas—and it worked. The F-104, with its small 3-½ per cent wing, together with the outstanding straight-winged *North American X-15*, which in 1967 exceeded 4,500 m.p.h. (7,000 km/hr), left the aerodynamicists far from sure of the way ahead.

In 1956 the British set up a Supersonic Transport Aircraft Committee to narrow the choice. Nobody in Europe was under-rating the American challenge. Fighters like the F-100 and F-104 were in service and there were reports of projects for a titanium Mach 3 bomber and a similarly fast US fighter—both deltas. These were to materialise into the *North American XB-70* and the *Lockheed A-11 (YF-12A)* interceptor and its reconnaissance version, the *SR-71 Blackbird*.

The A-11 was flying in 1961—just when the British and French were coming together on Concorde. Cool technical judgment was needed to decide, under the pressure of such impressive transatlantic sonic banging, to go for the conservative "knowns"—aluminium structure and Mach 2 speed—in a supersonic transport.

The Anglo-French decision was certainly based on plenty of data. The French in the mid-fifties put all their military bets on the judgment that the delta was the right supersonic formula. The wealth of experience gained from 1,500 Mirage IIIs has gone into Concorde.

Though it was not supersonic in level flight, the Hawker Siddeley (Avro) *Vulcan* delta bomber bequests to Concorde were low-speed, high inertia control and stability, and engine testing.

The Vulcan proved to be an ideal subsonic flying test bed for the engine chosen for the Concorde, the Rolls-Royce Bristol/Snecma Olympus 593. This engine was airborne many hours before Concorde flew. Turbojets bearing the name Olympus, all related to the engines of Concorde, were flying in British V-bombers in the early fifties.

Many and varied have been the machines and engines which have winged supersonic man through and beyond the speed of sound, mainly for military ends. All those supersonic years may be said to have endowed, directly or indirectly, the technical success of the Concorde civil supersonic airliner. The first supersonic farepaying passenger service is the climax, even the redemption, of all this work.

WILL IT PAY?

If Concorde operations by Air France and British Airways make a profit, few long-range international carriers will be able to remain subsonic. But if supersonic services prove to be uneconomic, assembly will most certainly cease before 16 production aircraft have been completed.

Although politics have always played, and will continue to play, an important part in the progress of the Western world's first supersonic transport, operating economics will be the final arbiter between success and failure. If Concorde, which was designed from the outset to make a profit, fails to live up to expectations it is highly unlikely that the Tu-144 will fill the gap. Not only will success spur the development of Concorde derivatives such as the proposed A1 and B versions, but there will be a second-generation machine. In any event, Concorde will have the Western market to itself for over a decade and, having re-established Europe as a supplier of long-range airliners, could pave the way for a follow-through Euro-American partnership.

The principle that Concorde will cost more per seat to buy and per seat-mile to operate than the present generation of wide-body transports is not disputed by protagonists of the SST. Cost is only one side of the economic equation, however. The key to Concorde's success will be its ability to attract high-fare passengers.

Although in very round figures Concorde will cost about twice as much to operate per seat-mile as a wide-body subsonic, first-class fares are typically twice the overall average. There will be no discount "advance-booking" and other fares on Concorde to dilute revenue. One of the problems for Air France and British Airways will be the desire of competing airlines to force a high supersonic surcharge to protect their obsolete fleets.

Concorde's high block speed means that in terms of seat-miles per hour its productivity approaches that of the wide-body tri-jets. It will have a greater impact on airline planning than its nominal capacity of 108 seats suggests.

In many ways, Concorde will be like any other airliner but more than most it will rely on efficient operation and professional standards of engineering and accounting. It might be fair, for example, to weight SST passenger costs to take account of the high standard of service, but unrealistic to allocate all indirect costs on a per flight basis when trip time is halved. Will it be fair to assign ticketing and reservations costs according to the price of the ticket when the real cost of booking a low-fare ABC passenger may be higher?

UTILISATION

Bearing in mind that some efficient airlines achieve annual 747 utilisations of 4,200hr, an operator with four or five Concordes should be able to return utilisations of about 3,600hr. This may involve some fairly sophisticated scheduling and maintenance, but the high cost

of the aircraft will probably provide the necessary incentive.

Just like all other airliners, the purchase price of Concorde has risen since the initial customers—Air France, British Airways, CAAC and Iran Air—signed preliminary purchase agreements. Each Concorde will probably now cost a new customer some $60 million. With spares, this represents a total investment of close to $75 million.

Concorde is essentially an aircraft for business passengers and there is no doubt that London-New York is the premiere business route—hence the battle for landing rights at Kennedy. The acid test of Concorde's profitability is therefore the 3,540-mile London-New York sector which will require an average block time of 3hr 30min.

Most airlines have a *depreciation* policy which writes down the cost of a new aircraft to a low residual value over a period of years. For a subsonic type a period of 12 to 14 years is not unusual; for Concorde 12 years to ten per cent residual would seem appropriate. This would give an hourly cost of $1,605.

Insurance rates vary from one per cent to more than three per cent of first cost;

two per cent would seem likely for Concorde–contributing $344 per hour. Flight-testing and route-proving by the prototype, pre-production and production aircraft have yielded *maintenance and overhaul* cost data to back up more theoretical studies. Direct and indirect maintenance costs calculated at a labour rate of $7.5 per man-hour result in maintenance charges of $420 per hour plus $911 per flight cycle. Total maintenance cost on London-New York would average $1,208 per hour.

Current levels of *salaries, expenses and training* suggest that each member of the flight crew would cost $140 per hour; Concorde will probably have a three-man crew costing $420 per flying hour. There is no doubt that the economics of the SST are sensitive to *fuel* prices and the 147,200lb of London-New York block fuel will cost $2,512 per hour at 40c per US gallon.

Total direct costs will therefore be $6,089 per hour.

Indirect operating costs include passenger service, aircraft servicing, landing fees, traffic services, reservations and sales, commissions, advertising and publicity, ground facilities and general and administrative costs. Using an updated version of the US Federal Aviation Administration indirect operating cost method devised by Boeing and Lockheed, realistic landing fees and 1975 levels of first-class cabin and ground-handling costs at a 50 per cent load factor, total indirects amount to $1,943 per flying hour.

TOTAL COST

Using the above results, the total cost of operating the 108-seat Concorde on London-New York is $8,032 per hour or $28,112 per flight. This gives costs of $7.94 per aircraft mile or 7.35 cents per available seat-mile (based on a flight time of 3.5hr, a distance of 3,540 miles, and a seat capacity of 108).

The London-New York first-class fare is $578. Allowing a five per cent dilution to allow for non-fare-paying passengers, the average yield per passenger is $549 per trip or 15.51 cents per mile. On this route, therefore, Concorde could break even at a load factor of 47 per cent–51 passengers–without a surcharge.

All the experience of the route-proving flights and the result of independently conducted passenger surveys indicate that there will be a big "slide over" of high-yield passengers from subsonic to supersonic services. Load factors of more than 60 per cent seem likely. These will give an annual profit per aircraft of some $7.7 million (at the same, existing first-class fares). One of the initial problems could in fact be how to provide sufficient capacity.

Concorde will have an important secondary effect on the Air France and British Airways route networks. Paris and London will be reconfirmed as the most important European gateways, particularly for high-yield passengers.

As noted above, other airlines are likely, through Iata, to insist on SST fare levels higher than first-class to protect their own subsonic first-class premium traffic. Air France and British Airways have agreed to settle for a 20 per cent surcharge above first-class on their initial Concorde routes. Assuming the same demand, this can only improve the supersonic economy.

Studies of other prime supersonic routes such as London-Rio, London-Sydney and Los Angeles-Tokyo show Concorde breaking even at load factors of 50 per cent or better at first-class fare levels. There is no doubt that Concorde can make money for the airlines.

RETURN ON INVESTMENT

If it is accepted that Concorde by itself is fundamentally profitable, two questions remain in the fleet-planner's mind. Will the aircraft provide a reasonable return on investment? What will be the effect on existing equipment?

More detailed calculations show that an individual Concorde is capable of providing an equal or better return on investment than a single 747. A mixed fleet of one-class Concordes plus economy/tourist/ABC 747s provides a better return on investment than a fleet of mixed-class 747s. The introduction of Concorde will release 747s for the carriage of payloads such as low-fare passengers and containerised freight, for which they are most suited.

The costs for London-New York give some guide to the likely rate for a Concorde lease, although the exact charge will depend on the length of the contract and the way in which the aircraft is supported by the lessee.

SUMMARY

- Concorde will cost about twice as much per available seat-mile to operate as a wide-body.
- Existing first-class fares will more than compensate for this, providing twice the yield of the average subsonic fare.
- Concorde will make a substantial profit on most business routes with a break-even load of about 50 passengers at first-class fare levels.
- A mixed fleet of Concordes and wide-bodies can make more profit and a better return on investment than an all-subsonic fleet.
- The efficient airline will take care to allocate its Concorde costs to allow for its higher yield and shorter flight times.
- Airlines will not be able to argue that Concorde has created surplus capacity; there has been adequate notice that air transport will not be the same after January 21, 1976.

ENGINEERING MASTERPIECE

Western Europe's two leading aviation countries joined together on November 29, 1962, to launch a supersonic airliner. This was not a political whim, a sop to President de Gaulle to allow Britain into the Common Market, as Concorde's opponents have often asserted.

Certainly the two Ministers had no idea of what they were letting themselves in for financially. They estimated the launching cost at £170 million. It has turned out to be more like £1,200 million. lion.

But technically this was no hasty project. According to one of the British engineers behind the Concorde, Sir Morien Morgan, the first design meeting was in February 1954–22 years before the first revenue customers would board the aircraft.

Time has vindicated the technically cool Concorde judgements, for which Sir Morien Morgan, Sir George Edwards and French colleagues like Satre and Servanty were responsible. That shape looks right and is right–from more than 1,000 m.p.h. down to below 200 m.p.h. It is a shape which has such good low-speed qualities that it has taught men the delights of hang gliding; and it will be right also for shuttling men to and from space.

Aerodynamics Transonic and supersonic flight introduces three major difficulties unknown at subsonic speeds: shock-wave drag, which in the cruise can be a third of the total; the rearward shift of the aerodynamic centre at transonic speed; and kinetic heating, or heating of the structure caused by skin friction–even in the thin, cold (-57°C) air of 60,000ft.

SLENDER DELTA

The best shape for optimum cruise performance at Mach 2 (twice the speed of sound) is a straight-edged slender delta twice as long as its span. But this shape has to be modified for stability and efficiency at low speeds, particularly on the approach and landing.

A bonus of the delta wing are the two vortex "sheets" which attach themselves at low speeds and high angles of attack, increasing lift. By rounding the straight-edged delta (thus producing Concorde's famous "Gothic" or "ogive" planform) the vortex sheets can be made to remain attached at high angles of attack.

One reason why a foreplane is not fitted to Concorde to assist pitch control, as on the Tu-144, is because it would disturb this valuable vortex sheet. Because of the high angles of attack needed to get landing lift and hence reasonable landing speeds, a drooped nose is necessary to give the pilot a better view for landing.

Concorde's beautiful shape was the result of 4,000hr of wind-tunnel testing. The wing is mechanically simple: it has no flaps or slats or lift devices, only elevons. These combine the functions of elevators and ailerons on conventional aircraft. Concorde's wing is also an ideal fuel tank, and a perfect mounting for the very long engines. Each powerplant is as big as a bus.

Judged on the aerodynamic criterion of lift/drag or L/D ratio, Concorde produces a figure of between 7 and 8 in supersonic flight, and between 12 and 13 in subsonic flight–not much less than the L/D of subsonic transports. This efficiency in the subsonic regime means that Concorde's miles per gallon are about the same subsonic as supersonic. Thus the range will not be too impaired when, because of the sonic boom, subsonic flights have to be made over populated areas.

Powerplant Concorde's Rolls-Royce/

Snecma Olympus 593 turbo-jets introduce into civil transport for the first time variable air intakes and nozzles. These ensure that the engines are running at their best throughout Concorde's uniquely wide speed and temperature ranges. The intake ducts swallow nearly half a ton of air a second in the cruise, and handle pressure differences of from 2lb/sq in to 20lb/sq in, reducing the speed of the free-stream air from Mach 2 to Mach 0.5 by the time it reaches the engines.

Thermodynamics Above Mach 1 heating caused by skin friction starts to make itself felt on an aircraft structure. Above about Mach 2.5 conventional aluminium alloys soften, and materials like steel and titanium—with their attendant problems of weight, fabrication and cost—have to be employed. Fuel nears the boil, and new materials for numerous details like seals, hoses and windows have to be developed.

Hence the choice of Mach 2 as Concorde's design Mach number. At this speed the general aerodynamic temperature is about 120°C, rising to more than 150°C on the leading edges (wings and nose). This is well within the strength and creep resistance of aluminium alloys.

A full-scale test-rig of a complete fuselage is under test at the Royal Aircraft Establishment, Farnborough, to verify Concorde structural fatigue strength. Unlike previous fatigue test rigs, this one actually simulates the heating as well as loading cycles of flight. The result should be an in-service life of 50,000hr.

Fuel for trimming The fuel in the wing tanks absorbs some of the cruise heat, but it also has another function: transonic trimming. As the aerodynamic centre of pressure shifts rearwards at transonic and supersonic speeds, a strong nose-down pitch has to be trimmed out. Instead of using "trim tabs" or other control surfaces, causing drag, trimming is done by changing the centre of gravity. Fuel is progressively transferred from front to rear tanks.

Testing Concorde is without doubt the world's most tested airliner. The flying programme alone has involved eight aircraft and a total of 5,500hr. (The jumbo jets each flew about 1,500hr.) Concorde testing has embraced wind tunnels, laboratories, simulators, mock-ups, test rigs and research facilities of industry, Government and universities on both sides of the Channel—all working to ensure that there will be no "unknowns" when the first revenue-paying passengers are carried.

Two complete airframe specimens have been used for structural testing, one for static loading at the Centre d'Essais Aéronautique de Toulouse in France, and the other for fatigue testing at the Royal Aircraft Establishment. Five major test rigs were constructed to simulate systems operation.

At Toulouse there are full-scale replicas of the flying control systems, together with the associated hydraulics and electrics. The fuel rig at Bristol is a reproduction of the complete aircraft system: mounted on a platform, it can be pitched and rolled to simulate flight attitudes.

Flight-testing has involved instrumentation and computers capable of monitoring and processing up to 3,000 measurements continuously, telemetering selected data to the ground. The trouble-free nature of the flight-test programme, and the reliability of Concorde during its proving flights, are attributed to the ten years of preparatory work.

The flight-test programme started in

March 1969 and led to certification at the end of 1975.

Various flight simulators and four research aircraft—a Mirage IV "analogue," the BAC-221 for high-speed and the HP-115 for low-speed research, and a Vulcan bomber with complete engine nacelle mounted under it—contributed to Concorde's trouble-free flight-test programme. Touching wood, after more than 5,500hr of the most extensive test programme in civil-aircraft history, there are no safety skeletons in the cupboard. The engineering objective now is low maintenance cost.

The extent of the technical achievement may not be easy for laymen to appreciate. The engineers of two countries, with different languages, units of measurement and engineering traditions, have built the most difficult aircraft ever attempted. It is now apparent that co-operation, though it takes longer, may produce a better end product. No single chief designer can say "It will be this way." In co-operation, proposals have to be debated, argued, tested and tested and tested again. This seems to produce better technical judgements and decisions.

History may judge Concorde to be a fine blend of the quick and decisive but sometimes rather rigid French style, and the common-consent but slower-moving British way.

Economy The continent which invented the jet transport—the de Havilland Comet and the Sud Caravelle—intended jet travel initially for first-class passengers and mail only. Everyone, including most engineers, believed that the mass markets, on long and short routes, would be retained by airliners powered by propellers. They were wrong. Jets have made possible £50 all-inclusive Mediterranean fortnight holidays for the masses. The marketing appeal and efficiency of speed changed everything in half a decade. So it will be again. By the 1990s the supersonic airliner will be within sight of becoming the standard method of travel on all long-haul air routes.

70 TONNES OF THRUST

While the aerodynamicists were thrashing out the optimum shape for Concorde, trying to strike a compromise between theory and practice, so the engine designers were faced with similar compromise decisions.

The engine for a supersonic aircraft needs two basic attributes. First it must offer high specific thrust (i.e. engine thrust divided by engine weight) to provide the required thrust for take-off and transonic acceleration. Concorde, which has no wing flaps to augment lift, has a high take-off speed and therefore has to accelerate quickly.

Secondly, the engine has to have low fuel consumption during the cruise to give optimum range. Three types of engines were available: a straight turbojet, a two-spool jet, or a high-bypass-ratio turbofan. The first type of engine would have demanded a turbine temperature too high for contemporary metals technology to produce the thrust needed. A high-bypass turbofan has a very high frontal area, increasing drag unacceptably. The middle of these three choices was therefore taken, and the result—the Rolls-Royce/Snecma Olympus 593—is a two-spool powerplant with good characteristics in both the critical areas.

Of the engines available or under development in the early 1960s the Olympus most nearly fitted the bill. Extensive modifications were necessary before it even met the basic Concorde requirements, and subsequently the engine has been further refined.

Bench-testing of the prototype 593D version began in mid-1964, almost five years before the first Concorde flight. A major contribution to the success of the programme was made by test flying an Olympus mounted in a representative Concorde nacelle under the fuselage of a Vulcan. This enabled subsonic in-flight problems to be identified and solved long before Concorde flew, reducing the cost of development and increasing the reliability and performance of the engine eventually installed in the aircraft.

Now, more than 11 years after the first bench trial, the Olympus has accumulated nearly 50,000hr of running, both in the test-cell and airborne. It has, with the aid of ground facilities such as the test-cells at the National Gas Turbine Establishment at Pyestock, been subjected to an environment far more hostile than it will meet in service.

A movable ramp inside the top of the intake mouth controls the shock-wave pattern at the front of the intake and the main 20ft diffuser section of the intake gets the flow down to a speed acceptable to the engine. The compressor inlet pressure is higher than on subsonic engines, and inlet temperature is also higher—120°C compared with -20°C for most subsonic engines. Other spill valves also double as supplementary intake scoops on take-off.

The very big intake-temperature range is the reason for the variable primary jet nozzle, which is the most efficient way of keeping the temperature of the engine within limits. It enables mass flow to be optimised for best fuel consumption in all flight conditions. The variable secondary nozzle keeps "afterbody" or base drag to a minimum.

The operation of these variable intakes and nozzles is the responsibility of an automatic electronic control system, with something like 850 electronic components per powerplant, the heart of which is a duplicated air-data computer. This receives all the inputs—not only temperatures and pressures, but fuel flow and temperature, yaw, sideslip, roll, pitch and the many other variables which have to be matched for optimum engine performance. This is absolutely critical, because on the Concorde there is a payload penalty of about 5 per cent for every 1 per cent increase in fuel consumption. The powerplant system could be managed by the flight engineer in an emergency.

The latest production version is known as the Mk 610. This variant, a development of the previous Mk 602 production engine introduced in 1971, has a revised low-pressure compressor and annular combustion chamber. These modifications virtually eliminate the exhaust smoke which was a characteristic of pre-production Olympus engines. The Mk 610 has a revised external design to ease maintenance, and now the engine is one of the simplest to overhaul.

One of the major successes of the Olympus 593 programme, along with the smoke-eliminating annular combustion chamber, is Snecma's TRA (thrust reverser, aft) nozzle. Braking performance of Concorde is enhanced, as on most commercial jet aircraft, by selecting reverse thrust on touchdown. Usually this is done by moving a deflector plate, or bucket, over the exhaust to turn the air flow forward. A similar arrangement is used in Concorde, where the original clamshell thrust reversers—positioned upstream of the final nozzle area—were replaced by the TRA assembly from 1971. The TRA nozzle, at the extreme end of the engine, not only controls reverse thrust but is used also in different positions during take-off and cruise. Take-off noise is reduced by diverting the exhaust air, the exhaust-gas pressure is controlled to optimise engine performance, and aircraft performance is improved by the lower weight of the TRA compared with that of earlier designs.

The 593 is now tuned to a pinnacle of performance and there is little that can be done, without a basic redesign, to improve any parameter significantly. In service it is proving reliable and is meeting every criterion. While a different arrangement might be chosen if a "cold start" on a supersonic engine were made today, the Olympus 593 remains a world-beater.

McDONNELL DOUGLAS STUDIES ADVANCED SST

While the Concorde landings issue remains unsettled and the commercial future of the Western World's first SST stays in the interrogative mood, the McDonnell Douglas Corporation is continuing studies of an advanced supersonic transport (AST). MDC-funded work on the arrow-wing developed by the US National Aeronautics and Space Administration has resulted in an "exploratory development study" of a Mach 2.2 aircraft capable of carrying 273 passengers up to 4,400 n.m., depending on powerplant. One possible engine, should the dictates of available technology and in-service time demand it, is a turbofan derived from the Concorde Olympus 593.

The study assumes that Concorde will attain a limited degree of success, that it will be accepted on world routes and that passengers will want to fly in it. The aircraft used in the study is fairly close in size and weight to the Boeing 2707-300 SST cancelled in 1971, but differs in its cruise speed (Mach 2.2, compared with Mach 2.7) and its wing planform. The net effects of these differences are longer range and lower direct operating costs (DOCs); the 2707-300 was expected to be comparable in DOC terms with the 707, and the AST is planned to be compatible with the DC-10-30.

The AST is designed to co-exist both environmentally and economically with the wide-bodies. The larger area and more extensive flappery of the arrow wing, coupled with new silencing technology, bring the AST within the FAR Part 36 limits for aircraft in its weight class.

The arrow wing is structurally more difficult than a delta of the same area, but MDC claims a lift/drag ratio up to 9.6:1 at Mach 2.2 for the AST, compared with 7.2:1 at Mach 2.7 for the original US SST design. (Refinement following prototype flight trials was expected to improve this to 7.95:1.) The moderately swept outer wings are fitted with leading-edge slats and conventional flaps and spoilers, but only the inboard section of the highly swept inner wing is slatted. The effect of this arrangement is to entrain vortices over the mid-span leading edge, rather in the manner of the gently curved Concorde root shape. Thickness-chord ratio at the root of the AST wing is 2.25 per cent, changing to 2.5 per cent at the leading-edge break and 3.5 per cent at the tip.

The arrow wing yields its best trimmed lift/drag ratio at an angle of attack of around 1°; this raises a slight design problem since ideal root incidence is 6° with respect to the wing reference plane. This results in a cruise floor angle of 7°, considered unacceptable from the viewpoint of passenger comfort and cabin service (widebody floor angle is 2°-3°). Limiting floor angle to 5° and root incidence with respect to the reference plane to 4° means a two per cent penalty in lift/drag ratio. MDC hopes, however, to improve lift/drag ratio by designing a nose-up pitching moment into the wing at cruise speed. The nose-up trim required in the cruise improves lift/drag ratio to

9.5:1, compared with the 9:1 obtained with the "baseline" wing.

McDonnell Douglas initially chose separate engine pods, with full-cone axisymmetric inlets, on the grounds of efficiency; improvement derives almost entirely from the intake design rather than from the four-nacelle layout. The axisymmetric intake is more demanding than the two-dimensional type chosen for Concorde in 1962, but a great deal of experience in intake management has been gained since that time.

The four-nacelle layout minimises the penalties of using physically larger engines, which can be prohibitive in dual installations. MDC has looked at four possible powerplants using the technology levels of 1978 onwards. The baseline chosen was a non-afterburning version of the General Electric GE4 developed for the Boeing SST, modified for a 1975 technology level and the lower Mach number, and fitted with a MDC-designed acoustically lined ejector nozzle to reduce noise at take-off by 12PNdB.

The first new-technology powerplant studied is a General Electric "mini-bypass" engine with a bypass ratio of 0.1:1. Using 1978 technology (although MDC estimates that development of a completely new AST engine would take at least eight years) the mini-bypass engine could offer eight per cent more range than the baseline engine. This would result from the lighter weight of the installation and supporting structure (being naturally quieter, a less complex silencer is required for the bypass engine) and the lower fuel consumption of the bypass cycle.

Pratt & Whitney has backed the duct-burning turbofan as the powerplant for AST. MDC calls the concept a "variable stream control engine"; the physical geometry of the engine is constant, but duct-exit temperature is held down to 400°F below design temperature on take-off and increases to the design value in the cruise. This produces increased mass flow on take-off, improving low-speed thrust and quietening the aircraft

to meet or undercut FAR Part 36 regulations with a simple acoustic ejector, while reducing mass flow in the cruise and improving specific fuel consumption. Pratt & Whitney estimates that the slow outer flow, with a 1.3:1 take-off bypass ratio, will provide an element of noise suppression. The duct-burning turbofan is the lightest of the powerplants evaluated, and MDC estimates that it requires a 1980 technology level. Installed in the AST, the duct-burning engine would give a range of 4,460 n.m.

Third option in the AST study was the variable-cycle engine (VCE), using mechanical variation of bypass ratio. The VCE has been hailed as the perfect powerplant for an SST, but the weight penalties are severe in the extreme. Compared with the GE4-derived baseline turbojet there is a 24,600lb increase in airframe weight with the VCE installed; this reduces range to 3,180 n.m.

The compatibility of the AST with a derivative of the present Concorde Olympus engine was not in fact studied in parallel with the other options and the engines are not easily comparable. Any derivative of the Olympus would be available rather earlier than an all-new powerplant. With a new engine, MDC observes, "an advanced-technology SST could not be certificated in the US before 1984, even if today industry was to select both the engine cycle and the engine size. It is hard to see how such a position could be established in the next two years."

Preliminary investigations of Olympus variants indicate that improvements in blade technology could, within six to eight years, yield a turbofan Olympus with a 1:1 bypass ratio. It would give the AST a 3,900 n.m. range and would meet FAR Part 36 without an exhaust suppressor.

The AST study may need some rethinking if the new and stricter noise standards now proposed in the US are accepted. At the moment it is unlikely that airlines are showing very much interest, but the MDC thesis—that Concorde may be successful and that a follow-up aircraft will be needed in that case—seems sound. AST would seem to be an obvious area for transatlantic co-operation in the 1980s; by that time the builders of Concorde will have learned a great deal about the task of operating and supporting SSTs.

THE SCISSOR WING TRANSPORT

Reprinted by Permission
AIRLINE PILOT (May 1976)

by Major William D. Siuru, Jr.
(USAF)

Ever since man started building airplanes it has been a universal belief that they should be symmetrical, the port side should be a mirror image of the starboard side. Perhaps this is because man got his inspiration for flight from the birds, or perhaps because a symmetrically arranged airplane is more aesthetically appealing. However, Dr. Bob Jones of NASA's Ames Research Center believes that non-symmetry has some distinct advantages. He feels significantly improved high speed transports can be obtained by using a radical departure in aircraft design—the "scissor wing" that pivots about its center.

In flight the scissor wing looks like a child's balsa glider that has had its wing knocked askew by an encounter with a tree. But according to Jones, it may be the solution to several key problems plaguing high speed commercial aviation. It could permit flights at sonic speed without the sonic boom, could give substantial savings in fuel usage and could provide reduced noise pollution in the vicinity of our airports.

The scissor wing, or the oblique wing as Dr. Jones likes to call it, consists of a single, straight and quite thin main wing that is attached to the fuselage at a central pivot point. For landings, takeoffs, and low speed flight, the scissor wing craft would look very much like a conventional airliner with the wing perpendicular to the fuselage, although the wing would be somewhat slimmer. For

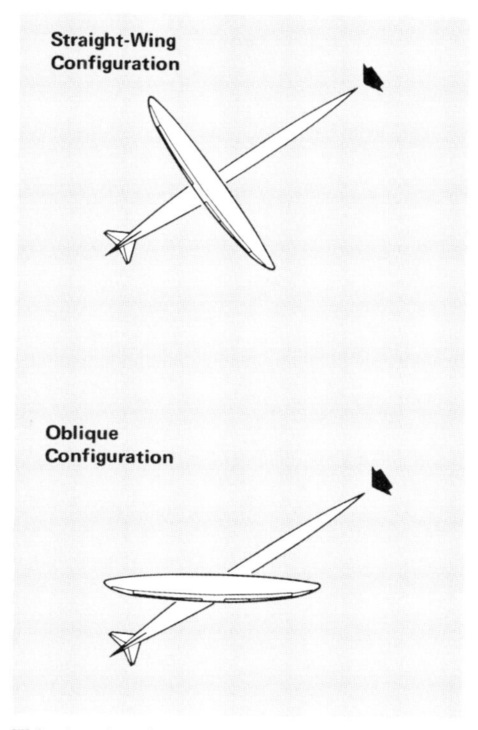

This drawing shows how the wing and fuselage of the new concept relate to each other like the two halves of a pair of scissors. The straight wing is mounted above the body.

high speed cruising, the wing would pivot with one half of the wing swung back and the other half swung forward. The wing could be pivoted as much as 60° from the straight position used on landings and takeoffs.

The theory for the oblique wing has been around for many years and aerodynamicists have known about its advantages for some time. However, the radical design has not been pushed, except by Dr. Jones and his colleagues at NASA and at the Boeing Commercial Aircraft Company who have pursued the concept vigorously with extensive design studies and wind tunnel experiments. Through these efforts Dr. Jones, who in 1945 was one of the fathers of the swept-wing that is a common feature of today's air transports, has the data necessary to back up his claims of the advantages of the scissor wing concept.

One of the main reasons we do not have a supersonic transport now operating within the United States and why our SST met with so much opposition is the sonic boom problem associated with such high speed flights. An airplane flying at altitudes of 35,000 to 40,000 feet can fly at speeds up to Mach 1.3 (about 850 mph) without a sonic boom occurring on the ground below. What happens is that while the speed of sound is about 650 mph at 35,000 feet altitude, the sonic speed decreases about 100 mph as the ground is approached. This is because the sonic velocity is related to the ambient temperature and the temperature on the ground is greater than at altitude. Because a shock wave can only exist in a supersonic situation, the shock wave from a plane flying at, say Mach 1.2 and 35,000 feet altitude would disappear before it reached the ground. Put in another way, the aircraft must fly at a subsonic ground speed to avoid a sonic boom at ground level. This subsonic ground speed still permits flight at altitude that is 200 mph faster than current jets.

The flight speed for which no sonic boom occurs on the ground varies somewhat with flight altitude, head or tail wind conditions, and ambient air conditions on the ground. Aircraft speeds considerably in excess of the 500 mph in practice today could be achieved with no effects being felt on the ground. The reason why a supersonic transport like the Concorde cannot fly at transonic speeds (near Mach 1) is because of its low efficiency in this speed range. The fixed-wing supersonic transport operates relatively economically at subsonic speeds or at speeds near Mach 2 (1,300 mph). At transonic speeds, however, it is so inefficient and its fuel consumption so high that its range would be reduced to unacceptable levels. By contrast, the scissor wing concept is designed to be efficient at these speeds and while it is not as fast as a true supersonic transport, it could reduce travel times rather significantly on passenger routes. For example, it is estimated that a transonic flight from New York to Los Angeles would be two hours shorter than the same flight today, while the reverse trip would take about an hour less.

Let us look at the reason why the scissor wing can operate economically at transonic speeds. A measure of the efficiency of a transport aircraft is its specific range, or the payload-miles per pound of fuel consumed. This is analogous to miles per gallon by which we judge the economy of our automobiles if we include in our comparisons the amount of load carried. The specific range is dependent on the lift to drag (L/D) ratio of the craft, in addition to several other factors. The higher the L/D ratio, the more efficient the aircraft design. The L/D of a typical aircraft at speeds less than Mach 0.7 to 0.8 is typically 15 to 20. However, at speeds greater than Mach 0.8 the L/D starts to drop off rather drastically. In the supersonic speed regime a typical L/D is in the range of only 7 to 10.

With the scissor wing concept the wing, when at an oblique angle with respect to the fuselage, is made to think that it is flying at speeds less than Mach 0.7 to 0.8 even when it is flying at speeds slightly in excess of Mach 1. When the wing is oriented obliquely, the airflow over the wing can be broken into two components. One component flows perpendicular to the wing, the other component flows along the length of the wing. This latter component has little effect. By keeping the perpendicular component below Mach 0.7 to 0.8 by increasing the angle the wing makes with the fuselage, the wing retains high L/D value needed for efficient operation even when flying at speeds that are mildly supersonic. For example on a wind tunnel model of the scissor wing, an L/D of 20 to 1 was attained at nearly Mach 1 with the wing pivoted to 45°. Studies have shown that the fuel consumption of a full-size airliner using a scissor wing would range from about that of today's jet liners when flying at 500 mph to about twice the fuel usage of current craft when moving at 900 mph. At this higher speed the fuel economy would still be twice that of the Concorde or the Soviet TU 144. And it all would be done without an associated sonic boom on the ground over which it passes.

The long and narrow wing of the scissor wing aircraft has another advantage. It has a very high L/D at the low speeds associated with landings and takeoffs and while in the holding pattern over an airport. At these speeds the wing would be at right angles to the fuselage and L/D's on the order of 20 or more to 1 would be achieved. This would lead to lower power requirements and thus economical fuel consumption. The low power would mean quieter operation and alleviation of a good portion of the noise that is becoming an ever increasing problem around our major airports today. In addition, the Concorde and TU 144 need an afterburner for takeoff because of their short wing span and thus result in takeoff noise that may cause them to be banned from US airports. The oblique wing craft could take off without the need of an afterburner.

The same spoofing of the wing to make it believe it is flying at a lower speed can be achieved by symmetrically changing the sweep of each wing of an airplane. The variable sweep geometry is now used on the Air Force's F-111 fighter. However, in this case, the pivot device for each wing section has to have massive bearings to carry the bending loads that are trying to break the wings off. These massive devices significantly increase the weight of the craft and reduce the payload it can carry. With the oblique wing, the entire wing can be built as one continuous section with a single pivot point at the center. Because the wing is continuous the bending loads are not absorbed by the pivot and the single pivot bearing device can be much lighter.

The asymmetrical appearance of the scissor wing initially makes one think it would be unstable in flight and would be hard to control. The wing looks like it would tend to pitch the craft violently. However, tests in NASA's wind tunnels have shown that this was not the case. Recently, NASA built and flew a radio-controlled scale model with an oblique wing that had a span of six feet. The model was put through all types of maneuvers including aileron rolls in addition to maneuvers more nearly akin to those that a normal air transport would see. While the model showed some unusual flying characteristics during the more aerobatic maneuvers (such as during loops it flew in a helical path) the plane was quite controllable. In the more straight and level flights, there was no noticeable difference between the scissor wing and a normal airplane.

The scissor wing may look funny and it would take some good advertising to convince the public that it is safe to fly in a craft whose wings don't look like they are securely attached to the fuselage.

However, it offers some real solutions to two major problems facing the airline operator—fuel economy and noise pollution—and it does this while increasing the speed of air transportation.

Major Siuru, *a Ph.D. in mechanical engineering, is currently serving as chief of the Supporting Technology Branch at the Air Force Rocket Propulsion Laboratory, Edwards AFB, Calif. His numerous technical and semi-technical publications include a soon to be published book on the F-16 fighter.*

ENERGY MANAGEMENT— THE DELAYED FLAP APPROACH

by John S. Bull
Research Scientist
NASA-Ames Research Center

Reprinted by Permission
SHELL AVIATION NEWS

The conventional jet transport instrument landing approach procedure requires high thrust settings for an extended time, with the accompanying community noise impact and relatively high fuel consumption. Significant reductions in both noise and fuel consumption can be gained through careful tailoring of the approach flight path and airspeed profile.

For example, the noise problem has been attacked in recent years with development of the Two-Segment approach, which brings aircraft in at a steeper angle initially, thereby achieving noise reduction through lower thrust settings and high altitudes. However, proposed implementation of Two-Segment approach procedures into routine scheduled airline service has met with some objections, such as those raised concerning wake turbulence vortex encounters.

In an effort to overcome these objections, the NASA-Ames Research Center is currently investigating the Delayed Flap approach procedure. In this case, the approach is initiated at a high airspeed and in a drag configuration that allows for low thrust. The aircraft is flown along the conventional ILS glide slope. Remaining at low thrust for reduced noise and fuel consumption, the flaps and landing gear are lowered at the appropriate times, so that the airspeed slowly decreases to

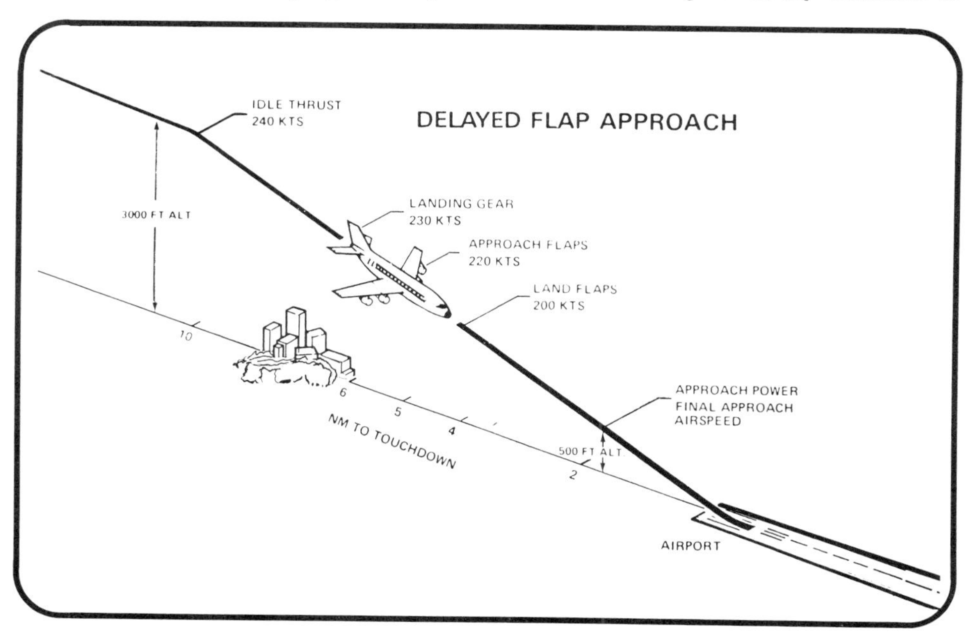

the desired value at the desired point from touchdown.

A flight test investigation of the Delayed Flap approach is underway using NASA's CV-990 aircraft. The CV-990 is a four-engine jet transport representative of the DC-8 and B-707 class. Fuel conservation and noise reduction benefits are being measured, and pilot acceptability is being evaluated in flight tests with guest pilots. Flight simulation investigations are also being conducted by the Boeing and Douglas Aircraft Companies under contract to NASA to determine applicability of the Delayed Flap approach profile to other aircraft such as the DC-8, DC-9, DC-10, and the B-727.

Figure 1 shows a typical Delayed Flap approach. In contrast to a Conventional approach, which is flown at a constant airspeed of about 150 kt and high thrust settings throughout, the Delayed Flap approach begins at a higher initial airspeed—240 kt—and decelerates at idle thrust through most of its duration.

The pilot intercepts the ILS glide path at about 10 n miles from touchdown and 3000 ft altitude. He then retards the throttles to the idle detent, and begins a slow deceleration. At about 6 n miles and 230 kt, the pilot is given a command from the avionics system to lower the landing gear. At about 5 n miles and 220 kt, a command is given to lower approach flaps, following which flaps are commanded to the landing position at about 4 n miles and 200 kt. The aircraft decelerates to final approach airspeed at 500 ft altitude, at which point the pilot advances the throttles to approach power and the last portion of the approach is flown at a stabilized airspeed similar to a conventional approach.

In headwinds, extension of landing gear and flaps is delayed; in a tailwind condition, they are commanded farther out on the approach. Thus, regardless of wind conditions, the aircraft is always stabilized for landing at 500 ft altitude, which is consistent with current airline procedures.

In addition to the normal instruments, the CV-990 cockpit features a Fast/Slow indicator, a message display, and a data entry keyboard. The Fast/Slow display, which is commonly found in most current jet transports, is a 'how-goes-it' display to tell the pilot how the aircraft is decelerating relative to the desired airspeed schedule. This is very similar to the way a Fast/Slow display is normally used—that is, to show the pilot his error from the reference landing airspeed.

The message display signals the pilot when to extend landing gear, approach flaps, and land flaps. The proper timing of signals is accomplished by a digital computer. In essence, the computer predicts the manner in which the aircraft will decelerate during the approach to land, taking into account the wind. Based upon this computed deceleration, the computer signals the pilot when the flaps or gear are to be lowered by flashing a command on the message display.

When the pilot has taken the required action, the display goes blank again until the next extension of gear or flaps is to be made. All this is accomplished such that the aircraft arrives at the final approach airspeed at precisely the desired point from touchdown. The data entry keyboard is used by the pilot to enter into the onboard digital computer the required navigation data, for example, field elevation.

The avionics that would have to be installed in a conventional jet transport in order to have a Delayed Flap capability would be simpler and less sophisticated than our flight research system used in the CV-990. Implementation of this pro-

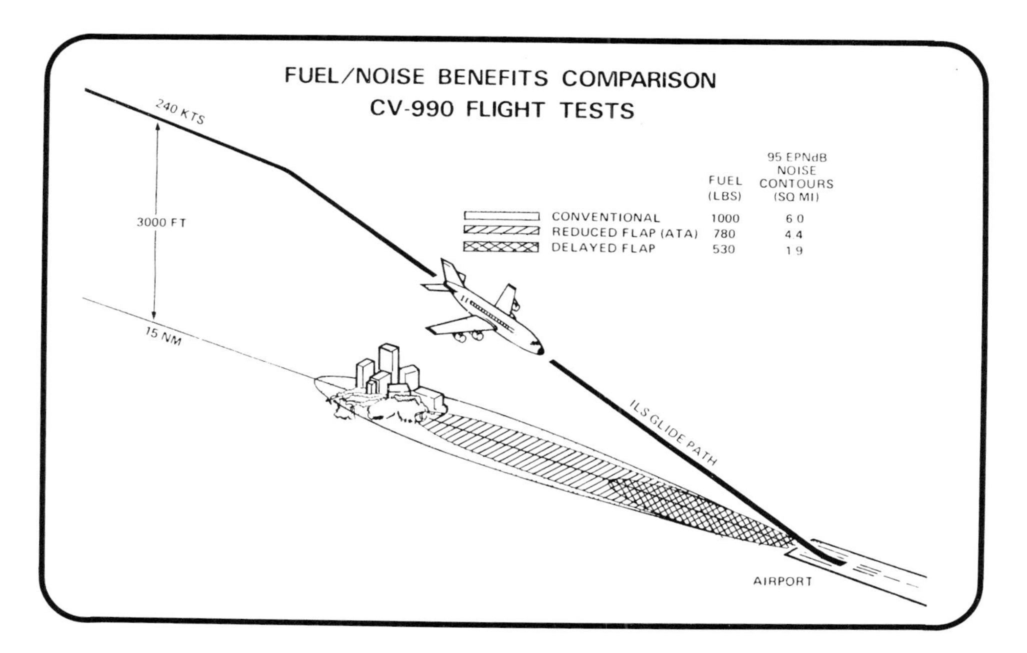

cedure in commercial service may require the addition of a DME navigation aid co-located with the ILS glide slope transmitter.

Guest pilots from the air transport industry were invited to fly the CV-990 in flight tests in November 1975 for the purpose of evaluating pilot acceptability of the Delayed Flap approach. Nine pilots participated from United Airlines, American Airlines, the Airline Pilot Association, the Air Transport Association, the Federal Aviation Administration, Boeing, Douglas, and Lockheed.

The Conventional approach—as already described—is one that is stabilized in airspeed and land flaps throughout most of its duration, and is representative of the way approaches were flown several years ago prior to efforts in noise abatement. The Reduced Flap approach is the current recommended Air Transport Association landing procedure; the first portion of the approach is flown at a reduced flap setting, and land flaps are not extended until 1000 ft altitude. The Delayed Flap approach uses a decelerating airspeed at idle thrust to 500 ft altitude.

The guest pilots were debriefed and given questionnaires to fill out after their flights. The Conventional approach was given an average rating of Excellent in pilot acceptability, the Reduced Flap approach was rated between Excellent and Good, and the Delayed Flap approach was considered to have Good pilot acceptability. The high initial airspeeds in the Delayed Flap approach had no effect whatever on pilot acceptability, and use of the conventional ILS glide path was a desirable feature. However, the guest pilots did voice concern over the problem of ATC compatibility that would be caused with the use of high initial approach airspeeds. In response to this concern, we are initiating work with the FAA.

Fuel consumption and noise reduction benefits for the three different types of approach were measured in flight tests at Edwards AFB. Figure 2 shows the results of these tests. In order to have the data represent as fair a comparison as possible, we were very careful to begin each approach under exactly the same initial conditions: 15 n miles from touchdown, 3000 ft altitude, 240 kt airspeed, with gear and flaps up. The fuel data shown is fuel consumption from 15 n miles to touchdown.

The first comparison that I would like to point out is the fuel savings and noise abatement that the airlines have been able to achieve with simple improvements in pilot procedures. The current airline procedure, the Reduced Flap approach, saves 220 lb of fuel over the Conventional approach, and has a 95 EPNdB noise contour only 74% as large. So the airlines have been able to show considerable gains simply with improvements in pilot procedure without adding new avionics to the cockpit.

The Delayed Flap approach offers about the same magnitude of benefits over the current airline procedure. It saves 250 lb of fuel over the Reduced Flap approach, and has a 95 EPNdB noise contour only 43% as large. The Delayed Flap approach does require additional avionics, but we expect that the cost of this could be recovered in a reasonable time by the savings in fuel.

On-going work includes an investigation with the FAA in the area of ATC compatibility. Also, applicability studies of the Delayed Flap approach to current jet transports will continue in areas of pilot acceptability, fuel/noise benefits, and avionics retrofit costs.

(This material originally appeared in the CONGRESSIONAL RECORD.)

YOUR CHANCES OF GETTING THAT AIRLINE JOB

Reprinted by Permission
AIR PROGRESS (July 1976)

by an Airline Captain

The author of this article, who prefers to remain anonymous, has 20 years experience as a pilot with a major airline. He has logged over 20,000 hours and is type rated in airliners ranging from DC-3s to DC-10s.

* * *

Whenever pilot vacancies arose, the airlines have sought as a rule applicants with the following qualifications: 23 years of age, a Masters degree in aeronautical engineering, an airline transport rating, a flight engineer's certificate, and 2,500 hours of pilot-in-command time in four engine equipment. Along with these requirements, they also preferred the clean-cut, good looking, square-jawed individual with the fabled steely-eyed look of the aviator. Of course, these demands were only a figment of the imagination in the mind's eye of every brand new personnel director. Realistically, the "perfect" applicant never applied since he was simply nonexistent. In contrast, the airlines have been known to employ precisely the exact opposite individual during desperate situations in the days of rapid expansion.

Today one may be Black, Polish and/or a female. Nonetheless, one must be persistent, determined and not easily discouraged. That is, one should resemble an image Hollywood might create of a dogged, determined, never-give-up type, aspiring movie star on the never ending quest of landing a major part.

Consequently, we have seen a number of excellent pilots with outstanding flying skills and phenomenal talent who threw in the towel after a number of unsuccessful interviews to pursue another way of life—to their regret in later years. On the other hand, we know of pilots with only bare minimums . . . where the ink on their commercial tickets hadn't even dried as they unexpectedly found themselves thrust into the copilot's seat on a scheduled trip—some on the very day of their first interview. Years ago, a pilot in executive status did all the pilot applicant screening and initial interviewing. Then and only then, a personnel director further screened a pre-hire who, if judged satisfactory, was given a check ride, usually by the chief pilot or some other pilot acting in his behalf. In the days of rapid expansion, promotion was so fast that I can remember one copilot tell me of how unprepared he was to fly as captain. He kept hoping he would fly into heavy weather and icing conditions pretty soon. He had never experienced this type of flying and was concerned because a captain's checkout was just around the corner. Here was this guy wishing he could acquire more experience before taking on a captaincy. Those were the days!

In the mid sixties, there was a period of such great pilot demand that it was the pilot applicant who became highly selective. It was *he* who screened the airline to decide where advancement would be most rapid. Many were so cocksure of themselves that they made it known to various airlines that it was *they* who were vying for *his* services and that *he* intended to insure himself of a good choice by seeking out the differences among airlines and only then making a well considered choice before signing on.

Through the years, pilot hiring concepts have seen radical changes. Stanine tests became the vogue. (That criterion was easily overcome by determined applicants who purchased the correct answers and went on to beat the system.) Today, the applicant runs the gamut by first talking with a receptionist before he can even come close to someone in personnel. Unless he has an inside contact, making direct contact with a pilot on the administrative level is normally the last step.

Historically, and still in practice, is the hiring airline's attempt at accurately forecasting an applicant's advancement. Most of these forecasts have proven incorrect for a multitude of reasons. A most recent example was the fuel crunch followed by an immediate set-back in the nation's economy whose rippling effect touched everyone.

In 1963, FAA Administrator Najeeb Halaby arranged a survey to learn if there was really going to be a pilot shortage in the next decade, as was forecast in various studies. This survey affirmed the possibility of a critical shortage in coming years, due to the retirement of a large number of World War II pilots and pilot aspirants with pessimistic attitudes resulting from dismal past employment attempts. The survey further revealed that the industry in the past had always enjoyed a surplus of qualified pilots.

In 1964 the Aviation Human Resources Status Board was activated. Its objective was to assess the long range requirements of commercial aviation for highly skilled and long lead time trained pilots. The board learned that available information, which airlines were reluctant to divulge, was hardly adequate. However, the report was completed and released six months later and with the cooperation of a few airline executives. Its prediction—a shortage! Yes, the report predicted that over 3,000 new pilots as a minimum would be needed due to natural attrition. In addition, it forecast a rise of over 4,000 additional pilots, which would expand the number of professional pilots to over 20,000. The board estimated that the new pilot hires for all airlines during the next ten years would amount to over 7,000. While statistics seemed to indicate a shortage, the availability of many more pilots standing in line to get on the airlines remained. We saw no shortage or spoke to anyone who did in this span of time. "Furlough" was on the lips of many junior pilots.

A sample cross section of a number of airlines two years ago predicted about 3,500 new hires would be required for pilots replacing retirees through the year 1982. However, a forecast of nearly 5,000 was predicted. Studies of the past ten years showed that medical retirements during the years 1969-1972 nearly equaled the mandatory age 60 retirement group, with this trend growing. The reasonable assumption: at least another 5,000 pilots would be needed. One airline executive boldly predicted an acute industry wide pilot shortage in the 1977-1982 time frame. *(This writer predicts it will never happen.)*

The last five years have been extremely tough on the junior pilots, many of whom have suffered furloughs and/or the new hire trying to get on.

There is no way an airline pilot can circumvent his mandatory retirement at

age 60. Nor is there any guarantee that that pilot will be replaced by a new hire filling the void. Many factors effect accurate predictability assessments of each airline. Always prominent in the forefront seems to be the deciding factor of basic economics.

Today, unless a small airline is sure of being awarded a substantial route award, the new-hire who is riding side-saddle (second officer) or copilot on a smaller jet, will most likely remain in that position for an infinite number of years, creeping up the ladder of advancement at a snail's pace. However, the job has its compensations. Besides flying the best equipment in the world, the pay is good, days off plentiful and the chow is excellent.

On some major airlines, a few pilots who may fly as captain actually choose a lower, first officer status which allows them a senior, choice run, perhaps overseas with only ten working days. This choice is often offered to the first officer too, who may bid "down" to the status of second officer. In either case, there is only a slight difference in pay as opposed to long and tiring working conditions in a higher status where they are unable to fly better trips owing to their seniority.

Currently, the fleets of major airlines are comprised of numerous advanced aged jets in the throes of retirement. They will soon require either replacement or, clinically speaking, be placed in cold storage in the name of economical demands. Either way, when they are replaced, chances are the ratio could be something like one (wide body) new airplane for every five taken out of service. With fuel and labor costs rising astronomically, you can bet airline management will keep the purse strings taut and employ high cost pilots only when absolutely needed. However, for those who will have the good fortune of getting "on" with an airline, there is some good news. Recent trends have allowed the new hire meteoric advancement heretofore unknown. That is, if he possessed the required drive and determination, along with his already accepted skills, was studious, neat, and worked like the very devil himself, he then could be a good candidate for the "office."

A number of airlines have selected relatively inexperienced pilots, quite low on the totem pole when it came to seniority, and have placed them in supervisory positions. This trend has proven convenient and satisfactory to the airline heads. At times so successful, even a few pilots have worked themselves right up to the high echelons of executive status. Word circulating among the line pilots was, and still is, "Be nice to your 1st or 2nd officer. He soon may be your next supervisor." It would never have happened if they were not in the right place at the right time.

Summing it all up: times will be tough for the new hire. The demand for airline pilots could well be compared to that of actors, entertainers and their like. Aside from proper credentials it all boils down to a matter of timing.

EMILY MAKES CAPTAIN

Reprinted by Permission
NBAA Convention News (September 16, 1976)

First woman airline captain in the nation, Emily Howell attained that rank when Frontier Airlines moved her into the left seat of a Twin Otter on its Denver-Cheyenne-western Nebraska route. Her promotion is as significant a break-through into the male bastion of airline piloting as was Jackie Robinson's into the exclusively white-skinned ranks of pro baseball when he went to play second base for the old Brooklyn Dodgers.

Women's voices are ringing with greater and greater authority in the aviation field these days, and nowhere is the sound more noticeable than at Denver-based Frontier Airlines.

Hired by Frontier in 1973 as the U.S. airline industry's first woman pilot, Emily Howell was elevated, on June 1, 1976, to the position of captain, becoming the first member of her sex in the nation to

command a scheduled carrier. The pioneering female aviatrix now flies Twin Otters (DeHavilland DHC-6s), between Denver, Cheyenne and western Nebraska.

Beginning as a second officer on Boeing 737s, pilot Howell moved up to first officer on Convair 580 propjets and Twin Otters before gaining her four stripes. She said that the reaction of fellow male airline pilots "has generally been positive, although, with a couple of captains, it took a few weeks before I became accepted as just another pilot."

Flying has long been a fascination, a fixation for this modest, serenely attractive woman in her mid-30s. A native Coloradoan, she recalled her first airplane ride from Denver to Gunnison, Colorado, in the late-fifties, "I thought I wanted to be a stewardess and that I should first find out if I'd get airsick. It so happened that I was the only passenger on the return trip so they let me visit the cockpit. I guess my enthusiasm showed, because one of the pilots suggested I learn to fly. That very weekend, I began taking lessons."

She worked as a department store clerk to finance her private ticket that same year. Within three years she had a commercial with instructor and instrument ratings. From then on aviation was her full career. It began with a secretary's job for a Stapleton FBO for whom she started to instruct in 1961 and eight years later she became manager of its flight school. In 1973 Frontier hired the nine most qualified pilots it could find—Ms. Howell was one of them.

Recalling the news stories in the Denver, Chicago and New York papers, and her appearances on the *What's My Line* and *To Tell The Truth* television programs, Captain Howell said, "I found it fun but a little embarrassing," but added, "on my first Frontier run, when United (Air Lines) and Air Force pilots got on the radio and welcomed me into the air, it was really great!"

"I am not a women's libber," insists Ms. Howell, divorced and the mother of a young son. "I am not striking a blow for anything. I expected to be treated just like any other pilot, and, after three years, I'm not disappointed. It's great. I really enjoy all of it."

In 1973, Howell received the Amelia Earhart Award as the year's outstanding woman in U.S. aviation and, later that year, she was presented Colorado's first Wright Brothers Memorial Trophy. In 1974, she became the first woman elected to membership in the Air Line Pilots Association.

Today, one of Frontier's 550 pilots, she has more than 9,000 hours in her logbook.

Not all Captain Emily's flying is on Frontier's routes. She's an avid aerobatic buff and is currently a partner in putting together a Skybolt biplane kit at Arapahoe Airport outside of Denver.

AROUND THE WORLD IN FORTY HOURS

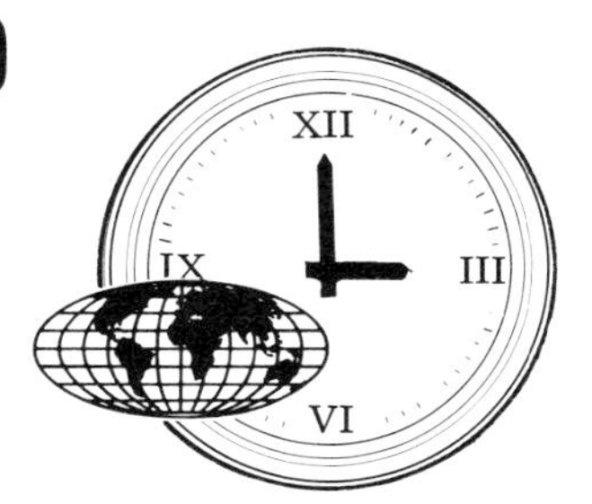

Reprinted by Permission
AIR LINE PILOT (August 1976)

by C. V. Glines

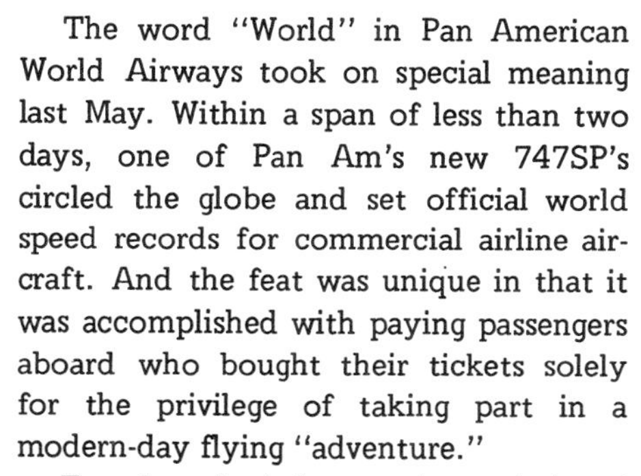

The word "World" in Pan American World Airways took on special meaning last May. Within a span of less than two days, one of Pan Am's new 747SP's circled the globe and set official world speed records for commercial airline aircraft. And the feat was unique in that it was accomplished with paying passengers aboard who bought their tickets solely for the privilege of taking part in a modern-day flying "adventure."

Even in today's jet age, the statistics of the epochal flight are awesome. The junior jumbo, christened *Clipper Liberty Bell* by First Lady Betty Ford in Indianapolis the day before, lifted off from Kennedy International Airport on May 1. Exactly 39 hours, 25 minutes and 53 seconds of air time later, it touched down again at JFK. With only two stops (New Delhi, India and Tokyo, Japan), the total elapsed time from takeoff to return was 46 hours, 50 seconds. The distance flown: 23,137.92 statute miles.

The previous record for round-the-world flight by a commercial aircraft was 62 hours, 27 minutes, 35 seconds established in November 1965 by a Boeing 707-320C that carried only a crew. Therefore, the Pan Am 747SP globe-circler was another in a long list of Pan Am commercial aviation "firsts" in that 98 passengers were carried. First class adventurers paid $2,912 for the privilege; economy class tickets were $1,838. Thus, the trip paid for itself and was not just a public relations gimmick that produced no revenue for the financially-troubled airline.

The idea could be said to be a by-product of the highly successful, pilot-led Pan Am Employe Awareness program that brought management and employes together in an unprecedented show of cooperation and solidarity. The improved communications, built up by the intense desire of both parties for the company to survive, encouraged many employes to submit worthwhile suggestions on improving company image and income. Among them was an idea submitted by letter directly to Chairman of the Board William T. Seawell last December by Captain Lyman G. "Mike" Watt, ALPA secretary. Would it be possible, Watt asked, that Pan Am might revitalize its image as a world airline by attempting an assault on the commercial round-the-world record and tie it in with the Bicentennial? The newly acquired 747SP's made it possible to circumnavigate the globe using only two normal Pan Am stops; certainly the old record of 62½ hours could be beaten easily. If passengers were taken, the trip would be fully justified cost-wise.

Chairman Seawell passed the idea on to his staff. Although Pan Am has established many flying "firsts" and set many records in the past, there had been no thought of record-setting in these days of penny-pinching and fuel conservation.

In the weeks following, a plan emerged

Captain L.G. "Mike" Watt, ALPA secretary and the man who suggested the around the world flight chats with passengers.

John M. Slattery, (right) official NAA observer, checks flight progress with Captain Al Frink (left) while Keith Ackerman, unaccompanied 11-year-old passenger and Beverly Harris, flight attendant, listen in.

that involved the entire New York headquarters staff at some point. The idea was not only feasible but judged an excellent way to put Pan Am back in the news favorably. While Watt has suggested departure from Washington with stops in Karachi and Guam, it was decided to depart New York and make stops at Delhi and Tokyo. Thus, the trip would consist of three nearly equal segments. Besides, the National Aeronautic Association (NAA) that acts as the official U.S. record-approving body for the Federation Aeronautique Internationale (FAI) requires that a world flight for the purpose of setting a record must be a minimum of 22,858 statute miles in length. To be assured that the minimum distance would be flown, most record-seekers deliberately plan to fly more miles than required in case their mileage computations are inaccurate. All distances flown must be verified by the National Oceanic and Atmospheric Administration (NOAA) as part of the record-approval process.

FAI rules also require that an NAA official monitor the flight to verify the route flown and elapsed times. John M. Slattery, newly-appointed NAA record-keeper and timer, volunteered.

While Pan Am operations and maintenance personnel worked out details in their respective areas of responsibility, the public relations and marketing departments geared up for publicity and sales. Ads were run in several leading newspapers and brochures were sent to Pan Am offices.

"How's your spirit of adventure? Your love of pioneering? Your desire to say 'I did it!' " read the brochure. There would never be another flight like it, the brochure said, and it would give a passenger something to talk about for years. Full-page newspaper ads announced: "On May 1, Pan Am will revise the *Guinness Book of World Records*. Would you like to be on board?"

With a spirit befitting an airline rejuvenated from within, the 98 passengers who signed up reported to the Pan Am Worldport at JFK in mid-afternoon on May 1st. Amid a large group of press, radio and television reporters, Chairman Seawell gave credit to Captain "Mike" Watt for conceiving the idea and his own staff for coordinating the arrangements around the world. At precisely 5:40 p.m. on May 1, the one-of-a-kind flight began with Watt as part of the crew. Captain Walter H. Mullikin, Pan Am's vice-president and chief pilot headed the 15-man cockpit team that would eventually bring the flight back home.

Clipper 200–*Liberty Bell Express*, as the flight was designated, headed east and flew through the brief night hours to Land's End, England, then proceeded across Europe via Paris, Munich, Istanbul, Ankara and Tehran to New Delhi, a distance of 8,081 miles–longest leg of the flight. Air time was 13 hours, 31 minutes–a city-to-city record. Average speed: 598 mph. Enough fuel remained to fly another 1,815 miles to Bangkok, Thailand, and still have the required reserve.

There was a two-hour fuel stop at New Delhi. Passengers, including Summer Bartholomew, Miss U.S.A., deplaned and were treated to a warm (104°F) welcome by sari-wearing women who gave each passenger and crew member a garland of flowers and a dot of red paint in the middle of the forehead. The handicraft shops were visited and a film of Indian hunting was shown.

The next destination was Tokyo but the route was not to be flown direct in order to assure the minimum mileage required. A newly embarked five-man crew took over the flight deck but the original five-man crew including Captain Mullikin stayed aboard since NAA rules require that the original crew must go all the way around in order for the record to be official. Two passengers bound for New York were added to the manifest.

The flight headed southwest toward Bombay then to Tokyo via Singapore, Jakarta and Manila. This leg was 7,539 miles and was covered in 14 hours, one minute. Adverse winds cut the speed to a relatively low 536 mph. However, another city-to-city record was placed in the record books.

Although only a two-hour fuel stop was planned at Tokyo, a work slowdown that had been in progress for some time among ground crews at Haneda International Airport nearly spoiled the historic attempt. In order to focus attention on their demands, drivers of ground vehicles parked them three and four deep behind the aircraft just as the captain signaled for the pushback. It was two frustrating hours later before the Pan Am negotiators were able to persuade the Japanese ground crews to allow the SP to depart. It was this two-hour delay that prevented Clipper 200–*Liberty Bell Express* from establishing an elapsed time record for all types of aircraft, set by three Air Force Boeing B-52's that circled the globe non-stop in 45 hours, 19 minutes in 1957 by refueling in flight.

Choosing the best winds available across the Pacific, Clipper 200, now being flown by its third crew, headed east and made landfall at Vancouver after dawn. Through FAA cooperation, the crew was able to obtain a nearly direct route using the INS from there to the New York area. Speed for the 7,517 miles was calculated at 641 mph. Air time: 11 hours, 53 minutes–still another city-to-city record. Ten more New York-bound passengers had been added to the manifest at Tokyo for the final leg.

During the last two hours of the flight, another aviation "first" took place when U.S. Customs and Immigration officials who had come aboard in Tokyo went through the aircraft and processed all passengers and crew members. According to the officials, this was the first time to their knowledge that the normal customs declaration and immigration clearances had been granted during flight in a commercial airliner. The only exception they knew of was Presidential aircraft that are cleared in this manner to allow members of the President's staff to depart the airport immediately upon arrival in the States following an overseas trip.

After docking at JFK, Pan Am's new president, F.C. Wiser, came aboard and welcomed the world travelers. Debarking, they were beseiged by press, radio and television interviewers who wanted to know what it was like to circle the globe in record-breaking time. The answers were unanimous. It was wearying but fun. Together, they had set new marks in the jet age and their names had gone down in the history books as the first public group to accept an invitation to participate in such a one-of-a-kind trip.

Captain Walter Mullikin, 25,000-hour veteran with Pan Am, praised the passengers for their spirit, enthusiasm and durability. "I have never seen such a team-spirited group of passengers who were so keenly interested in aviation and in the flight of the *Liberty Bell*," he said. "In all my years with Pan Am, this flight is the high point of my career."

The band of passengers who made the entire trip consisted of 84 men and 14 women and included four husband-wife couples, three father-son couples, one grandfather-grandson couple, and one lad, age 11, flying solo. The youngest was the grandson, seven years old, and the senior

The arrival at Delhi, India, was a high point of the flight and all hands were decorated with garlands of flowers and a red dot on the forehead. Shown is Summer Bartholomew (Miss USA) and other passengers accepting good wishes from the greeting party.

citizen was an 80-year-old aviation consultant whose list of personal aviation firsts included being the first passenger to fly around the world on a Pan Am 707 in 1959. Another veteran globe circler was a woman writer who had flown around the world in 1953 when it took four airlines, six plane changes, landings at 19 airports in 15 countries and 91 hours of flight to make the trip. Still another was retired Air Force Colonel Tony Story, former pilot for General MacArthur, who completed his 24th round-the-world flight on this trip.

The saga of Clipper 200—*Liberty Bell Express* came off without a single malfunction of any kind—a tribute to the technical superiority of the aircraft and its flight and cabin crews. It is a safe bet that the four records it set will remain on the books for a while since no other type of aircraft now available could duplicate the flight with only two stops and carrying paying passengers. The SP's main selling point—that it can fly farther, faster and higher than any competition—has been proven. It may take a whole new generation of transport aircraft to overcome the advantages that the SP seems to offer, especially for transoceanic travel.

AND THAT'S NOT ALL . . .

Reprinted by Permission
AIR LINE PILOT (August 1976)

The round-the-world flight of *Clipper Liberty Bell* was not the only newsworthy Pan Am event of last spring. On April 25 and 26, commercial aviation's version of the "Orient Express" took to the air when Pan Am inaugurated the first Los Angeles-Tokyo nonstop flights with its newly-acquired 747SP's. Service is daily to and from Los Angeles, thrice weekly out of New York.

Scheduled flight time for the westbound 6,754-mile trip out of New York is 13 hours, 40 minutes, while the return trip is advertised at 11 hours, 30 minutes. Three hours, 45 minutes are lopped off the one-stop westbound service which stops for refueling in Alaska.

The L.A.-Tokyo SP flight cuts three hours, 15 minutes off the normal 747 transit time. A total of 266 passengers can be accommodated—44 in first class seating and 222 economy passengers. The upper lounge of the SP has been slightly enlarged and main deck galleys have been relocated away from traffic areas to permit unrestricted passenger flow.

On the first New York-Tokyo flight, six flight deck crew members were aboard and took informal turns at their respective positions throughout the flight. An FAA inspector was also aboard. Two bunks, one above the other, were provided for resting crew members. They did not prove popular because of their narrow design and location where the noise of the cabin door prevents adequate rest.

On the return flight, only five flight deck members were required. The sixth deadheaded.

Although a few passengers complained of the long confinement, most seemed glad to complete the flight and save nearly four hours of total enroute time. The most disconcerting element proved not to be in the air but the inadequate and crowded customs and immigration processing upon arrival at Haneda.

There was one adverse physical aspect noted by *Air Line Pilot* during both prolonged flights—dehydration. While pilots of commercial jets have long known of this, few passengers seem to be aware that prolonged flights at high altitudes up to 45,000 feet tend to dehydrate the body. Symptoms are a dry, sore throat and, in some cases, a nose bleed may develop. If alcoholic beverages are consumed, the dehydration process is enhanced. The easy remedy is to drink liquids that quench thirst such as water or soft drinks.

Although it has now been proven that the SP can easily make the trip from New York against the wind to Tokyo and still have the required fuel reserves, Pan Am must accept an average 60-passenger penalty westbound. Depending upon wind conditions, about 206 passengers can normally be accepted on the 266-seat jumbo. Passengers are asked their weight before boarding so maximum fuel can be carried. Little, if any, freight will be carried westbound. However, there are no restrictions on the eastbound return leg or in either direction out of Los Angeles.

Despite the westbound limitation out of New York, Pan Am is betting that the SP will fill a need and be a profitable venture, especially when it is run only three times weekly and is supplemented the other four days by regular one-stop 747 service. Other U.S. airlines are adopting a wait-and-see posture. Meanwhile, Pan Am jots down more aviation "firsts" in the history books as it has done in the past half-century.

NOTAMS

This new section of the JEPPESEN SANDERSON AVIATION YEARBOOK was set up for those outstanding stories or photographs that would not fit into one of the four established categories. In this section will appear varied subjects: aviation pioneers, unique events or photographs, those items that touch more than one area of the world of aviation.

This year's NOTAMs section contains such a varied mix: biographies of two aviation pioneers who died in 1976–Howard Hughes and Grover Loening; photo coverage of the National Air and Space Museum; one of the few presently available photographic studies of the Navy's rigid airship, MACON; and what may become one of the classic aircraft accident photos of all time.

The NOTAMs section is planned to contain special articles covering many different areas of aviation. We hope that you find it so.

Buffeted by crosswinds, a rescue plane bringing aid to victims of the Guatemalan earthquake crashes into a truck while trying to land on a mountain highway near Sanarate, northeast of Guatemala City. Miraculously no one suffered serious injury. The two men running at the left leaped from the truck just before impact. (6/4/76)

Photo by Robert W. Madden
National Geographic Society

TRIBUTE TO HOWARD HUGHES—PILOT

by Jerry Phillips

Reprinted by Permission
GENERAL AVIATION NEWS

Howard Hughes; the name was better known than the man, especially after his self-imposed exile to places other than his native land. From Las Vegas to Nicaragua, then to Nassau, to England, the Grand Bahamas, and finally to Acapulco, Mexico. It was from here on the 5th of April 1976 that Howard Robart Hughes, age 70, took off on his last flight. In the element of his greatest love, on the way to Houston, Texas for medical treatment this unusual man died in an airplane at 30,000 feet. Howard had held forth incommunicado to the public and to the news media that was so unkind to him in his early active aviation years. The press and the radio always played up the sensational aspects of Hughes life; the millionaire playboy, the glamour girl escort, the big shot movie producer, the buyer and seller of airlines and in later years the eccentric recluse with a germ phobia, who sold TWA and bought Las Vegas.

Very seldom did they give him credit for his many good, constructive accomplishments in the art of flying and his many aircraft design innovations and improvements. So to you Mr. Hughes, in the great beyond, we would like to toss a bouquet of recognition and give a belated pat on the back for your great aeronautical accomplishments and your personal contributions to aviation in America. These are things we oldtimers vividly remember, but perhaps many newly arrived aerospace pilots never heard of. Therefore, a brief resume' of Howard Hughes, the pilot and his aeronautical accomplishments, is in order.

Howard Hughes in his early 20's.

Howard Hughes was a pilot not by profession, because he never flew for money, but for the love of flying and an intense desire to be good at it. He was an aircraft designer but not an aeronautical engineer. In fact, he had no degree and no formal education. At nineteen he was a dropout from two well-known prep schools. Yet, this self-taught engineering genius helped design a racing plane that broke all world records, a flying boat that was the largest aircraft ever built, a fighter plane of radical design and super performance. Then he flew them all on their first test flights. He insisted that because the concept was his idea, he would be the one to test fly it.

Young Howard Hughes shown with his Boeing Pursuit.

His real flying career started in the mid-twenties. With ability backed by enthusiasm and stubborn determination, his efforts in the cockpit brought him and America many honors and many aviation firsts. Although he had never flown competitively, in 1934 at the All America Air Meet in Miami, Florida, he entered the unlimited closed course race and won first place and the coveted "Sportsman Cup." He set many speed and long distance records and established numerous firsts in point to point flying. He was the first airplane builder to incorporate flush riveting and a retractable landing gear system in an aircraft design. He was awarded the Harmon Trophy in 1937 for "The most outstanding contribution to aviation in 1936," a special congressional medal for achievements as a pilot in 1938, and the Collier Trophy in 1939 for "The greatest achievement in aviation in America, the value of which has been thoroughly demonstrated by actual use during the previous year." In addition, many foreign and local honors were awarded this man for his outstanding flying ability.

In his long and varied flying career he put his life on the line many times. His confidence in his ability usually outweighed his fear. But Howard did have three major crackups, all serious enough to have easily resulted in his death. The first occurred in a TM Scout, a single place World War I observation plane, while we were filming early flying sequences in his wartime aviation movie "Hells Angels". Howard wanted a scene where the plane would approach the camera at a very low altitude, then bank sharply to the right turning away from the camera. We told him that due to the torque of the rotary engine, a sharp bank at low altitude would be extremely dangerous. He minimized the danger, and climbing into the cockpit of one of the TM Scouts he said, "I'll show you how I want it done." He took off, zoomed up in a steep climb and came back at the camera a few feet off the ground. Had he been a foot higher he would have made it

Howard Hughes and the H-1 prior to takeoff for the world's speed record.

Howard Hughes at the controls of the Lockheed Lodestar at the New York World's Fair, 1939.

OK, but the torque speeded up the applied bank and the wing tip dug into the ground causing the aircraft to cartwheel and crash. Before the dust cleared we were pulling a bloody Howard out of the tangled mess. When Frank Clark, our chief stunt pilot, saw that Howard was still alive he anxiously asked, "How is his right hand?" "His right hand? It's OK I guess, but why do you ask?" questioned the first aid man. "Well, you know today is payday and he has to sign those checks!"

The second serious crash was in his Sikorsky amphibian while making certification flights at Lake Mead with two Civil Aeronautics Administration pilots and his mechanics. Something went wrong on one of the landings, a wing dug into the water, the plane ripped apart and one of the CAA pilots was killed. Howard suffered lacerations and severe internal injuries and had nearly drowned before he was rescued. The plane sank almost immediately. He later recovered the plane from the bottom of the lake, completely rebuilt it and flew the certification tests.

The third near-tragic crash, and probably the closest he came to losing his life, was on the initial test flight of the XF-11, a twin engine, super performance fighter aircraft Hughes had designed and built for the air corps. Against counsel from the military and close friends he chose to make the first test flight himself.

On July 7, 1946 in the early morning hours Howard roared off the runway at his Culver City plant and soon disappeared out over the Pacific Ocean. As he turned back over Santa Monica at 400 mph the plane suddenly started to yaw to the right; no amount of opposite rudder could correct the situation. He was losing air speed and altitude rapidly. Ahead, on the edge of Beverly Hills, he could see the Los Angeles Country Club with its long open fairways. If he could stretch the irregular glide he could make it to that open green area. His sink rate was greater than he expected and on the edge of Beverly Hills the plane tore thru the roof of one house, crashed into and demolished another ending up in the middle of the debris and bursting into flames. He was pulled from the burning wreckage by a Marine Sergeant and rushed to the Good Samaritan Hospital. The doctors gave him less than a fifty percent chance for survival.

Despite a broken collarbone, broken ribs, a collapsed lung, many internal injuries, a fractured skull and deep lacerations on his face and head, his first concern after being informed there were no other casualties was what caused the crash. Because he suspected malfunction of the propeller pitch control he ordered a detail investigation of that mechanism. This later proved to be the source of trouble. In true Hughes fashion, less than six months later he vindicated his faith in the aircraft by test flying XF-11 number two in a perfect flight that led to government acceptance of the plane.

For the benefit of those who do not know or remember the magnitude of Howard Hughes' aeronautical achievements the following chronological sequence of his outstanding flying accomplishments is noted.

1927-1930: Howard Hughes produced "Hells Angels," the most spectacular and authentic aerial moving picture ever made, in which he flew and directed many of the flying sequences and at one time assembled forty pilots and forty planes for the mass aerial dogfights.

January 1934: Won the coveted "Sportsman Cup" at the All America Air Meet in Miami, Florida by flying his re-designed Boeing pursuit to first place in the unlimited closed course race.

Howard Hughes at the controls of the H-4 Hercules at Long Beach, Ca., 1947.

September 1935: Broke the worlds speed record for land planes by flying his H-1, a plane designed and built under his personal supervision, to a speed of 352.39 mph at Santa Ana, California. This flight broke the French record of 314 mph and brought the honor of the worlds speed record back to the USA.

1936-1939: Over a three year period he bought controlling interest in Trans-continental and Western Airline, the largest commercial airline in the US. He immediately set about getting new air-craft, acquiring new routes and imple-menting many new flight operation features in communications, navigation and safety. He met with Lockheed engineers and largely influenced the design of the Lockheed Constellation.

January 1936: Established a new transcontinental flight record by flying a Northrop Gamma, re-designed to his own specifications, from Burbank, California to Newark, New Jersey in nine hours and twenty-seven minutes.

April 1936: Broke all records for flying time between New York and Miami by covering the distance in four hours and twenty-one minutes.

May 1936: Established an east-west record between Chicago and Los Angeles of eight hours, ten minutes in the face of extreme head winds.

January 1937: Broke all land plane distance speed records by flying his H-1 from Los Angeles to Newark in seven hours, twenty-eight minutes averaging over 332 mph for the 2,500 mile trip.

January 1937: Awarded the Harmon Trophy "for the most outstanding con-tribution to aviation in 1936." The award was personally presented to Hughes.

July 1938: Set a new round-the-world flight record of three days, nineteen hours and eight minutes. Flying his Lock-heed Lodestar, redesigned to his own specification with many new navigation and communication aids, he flew com-pletely around the world being the first to circle the globe in a transport type aircraft. This flight also established a new New York to Paris record of sixteen hours, thirty-five minutes.

1938: The National Aeronautics Association named him "Aviator of the Year" and the Congress of the United States awarded him a special congressional medal "for achievements as a pilot."

December 1939: Hughes' name was officially added to the list of aviation immortals when he was awarded the Collier Trophy "for the greatest achievement in aviation in America, the value of which has been thoroughly demonstrated by actual use during the preceeding year." The presentation was made to Hughes personally by President Roosevelt.

September 1942: Hughes, with Henry J. Kaiser, signed a contract to build the worlds largest flying boat. Kaiser later dropped out and Hughes continued on his own. When he could not get aluminum, he went ahead and built the super flying boat out of plywood. Its size dwarfed even modern day jets with a wingspan of 320 feet, length of 218 feet and height of over 79 feet. The government put

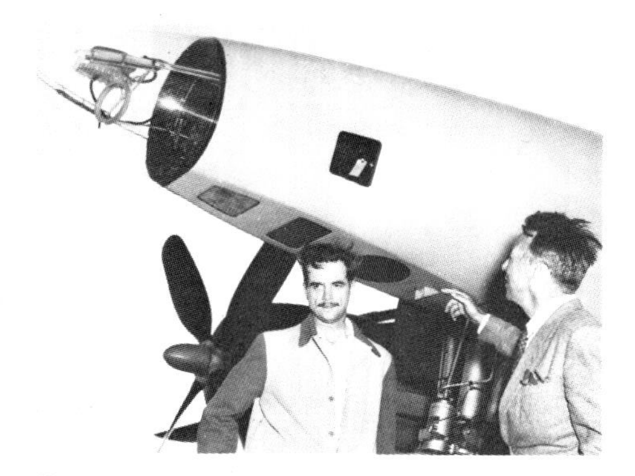

Howard Hughes and Sen. Harry Gain at the XF-11, No. 2.

$18,000,000 into the project and Howard put in more than twice that amount of his own money. There were many delays and much criticism from government authorities. Not until 1947 was the much belabored seaplane completed.

April 1944: He took delivery of the first TWA Lockheed Constellation, loaded it with newsmen, movie stars and friends and personally flew it on an inaugurating flight from Los Angeles to Washington, D.C. in the record time of six hours.

The Hughes H-4 "Spruce Goose" Hercules on first flight in Long Beach Harbor, November 2, 1947, with Howard Hughes at the controls.

The Hughes racer.

July 1946: Made the first test flight of the XF-11, a super performance, high speed fighter plane. Ignoring the suggestions and pleas of his associates, he reaffirmed his stated policy, "I always test fly the planes that I have built." On a Sunday morning, with only a few military and press personnel present, he took off on the flight that ended in a fiery crash and nearly cost him his life. Less than six months later he again elected to test fly XF-11 number two, having determined the cause of the crackup of number one and having taken steps to correct it, he was sure the plane would perform OK. The flight was a success from start to finish, he met all the requirements of the Army Air Corps and the plane was accepted.

November 1947: Test flew the Hercules, the worlds largest aircraft, on its one and only flight. After much criticism for delays and pressure to complete and fly the aircraft, he told the government officials that the aircraft would fly by November 1947 or he would leave the country and never come back! On November 1, 1947 he set the stage for a dramatic event. The flying boat was to fly! After a few taxi runs in the Long Beach harbor with press and dignitaries aboard, he unloaded all but the crew, then taxied out, opened the throttles on the eight engines and took off. Immediately after becoming airborne he levelled off at about 75 feet, flew about one mile, then landed and taxied back to the hangar. He kept his word, the "Spruce Goose" (as it was affectionately called by his employers) had flown and Howard Hughes had flown it, altitude 75 feet, distance one mile! Back into the hangar it went to be maintained and guarded for over 28 years but never to fly again. (The latest rumor is that The Summa Corporation, one of Howard's holding companies, announced it would junk the flying boat, and immediately a group of aviation enthusiasts started a movement to save the aircraft intact and put in in a museum.)

With the flight of the "Spruce Goose" Howard's flight activities gradually tapered off but up to that time one will have to admit that this man made quite a contribution to the development and progress of aviation. So again we would like to say in belated recognition to Howard Hughes, a pilots' pilot, congratulations for a job well done.

40 planes and 40 pilots on location at Oakland, Ca. for the filming of Howard Hughes' "Hell's Angels."

End of Howard Hughes' test of the XF-11, No. 1 in Beverly Hills, Ca., July, 1946.

THE FOUNDING FATHER

by George C. Larson

In his capacity as president of the National Pilots Association, Michael Loening once requested of his father, a fellow NPA member, that he contribute as a consultant to a study the NPA was doing concerning general aviation. As Michael recalls it, his father, Grover, winner of nearly every award devised to recognize great achievement in aviation, concluded his report with a neatly penned: "Respectfully submitted, Grover Loening." It was one of the last pieces of work he completed before he suffered a progression of illnesses that took his life on February 29, 1976. To Mike, that typical touch of humility at the conclusion of his father's report epitomized the style of his entire life.

Grover Loening was more than merely an engineer and designer. From the very beginning, with his Flying Yacht, he demonstrated that his ideas were to be far ahead of the prevailing way of doing things. In the case of the Flying Yacht, the first strut-braced monoplane ever built—and on a flying boat, to boot—Loening's vision leaped too far ahead for his contemporaries' narrow-minded views of proper aircraft design. Confronted with official resistance to a single wing, he was forced to build his best-known design—the tractor-engine Amphibian—as a biplane, to overcome the anti-monoplane prejudice. In his book, *Amphibian*, Loening wrote:

"All through its life, even though it was faster and climbed higher than the DH-4 with the same power, the Loening Amphibian was handicapped by 10 miles an hour less speed . . . 2,000 feet less ceiling . . . 200 pounds more weight, than if its designer had kept faith and made it a graceful monoplane."

Loening was even harder on himself for giving in than was the aviation world for forcing him to do it.

One of the principles that Loening never abandoned was his belief in the superiority of amphibian craft to land-based aircraft as a flexible addition to the total transportation system. Even in this age of jumbo jet air travel, he continued to insist that we would all be better off with similar aircraft equipped to land on large bodies of water. The difficulties our airports are having with community relations these days certainly make it appear that once again, Grover Loening was correct and that the world missed another opportunity when it ignored him the second time.

Loening is also credited with the first practical design for a retractable landing gear; the device appeared first on his Amphibian. Although the design resulted from the Amphibian's need to land on both runways and water, the idea of folding the wheels aboard would open up realms of performance and that would have been unapproachable without Loening's innovation.

Son Michael recalls most vividly his father's strict sense of business ethics and the fundamental honesty of his dealings in aviation. Perhaps that is what drew so many people to this brilliant man and created so much demand for his advice and counsel. After learning the ropes with the Wrights in Dayton, Ohio, he became chief engineer of the Army's aviation section in San Diego. He held the first aeronautical degree from an American university—Columbia, in 1910—and his decision to cast his lot with the budding industry later took him to its heights. He was one of the directors of Pan American Airlines, a Medal of Merit winner for his work as an advisor to the War Production Board during World War II and the chairman of the Helicopter Advisory Committee on Aeronautics. His Flying Yacht won the Collier Trophy in 1921, and his Amphibian became the standard utility surveying aircraft for the Army, Navy and Coast Guard.

What was most important to general aviation was his creation of aircraft that appealed to private owners. Initially, his Flying Yacht was affordable by only the very wealthy, but it established a precedent in this country that eventually made the U.S. the leader of the world in the size of its privately owned lightplane fleet. He also was instrumental in promoting the helicopter for military use, and later for civilian transport.

Loening lived for 87 years, all but 15 of them steeped in the study and creation of better aircraft. The long list of his awards and prizes must have pleased him, but certainly no more than the pleasure of looking skyward and seeing so many of his ideas and designs hard at work aloft.

BARNUM WOULD HAVE LOVED IT

by Jim Snyder

Edited for Space

Reprinted by Permission
AIR LINE PILOT

You could call P. T. Barnum a lot of things, but most of all, you would have to call him a master showman. That is why Barnum would have been right at home in the new National Air and Space Museum. Barnum used razzle-dazzle and a slight touch of chicanery to sell his product; the people at the Smithsonian Institution have packaged theirs in light, electronics and sentiment.

When you come to see the museum, plan to spend at least a day. For the aviation buff, the museum is the ultimate treat—the kind of affair the Smithsonian is noted for providing.

Start your tour by making a circle of the building from the outside. If you're staying in a Washington hotel, take a cab to the Mall entrance. Cabs are cheap in Washington. If you're driving, try to come early enough to get a parking space in the building's basement garage. Entrance is on the west side of the museum, from the street nearest the Washington Monument.

Whether you "cab" or drive, arrange to start your outside circuit at the western-most front window facing the Mall. This is the Air Transportation gallery featuring one of the most spectacular arrays of hanging aircraft ever seen in a museum. It includes a DC-3, suspended in flight.

Go to your left, past the main entrance with its collection of famous machines representing the various "Milestones of Flight." Continue to the easternmost gallery window. This is Space Hall. Up front in the window are Apollo and Soyuz spacecraft in rendezvous position as they were in space last July. Behind them there is the backup Skylab orbital workshop and a startling array of missiles.

Now, if you have the patience, go on around the building and scan the other windows. If you've seen enough from the outside, backtrack to the center gallery and go through the main entrance. Here the fun begins.

Aircraft in the central gallery were chosen to represent various stages in the development of American aerospace technology. As you would expect, the Wright Brothers' *Flyer* of 1903 is hanging in the center of the hall. This is the original airplane, flown by Orville and Wilbur Wright on that first day, 73 years ago. It

The Bell X-1, GLAMOROUS GLENNIS, cruising over the Wright Brothers' FLYER

A Macchi 202 headed for a Boeing B-17 mural

has only been at the Smithsonian since 1948. It had been at the Science Museum of London since 1928. Before that, the *Kitty Hawk*, as it has become known, was on general exhibition at air shows around the United States.

In the far upper right hand corner is Lindbergh's *Spirit of St. Louis*. The *Spirit* was freshly trimmed and cleaned before being hung in this gallery. However, as with every specimen with a precise place in history, nothing was done to mar its authenticity. When you go to the upper balcony you'll be able to get a good look at the airplane's instrument panel through the open cockpit door.

Also in the central gallery is Samuel Langley's Aerodrome model; the first X-15 rocket research aircraft; the Bell X-1, *Glamorous Glennis*; Robert Goddard's rocket; John Glenn's Mercury spacecraft and the *Apollo 11* Command Module which Museum director Michael Collins flew to the moon with his companions, Neil Armstrong and Buzz Aldrin. There is also one of the rocks which came back with them in that ship. This gallery, therefore, serves as a quick sketch of the major steps in American aviation. Let's move on to the nitty-gritty.

Go straight ahead toward the rear of the building and pause a moment to look at the two wall murals which set the building's tone. On the west wall, Eric Sloane has pictured a single airliner, traversing a lonely but not barren stretch of the American West. All the flight elements are there, including a very convincing storm.

On the east wall, Robert McCall has done an awesome piece of work in his "Space Mural—A Cosmic View." There are larger-than-life astronauts walking the face of the moon surrounded by galaxies, nebulae and planets. The earth serves as the centerpiece. These murals make a dramatic statement about the museum and its contents.

While you're in this area, be sure to buy the guidebooks prepared by Smithsonian staffers. One book deals with aircraft, the other with spacecraft now on display. The books are full of history and pictures and will make your visit more interesting besides furnishing good reading when you return home.

Now go west. The first large gallery ahead is Air Transportation. It is the one you saw first from the outside. This is one of the strangest fleet formations

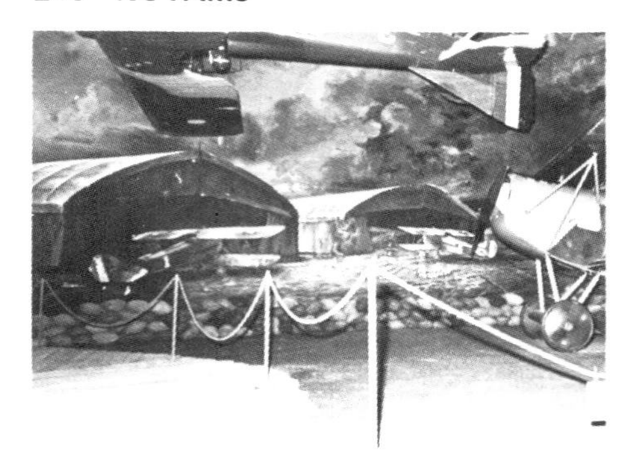

The World War One diorama with a Spad VII overhead

you'll ever see. There is a Ford Tri-Motor from American Airlines, a DC-3 from Eastern, a Boeing 247D from United, a Northrop Alpha from TWA, a Fairchild FC-2 from Panagra and a Pitcairn Mailwing, once owned by Eddie Rickenbacker.

All these airplanes are hanging from exposed ceiling members which were specifically designed to support heavy machines. A company which specializes in erecting high structures contracted to hang the airplanes. Its chief of operations told me, "While I take a lot of pride in this job, it sure makes me nervous to think about scratching up some history." He didn't, but just seeing the DC-3 hanging there is a heart-stopper.

Across the hall, drop in and drool over a collection of exhibition aircraft, including the prettiest Pitts Special and Bücker Jungmeister you'll ever see. Next to them is a gallery of some of the best in current-day private aviation.

If you're a helicopter buff, the next room on this floor contains a good representation of rotary wing aircraft.

Additionally, the room contains one of the Army's seven experimental autogiros, a Hiller-Copter and one of those machines designed to change from helicopter to car.

The interior of a Martin B-26B Marauder

Go upstairs and turn right. You're now on the hangar deck of an American aircraft carrier, looking at some of the Navy's most famous fighters. There is a Dauntless, a Skyhawk, an F4B-4 and a Wildcat.

The Wildcat was restored for the museum in 1974 by Grumman Aerospace Company volunteers. Some of them had worked on the Wildcats when they were being turned out in World War Two. The only thing missing at the time of restoration was the engine ring-cowl. The Marine Corps Museum loaned the one now on display. The cowl was taken in 1965 from the Wake Island Memorial where it had been displayed as a reminder of the great battle there in 1941. Although the outside of the cowl looks good as new, the inside still bears the scars and continues to serve as a memorial.

As you go along, you'll notice a variety of films and operating dioramas which illustrate the history and operation of the machines on display. There are 85 audio-visual setups which are controlled by a central computer in the basement. It chugs away, adding a dash of dazzle and merriment while pointing up what the artifacts did in their operational lifetimes. Be sure you take time to watch and listen, because they'll add considerable depth to your visit.

Go across the bridge to Gallery 205: "World War Two Aviation." In here, you'll find some of the museum's special prizes.

First, look at the Messerschmitt Bf109. It was brought to the United States with a group of other German aircraft at the end of the war because Air Force experts wanted to see what made them tick. After the evaluation was done, the Messerschmitt was stored with the others at O'Hare Field in Chicago. Later, it was moved to the Smithsonian's restoration facility at Silver Hill, Md. It has been totally restored inside and out and is probably one of the best preserved of its kind in the world.

The Messerschmitt's first major opponent, the British Spitfire, sits nearby. It came to the museum in 1949, after going through the same evaluation process as the German.

Then there is the Japanese Zero. It is a very rare specimen, brought to the United States from Saipan in 1944. Its origins are unknown because identifying numbers were removed after capture.

Not to be overlooked is a P-51 Mustang in the unit markings of the 351st Fighter Squadron, 353rd Fighter Group, 8th Air Force. In actuality, this airplane did not fly with that unit, but has been decorated to commemorate the fighter groups that escorted the bombers that flew from Britain.

The museum staff has gone to great pains to maintain the authenticity of each plane in the building. While the Mustang does not have its original markings, there is a plaque inside the fuselage which gives complete details of its origin for future reference. Every machine in the museum has been similarly treated.

Turning east, go to the next hall: "Balloons and Airships." Here, you'll get a special appreciation for the folks who first thought it would be fun to fly "lighter than air." There is a special display of Montgolfier balloons next to a mockup of the *Hindenberg* control cab. The mockup was used in filming the recent motion picture. It was donated by the production company.

ATC gallery

Next door is a gallery which will be near and dear to the hearts of cockpit crews. This is a history of air traffic control, complete with a reproduction of the first federal center in Newark, N.J. Listen to the voices on the radio and see how operations then compare with operations today. The audio-visual process plays a heavy part in making this exhibit come alive. Going through the room, you'll probably recognize the face of actor Cliff Robertson as he portrays an FAA radio repairman. The face, projected onto a mannequin, grows older as the repairman discusses with his wife the progress taking place over a period stretching from the late 1920's to the early 1970's. The airplane featured in this gallery is Amelia Earhart's Lockheed Vega.

"Air Traffic Control" is one of several exhibits which were first tried out in the old building. After several months of debugging, the exhibits were packaged and set aside to await placement in the new museum. This kind of planning has

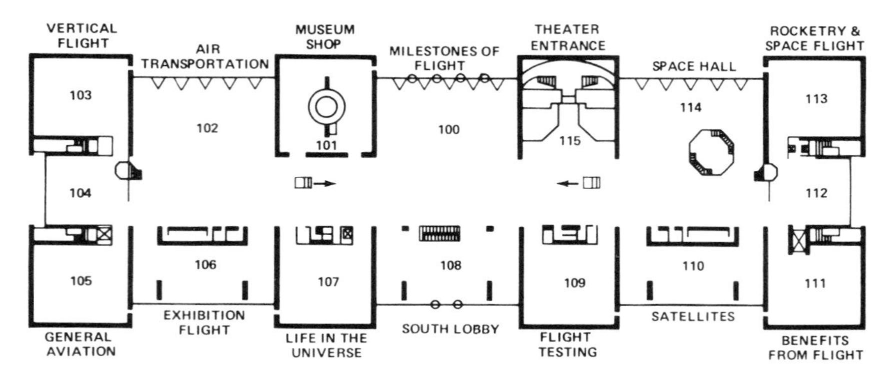

FIRST FLOOR PLAN

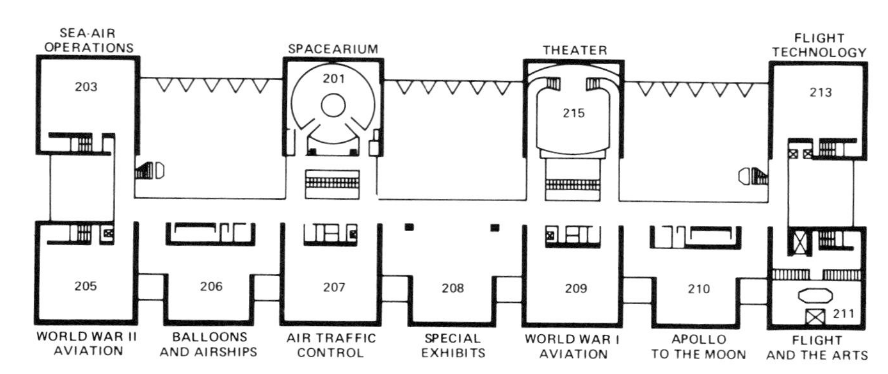

SECOND FLOOR PLAN

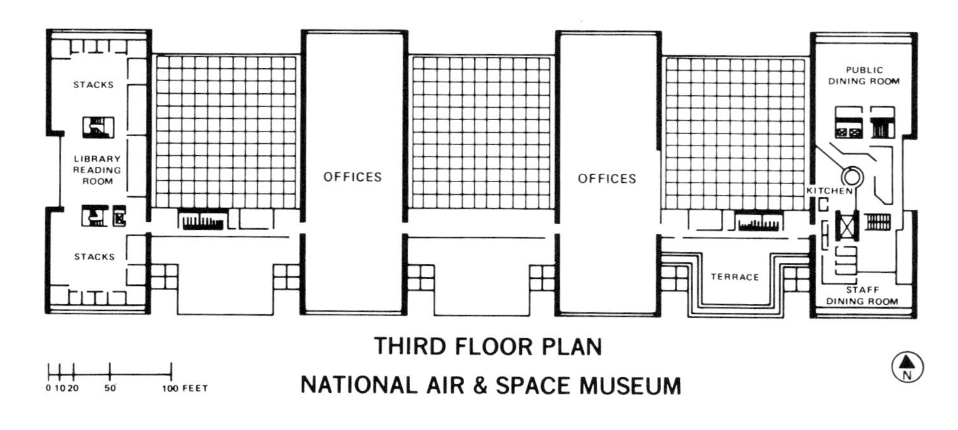

THIRD FLOOR PLAN

NATIONAL AIR & SPACE MUSEUM

To accommodate the movement of people through the museum, the circulation system, as noted on the adjacent floor plans, runs parallel to the Mall on the ground floor and second level. The pattern allows the visitor to quickly understand where he can go to view the various displays. All major exhibit areas, housed in a series of enclosed galleries and three open, skylighted galleries, are located on the first and second levels and are linked directly by the linear circulation path.

kept the museum ahead of schedule and, happily, within the budget approved by Congress.

There are 64 aircraft in the museum out of a total collection of some 250. Some of these planes, such as the prototype Boeing 707 and the Curtiss NC-4, are on loan to other museums. Most, however, are available to be circulated from time to time into the next gallery. Right now, in the room marked "Special Exhibits," you'll find Lindbergh's Sirius, the Douglas World Cruiser and the Wright *Vin Fiz* flown by Cal Rodgers on the first transcontinental flight.

The gallery after that is right out of World War One. It is an aerodrome in France, down to the mechanic's shop and strewn airplane parts. In this set, you'll find the Smithsonian's SPAD's (one was flown by Billy Mitchell), and a Fokker D-7. The scene depicts the capture of the Fokker as it really happened. Go to the "Ready Room" and listen as someone questions the German pilot. This diorama is another package which was first tried out in the old building and moved to the new.

At this point your stomach may be growling, so take the elevator on the east end up to the third floor, where a public cafeteria is available.

Now it's time for a decision. There are two presentations in the museum which require an entrance fee. Which do you want to see first? On your left as you face "Milestones" and across the hall as you leave the elevator is the "Spacearium." It's a big planetarium whose chief piece of equipment was donated by the West German government. Using the Zeiss Model VI planetarium instrument and numerous auxiliary projectors, the museum presents a realistic simulation of the heavens. Burgess Meredith is the narrator of a program which takes the viewer through the changing concepts of the universe.

The other admission gallery is a 485-seat theater. The screen is especially large and close to the steeply inclined audience. The projector is one of the new IMAX systems which gives the audience a very large, bright, flicker-free image that is so convincing that the viewer feels he's seeing the pictured objects first-hand. The films detail the history of flight, taking you coast to coast in everything from balloons and early open cockpits to high altitude jets to spacecraft.

Across the hall from the theater is "Apollo to the Moon." There are four types of lunar samples to see. Mercury, Gemini, Apollo spacecraft and their related paraphernalia are set for close-up study. The equipment is pretty familiar stuff to most Americans by now, but in this setting you begin to get a feel for the conditions in which early astronauts had to perform. Take your time here and you will learn a few things they didn't tell you on the radio or television shows. Watch the window displays as you pass through this exhibit and make sure you enter all the booths. There are surprises around every corner in this gallery.

If you appreciate art, the Smithsonian has a fine collection of paintings dealing with the aerospace environment. There are Norman Rockwells in the gallery next door to Apollo and some originals done for the movie "2001".

After you've browsed awhile, cross the hall to "Flight Technology" and see a super star. Here, in a very special setting, is Howard Hughes's H-1 Racer. The H-1 was considered an advancement in the state of that day's art. With it, Hughes flew from Los Angeles to New York in 7 hours, 28 minutes and 25 seconds. The date of that event was Jan. 19, 1937.

Take a very careful look at the Racer's construction. Notice the smooth fit of the landing gear against the wing, the flush rivets and joints, the close-fitting cowl. Even the screws in the wing fillet are flush and the slots are turned into the airflow. Hughes used a system which caused his ailerons to droop 15 degrees when the flaps were fully extended, giving the airplane better lift along the entire length of the wing. Anything sound familiar? Remember, this airplane flew in the mid 1930's.

Look at the exhibits around the hall. There are special tributes to the great men of aviation technology and there is a puppet show.

The puppets are another of the basement computer's responsibilities. The characters are a team of industry people, including a designer, an engineer, a manager and a test pilot. As they talk about designing and building a new airplane, you are subtly guided through a short course in what it takes to design an airplane. The show ends with two of the characters in old age discussing what they've seen happen in their lifetimes. The puppets are about the size of small children. They were done by no less an expert than Bil Baird.

As you come this direction, you'll be looking down from the balcony into the massive front galleries. Go down the stairs from here for a close look at a real Lunar Module.

This is one of the most underestimated machines in the space program. It was the first true only-for-space craft to be designed for manned flight. There was no need for aerodynamic shapes, so everything just sort of hung out. In addition, the LM looks flimsy because of its lightweight outer skin. But there is a tiger underneath. The LM has a full aircraft-strength framework and a computer brain just like Apollo's. As you stand in front of the machine, you'll be surprised by how big it is. It is one of those critters that photography "never did justice."

An astronaut climbs aboard the Apollo Lunar Module after placing a U.S. flag on the moon.

'You are there'

Imagine yourself crawling out on the LM's "front porch" and looking around at a strange lunarscape. Swathed in your massive spacesuit, turn around and back down the front leg's ladder till you get to the forward landing pad. Now, you're ready to step onto a totally alien world. Just as your foot swings out to the lunar sand, you hear a small voice behind you say, "I want a drink of water." The bubble has burst, so let's move on.

If you're a rocket buff, don't miss "Rocketry and Spaceflight." It's the room directly to the left of the LM. If you're looking for new examples of how aerospace has impacted on society, go to the right to "Benefits from Flight."

Next is Space Hall itself. Here stands *Skylab*, stretching to the ceiling with one of its solar wings unfurled. This machine is so large it gave the museum's people palpitations. It cleared by one-quarter of an inch when they brought it in to assemble.

This is actually the backup *Skylab* hardware that stayed behind when the flight model was launched. There were a few moments when NASA thought this one might also be launched as a replacement for the orbital ship when an accident on liftoff caused the solar wings to remain closed. Astronauts proved once and for all that there is a place for people in space when they simply went outside the spacecraft and repaired it. So as it turned out, this little *Skylab* stayed home.

In the middle of the room, you'll find a missile pit, complete with several launch vehicles in place. Hanging from the ceiling is a certain lifting-body spacecraft which any kid who sees the *Six-Million Dollar Man* will recognize immediately. See what yours say when they see it.

There are two more galleries to visit before you can say you've been through the whole place. There is one dedicated to space satellites and there is one on "Flight Testing."

An official description of "Flight Testing" says it "highlights the historical evolution of flight research aircraft and examines the interaction of flight testing, ground testing and research vehicles. Artifacts include *Winnie Mae*, the Bell XP-59A Airacomet, the Hawker Siddeley XV-6A Kestrel and a Lilienthal glider. What all that means is that this room holds some examples of how far Man will stick out his neck to stretch his capabilities.

Take a hard look at Wiley Post's *Winnie Mae*. This beautiful 1930 model Lockheed Vega was modified by Post several times to fit his changing needs. Since he was not a wealthy man, Post relied on *Winnie* to carry all his experimentation gear rather than spreading it among several specialized planes. She is certainly not the same machine she started out to be at the Lockheed factory. Notice that each cylinder has its own exhaust; that there are supercharger cooling radiators in her cowl. Look at how Post mounted the inlet for his 10:1 supercharger directly in front of the windscreen. Look at the belly-skid which was installed for wheelless landings. The wheels were dropped away on high altitude flights to reduce drag. Look at the different colors painted on the inside propeller tips. They were put there so Post could crank over the prop to the best horizontal position for a belly landing. Getting the colors in the right positions meant the pistons would be at the right parts of their strokes to prevent internal damage to the engine.

A Grumman Gulfhawk II

Winnie is a beauty that deserves some extra attention if you have the time.

Time is what you ought to have here. This new museum took a long time to build. It was put off by Congress for an awfully long time, much to the exasperation of people who knew of the Smithsonian's vast aircraft collection and of the truly magnificent things that aviation has done in America.

The museum is an elegant storehouse for these treasures. Doing what good museums always do, it is a fine research facility as well.

There is an impressive library for the serious student. If you wish to use it, phone ahead and let them know you're coming. The library contains any number of books no longer in print to help you write your new one or to simply help you prove a point. There are films, still pictures, letters and documents available for inspection. Some Xeroxing can be done at the facility for a fee. The staff is more than helpful.

The new museum is worth a special trip to Washington. It is a "gee-whiz" place, a flag-waver, and "American Aviation's Bicentennial Bonanza." It says, "Look what's been happening here."

Barnum would have loved it.

GALLERY 102: THE ULTIMATE HANGAR

by Stuart Nixon

Edited for Space

Reprinted by Permission AIR LINE PILOT

If you had a large structure in your backyard—something roughly 100 feet square by 50 feet tall—in which you could display a group of airplanes to commemorate the history of U.S. commercial aviation, which planes would you pick and how would you arrange them in the building?

That, simply put, was the problem facing planners of the cavernous new National Air and Space Museum, which contains not just one but 26 spaces (called galleries) in which aircraft can be displayed, including a space 115 feet long, 103 feet wide and 52 feet high devoted to "Air Transportation."

In their particular case, the Smithsonian designers had one factor going for them that you might not if you were doing the same thing behind your garage—enough money to build a superstrong roof. That meant they could suspend the planes in the air, where they belong anyway, enabling visitors to see them easily and compare their sizes, and freeing up valuable floor space for secondary exhibits. The only catch to this scheme was figuring out which planes to hang from which beams, and in what order to install them so that each could be raised into place without hitting one that had already been suspended.

The man selected to worry about logistical headaches was Walter J. Boyne, NASM curator of aeronautics, who was responsible for coordinating the restoration, transportation, reassambly (following delivery to the museum) and suspension of the planes in the gallery.

"They called me the 'Move Operations Officer,' " Boyne says.

Figuring out which aircraft to include in the Air Transportation gallery was not as simple as it might sound, Boyne explains, since, first of all, only a limited number of candidates were available to the Smithsonian to choose from. Also, the museum decided to display only "authentic originals, not reproduction airplanes," further narrowing down the field. These constraints necessitated a few trade-offs, which Boyne admits have already stimulated some complaints.

Those planes that did win a spot in

Afternoon sunlight floods the airspace of Gallery 102, washing the polished skins of the DC-3 (top) and Northrop Alpha. Dummy pilots, complete with authentic uniforms, have been placed in the cockpits to help museum visitors better understand the relative size of each plane. Visible through the gallery's massive windows are the Washington Mall and National Gallery of Art.

Gallery 102 are six famous workhorses whose names are second-nature to students of aviation history.

The oldest (insofar as records indicate), and also the lightest, of the six is the original prototype of the Pitcairn Mailwing (PA-5, Serial No. 1), which the Smithsonian obtained from Eastern Airlines in 1957. This plane saw service with EAL's predecessor, Eastern Air Transport, beginning in 1927, and was restored for the museum by a retired EAL pilot, Captain Joseph F. Toth, who did the work partly at his Miami home and partly in a rented shop next to Opa Locka Airport.

Second of the six planes is a Fairchild FC-2, one of four such ships flown by Panagra (Pan American-Grace Airways) in South America in the late 1920's. The Smithsonian acquired the aircraft from Panagra in 1949 and rebuilt it at the institution's restoration facility in Silver Hill, Md.

The third bird is a Northrop Alpha (4A) that was first owned by the Department of Commerce, then later sold to Transcontinental and Western Air (predecessor to TWA), which used it to carry mail and cargo in the early 1930's. Eventually, the plane passed to the Experimental Aircraft Association, which donated it to the Smithsonian in 1973. It was restored in Kansas City by TWA volunteers before being delivered to Washington.

American Airlines donated the fourth aircraft on view in Gallery 102, a Ford Tri-motor (5-AT-B) whose storied career began in 1929 when the plane entered service with Southwest Air Fast Express (SAFE), a company bought out by

American the following year. AAL retired the plane in 1935, then reacquired it some 20-odd years later from a man in Mexico who was using it as a house (complete with a chimney through the roof). It was restored by the Smithsonian at Silver Hill, under a grant from AAL, and flown on a series of public relations tours for the airline, which used it to make the first regular flight that departed Dulles Airport in 1962 right after the airport had opened.

Number five is a Boeing 247D, presented to the Smithsonian in 1953 by the Commerce Department, which had dubbed it "Adaptable Annie" because of its performance in various safety experiments. The plane first went aloft in 1934 when racing pilot Roscoe Turner leased it from United Airlines for the MacRobertson London-to-Melbourne Race, in which it finished third. It was operated in regular service by UAL until 1937 and later restored for the museum by CNC Industries with a UAL grant. In its present form, it carries two sets of markings: the left side as it appeared when Turner raced it, and the right side as painted by UAL for commercial flight.

Last on the list is the one plane whose inclusion in the gallery was never in dispute: the venerable DC-3 "Gooney Bird" from Douglas. The Smithsonian's specimen, donated by Eastern in 1953, ran up more hours (56,782) than any other member of EAL's DC-3 fleet. It was flown in service for 15 years (1937-1952) and prepared for display at Silver Hill by volunteers from EAL. It is both the heaviest of the six planes (17,500 pounds) and the heaviest object suspended in the museum.

Given the dimensions of the gallery and the various sizes of the six aircraft (whose wingspans range from 33 to 95 feet), it was immediately apparent to Smithsonian technicians that some imagination would be required to juxtapose the planes in the available space so that each would be easy to see from the gallery floor and from the second-story balcony that overlooks it. Using a three-dimensional model of the hall, the technicians first arranged the planes at angles to the side walls, then discovered the planes could not be suspended in that position due to load limitations in the room's five roof trusses and the finite number of pick-up points on the planes themselves. Going back to the drawing boards, the planners decided to align all six craft in one direction—toward the center of the museum—both because this would work from a practical and esthetic point of view and because it had the added advantage of carrying through on the museum's

symbolic concept of pointing virtually all aircraft in NASM's five open bays (of which the Air Transportation gallery is one) toward the Wright Brothers' *Flyer* in the central lobby. This last factor—not likely to be noticed by most visitors to the museum—is just one more example of the devotion to detail that went into the museum's design.

All of the aircraft were brought to the Smithsonian by truck, accompanied in most cases by airline employees and other volunteers who had helped restore the machines and prepare them for their final journey. Deliveries were spaced out over several months to allow time for reassembly and suspension of each plane. Access to the museum was through three large doors at the west end, specially designed for that purpose.

When it came time to do the actual hanging, Boyne recalls, the customary wisecracks of construction workers gave way to a nervous silence. Hoisting a DC-3 inside a building was not, after all, the kind of event you see every day. Even though the rigging company was insured, it wouldn't have made much difference in the event an irreplaceable, 10,000-pound aircraft had dropped 30 or 40 feet onto the gallery's floor. As each plane was lifted into place and safely secured, the release of tension among on-lookers was audible. On more than one occasion,

BELOW: Constrained by wires to fly an endless holding pattern in Gallery 102 are (clockwise from 12 o'clock) the Ford Tri-motor, Northrop Alpha, Boeing 247D and Douglas DC-3. Not visible are the Fairchild FC-2 (behind the DC-3) and the Pitcairn Mailwing (below the Alpha). Note the symmetrical positions of each plane's propellers. LEFT: Shown being hoisted into place are the Alpha (left) and Mailwing (right). "United" on the workers' jackets identifies the rigging company, not the airline.

Boyne says, spectators broke into applause.

The wires holding the planes are designed to last as long as the museum—and longer. "We are very conscious of the fact that airplanes are hanging over people's heads," Boyne confides.

The highest of the six aircraft is the Ford Tri-motor, the first to be deployed. Beneath it is the DC-3, then the other two passenger transports. Lowest are the two mail planes.

The condition of the planes varies; strictly speaking, none is airworthy. "All of them would require substantially more than an oil change and new plugs to go out and fly," says Boyne.

All of them, of course, were hung "dry."

As to what's inside each aircraft, Boyne reports that the DC-3 "looks exactly as it did on a flight . . . right down to the doilies on the seats." Likewise, the FC-2 retains most of its original cabin appointments. The Ford, on the other hand, is outfitted as it was during the latter part of its career after American had restored it, not as it was in first-line service. The 247 remains in its "Adaptable Annie" format, mainly because the Smithsonian didn't have time to convert it back to the passenger configuration. All of the planes, except the Mailwing, are lighted inside to facilitate viewing their interiors.

In addition, mannequins have been placed in each aircraft's cockpit, complete with uniforms and flying gear, to give viewers a better sense of proportion and lend further authenticity to the display. No mannequins are used in the cabins.

Even though the six aircraft dominate the gallery and serve as the natural focus of interest, visitors who get tired of looking up will find a room full of other historical items displayed on the floor, including engines, propellers, uniforms and other artifacts. There is also a short film on air transportation, a three-dimensional mural on airmail and a series of scale-model planes.

For pilots of the present era, and for everybody else who has a chance to visit the museum, the gallery will serve as a fantasy trip through the earlier years of this century when America discovered what aviation pioneers already knew: flying is a lot more than leaving the ground—it is a new way of looking at the world. No one in those days could have imagined that a place as spectacular as NASM would be created to capture the romance of air travel and hold it fast for later generations. But today, thanks to the vision of Smithsonian artists and craftsmen, the ultimate hangar has been built, in which six extraordinary machines will forever wing their way into the memories of men.

AIRSHIP

Reprinted by Permission
NAVAL AVIATION NEWS
Edited for Space

George Carroll was one of the Navy's first aerial photographers and produced the USS MACON pictorial on these pages. In the early 1930s he was ordered to the Sunnyvale Naval Air Station (renamed NAS Moffett Field in 1942) to film all aspects of ZRS-5 operations. Earlier airships, SHENANDOAH and AKRON, had been lost. Carroll was given free rein while aboard and took pictures from every imaginable angle. He even rode in the "spy car." This one-man vehicle was designed so that it could be suspended by cable below a cloud layer for scouting endeavors. It was towed by the airship which remained concealed by the clouds. After a ride suspended 500 feet below the mother ship, Carroll admitted that "I only did it once and that was enough."

Sunnyvale was commissioned in 1933, and featured a huge airship hangar with orange-peel doors at either end. Carroll photographed the air station from the non-rigid airship J-4, the first to fly from the California base. USS *Macon* arrived at the south mooring circle on October 15 after a 73.3-hour journey which began at NAS Lakehurst, N.J. Commander Alger Dressel was in command on the flight. The course took the airship south across Atlantic coast states then westward over Georgia, Alabama, Mississippi, Texas and Arizona. *Macon* swung north near San Diego and followed a Pacific coastal route to San Francisco Bay. The only trouble spot along the way was San Angelo, Texas, where Dressel and crew orbited for several hours waiting out lightning-streaked thunderstorms.

Ground crews waited on the mooring mast as *Macon* approached. The mast cup resembles a doughnut when viewed from the forward service platform door aboard *Macon*. The nose cone was locked into the cup. The lower vertical fin was gently secured onto the stern beam. This beam was connected to the mast unit by solid couplings which were supported on flat-bed railroad cars. These allowed the airship, mast and stern beam to be slowly moved from the mooring circle as a unit with little or no stress on the airship's structure.

The highly trained ground crew at Sunnyvale consisted of five officers and 150 enlisted men. *Macon*'s personnel complement included ten officers and 50 enlisted men assigned to lighter-than-air related duties. Four officers and 15 sailors aboard had heavier-than-air duties.

The airship was 785 feet long and had a 144-foot maximum width. It was made of duralumin 17-SRT metal and was covered by 33,000 square yards of cotton cloth. The cloth was treated with four coats of clear dope and two coats of dope treated with an aluminum pigment.

Macon had 12 main frames 74 feet apart. Thirty-three intermediate frames were spaced 16 feet from each other. There were 36 longitudinal girders and three keels—port, starboard and upper. Hollow-hull space measured nearly 7.5 million cubic feet. Twelve helium gas cells, made of gelatin latex, provided lift. The airship's dead weight was about 242,000 pounds but it had a gross lift for normal operations of 410,000 pounds with helium cells filled to 95 percent capacity. The gross lift figure included 168,000 pounds for the crew, food, gasoline, oil and ballast water.

At the forward section of the lower vertical fin was the auxiliary airship control station, an arrangement similar to that in surface vessels. It was equipped with rudder and elevator control wheels, altimeter, airspeed indicator, rise and fall indicator and a telephone. The station was manned during mooring and unmooring operations.

The author spent much of his airborne time here. For photographers this was an ideal vantage point. There was an unrestricted view forward and to either side of the airship. Despite vibrations, and the swaying sensation of structural movements, plus singing tension wires, it provided an excellent platform for photography.

Orange peel doors at the north and south ends of the hangar allowed the airship to be moved out of or into the selected aperture depending on prevailing winds. The hangar was almost four football fields long, over 300 feet wide at ground level and nearly 200 feet high.

The south area was utilized as often as possible. The Bayshore Highway near that end afforded the public a grandstand view of airship-handling activities. Cars and spectators jammed the area for several hours before and during flights. Many had cameras which led Carroll to assert that *Macon* became the most photographed airship in history.

Looking like vertical rows of windows were finned radiator-type units stacked above *Macon*'s eight engines, four on either side. These units condensed the exhaust into a contaminated water mixture which was collected in ballast bags inside the airship. About 100 pounds of water could be recovered for each 100 pounds of gasoline burned. Equal load distribution in flight was achieved this way.

The port keel walkway and duralumin girder framing seemed like an endless maze of metal. Holes in the girder structure reduced weight without loss of structural strength. Round object was one of 110 gasoline tanks dispersed equally on either side of the airship. They held 126,000 pounds of fuel. Quilt-like fabric overhead was part of helium gas cell.

Macon's water recovery system was connected to the water ballast bags which were positioned inboard of the port and starboard cells adjacent to the gas tanks. In flight, crewmen labored continuously pumping fuel and water ballast to ensure proper distribution of weight throughout the huge craft, a critical requirement.

There were about 30 bunks in the enlisted crew's sleeping compartment. Two members alternated using a single bed. Crews were divided into two duty sections, worked four hours on watch, followed by four off. A six-foot by six-foot compartment, about seven feet high, adjacent to the bunk room, was the airship's smoking room. It was the sole locale where smoking was permitted. A maximum of six men were allowed in at one time but ventilation was so poor that a light smoker had only to step inside, inhale a few breaths, and his smoke break was complete.

The helmsman swung his wheel to operate the rudder sections on the vertical upper and lower tail fins. Virtually a house of glass, the cabin resembled a ship's bridge and was impressive in its neat and functional design.

View on the right shows cabin's port side, looking aft. In upper left corner were emergency helium release controls. In the upper right area the water ballast emergency dump controls were rigged. Twelve tons of gasoline and water could be quickly released. Elevator wheel in center controlled elevator sections on the horizontal tail fins.

Picture on the left was taken looking aft on the control cabin's starboard. Annunciator signal units were connected to the eight engine compartments. Telephone was one of 16 automatic-dial types which functioned as *Macon*'s intercom.

Macon was powered by Maybach VL-II, 12-cylinder V-type, German-built engines. Each was rated at 560 hp at 1,600 rpm. The engines were connected to the three-bladed propellers, 16 feet in diameter, through Allison gear. The engines could develop maximum horsepower when operated either in clockwise or counterclockwise direction. Motor mechanics in the engine compartment maneuvered the adjustable pitch propellers upon command for forward, backward, upward or downward thrust through the annunciator signal system.

The galley featured a small propane gas range where a cook prepared food for the entire company. In the adjacent dining compartment 12 men could eat at one

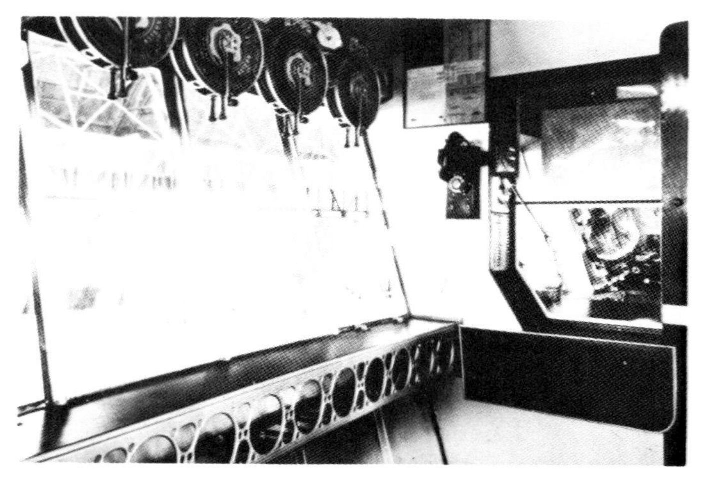

time. Dinnerware was made of lightweight durable plastic. Silverware was also lightweight, and was manufactured from chromeplated aluminum.

Five F9C-2 *Sparrowhawk* fighters were assigned to *Macon* and flew vector search missions during fleet maneuvers. The Curtiss-built planes would approach the airship's trapeze from below, then climb slightly at a slow closure rate as *Macon* achieved its top speed of approximately 85 mph. The *Sparrowhawk*'s hook would "catch" a trapeze and then be hoisted up and into the aircraft. It was then transferred to a trolley along a monorail. Four aircraft were secured in a hanging position in the hangar space. The fifth stayed on the trapeze.

N2Y-1s also made hook-on landings aboard *Macon*. The two-place trainer was used for ferrying passengers from the ground to the airship and vice versa—shades of *Star Trek*'s transporter device.

The airship cruised at 55 knots, had a max speed of 75 knots and could remain aloft for more than 100 hours covering 6,000 miles. From April 1933 to February 1935 *Macon* made 54 flights and logged 1,798 hours.

Macon's longest flight occurred in July 1934. It lasted 83 hours, and covered 4,000 miles. The airship intercepted USS *Houston*, a heavy cruiser, 1,500 miles west of Central America. Aboard *Houston* was President F. D. Roosevelt.

Tragically, *Macon* was to go the way of her predecessors. Two men perished, the remainder were rescued when she crashed in the Pacific.

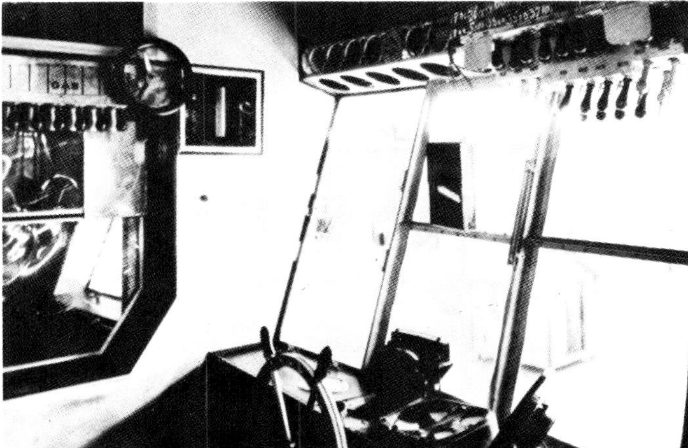

It was February 12, 1935. LCdr. Herbert V. Wiley was in command and in addition to 83 personnel, there were four *Sparrowhawks* aboard on exercises off the California coast.

As *Macon* neared the coast on the way home she encountered turbulence. Wiley tried to keep *Macon* under the clouds and descended to 1,250 feet. Despite the weather all seemed well and the cook was preparing a roast chicken dinner. Mooring was scheduled for 7 P.M.

At a little after 5 P.M., three miles west of Point Sur, a sharp jar was felt throughout the airship. The elevator wheel spun out of the controller's hands and *Macon* nosed up. Within moments a gas cell located aft was reported to have deflated. Wiley, who was one of three survivors of the *Akron* crash, ordered water ballast and fuel dropped from the ship's after section. He directed other adjustments to try to level the airship, releasing helium from the forward cells. Rear port and starboard engines were set at full rpm in an effort to level the airship and restore trim and control.

Chief Radioman Ernest Dailey began sending SOS messages which he did continuously in the next perilous moments.

Part of *Macon*'s outer covering had been torn away. It soon became evident that serious structural damage had been sustained. *Macon* was literally out of control, her nose tilted 25 degrees up and she was rising rapidly.

Wiley ordered all available personnel to the nose. Sixteen managed a monkey walk along the girders scrambling uphill for about 325 feet. Their weight up forward failed to help.

AMM first class K.H. McArdle was perhaps the coolest member of the crew at this moment. As he passed the galley on his way forward, McArdle grabbed a whole roasted chicken, stuffed the bird into his flight jacket and pulled up the zipper.

A ship's officer passed McArdle, who had sat down on a catwalk to rest and eat his chicken.

"McArdle," asked the stunned officer, "how can you sit there eating chicken when our ship is going down?"

"Sir," replied McArdle, "if I have to meet my Maker I want to meet Him with a full stomach."

Automatic relief valves had opened as *Macon* passed 3,000 feet in its ascent. Since an excessive amount of helium had been released during the climb, the airship peaked at 5,000 feet. It began to fall, tail down, at a rate of 750 feet per

minute. Halfway to the waiting ocean the rate was slowed to 300 feet per minute through careful managing of gasoline and water ballast. The descent rate was further reduced to 150 feet per minute but collision with the sea was inevitable.

At 5:39 P.M., some 33 minutes after the jolt which shook the entire airship, *Macon*'s tail struck the water. She sank slowly, stern first, the framework collapsing as sea swells struck her. At 6:20 P.M., about 12 miles from Point Sur, *Macon*'s nose disappeared and the airship, which was nearly the size of an aircraft carrier, was swallowed up by the sea.

Chief Dailey, and Florentine Edguiba, a mess attendant, lost their lives. The 81 survivors were rescued by Navy cruisers. The last was hauled to safety at about 7:40 P.M.

MILITARY AEROSPACE

AVIATION YEARBOOK

INTRODUCTION

The major question in the U.S. military/aerospace field as 1976 drew to a close is what effect President Jimmy Carter's administration will have on it.

Perhaps the biggest question mark hovered over the Air Force's B-1 bomber. Early in the primary races last spring, as a relatively obscure candidate, Carter had said he favored scrapping it. Later, as President-elect, after being briefed on diplomatic and military issues, he said he would reserve judgment until after he took office.

Meanwhile, early in December, the Defense Department approved funds extending B-1 production to early summer 1977 and authorizing procurement of additional tooling and long-lead time items running into 1978-79. Some observers question whether a lame-duck defense team would have taken that step without Carter's tacit approval.

In the past, incoming Presidents, faced with the full responsibilities of office, have found it impossible to carry out all their campaign promises. Their programs, too, are subject to Congressional approval or rejection.

Certainly no U.S. military decisions can be taken without regard to the actions of the other major world powers—especially the USSR. The best shot in the arm for U.S. air defense in 20 years is the USSR's Backfire bomber. Is it a threat to the U.S., or is it, as the Soviets insist, only a peripheral bomber limited to attacking targets in Western Europe and mainland China? Because of the impact of Soviet views on U.S. defense policies, we lead off this section with an interpretation of Soviet attitudes toward war and peace.

There is a brief story also on economic and technical factors that affect military aviation on both sides of the Iron Curtain.

Aside from the election results, it would be difficult to reach any real consensus in rating the other top stories on military aviation and space projects in 1976. There were enough major events to draw votes from a wide range of disciplines.

Entebbe, for instance. For courage, skill, and flawless planning, the rescue by Israeli commandos at Entebbe airport in Uganda of more than a hundred hostages from a hijacked Air France A300 Airbus must rank high on anyone's list.

For sheer technical brilliance nothing tops the exploration of Mars by two Viking landers and the incomparable color photos they transmitted back to earth—although scientists may be pondering for years the significance of the data they obtained.

Then there's the story of Soviet Air Force Lieutenant Victor Belenko and the MiG-25 Foxbat he flew to Japan, giving western technical experts an unprecedented opportunity to evaluate the state of the art in Soviet military aviation.

You would certainly get votes for the rollout of the Rockwell International OV-101 orbiting vehicle, or Space Shuttle, that promises to inaugurate a whole new age of space exploration. Although the Shuttle itself will operate barely beyond the earth's atmosphere, it will be capable of launching probes to distant planets and building large space stations by hauling up sections, piece by piece, to be assembled in space.

To speak of carriers, we have a report on the Soviet's new carrier-based VSTOL fighter, tentatively designated the YAK-36. It operates from the KIEV, a second generation vessel in the Soviet Navy's progress toward a carrier fleet, following their helicopter carriers of the MOSKVA class.

Our potpourri includes pilot reports on two new U.S. planes at opposite ends of the military aircraft spectrum—one a prime candidate for the ugliest design of the decade, the Fairchild A-10 antitank attack plane, nicknamed the Wart-hog; the other probably one of the most photogenic—and deadly—the Rockwell International B-1.

Canada made these pages by initiating what promises to be the most expensive military aircraft program in its history: a program to co-produce 18 Lockheed P-3 Aurora long-range patrol aircraft costing an estimated $750 million.

At presstime, the U.S. Army had announced the winner of one major helicopter contract – the Utility Tactical Transport Aircraft System (UTTAS) for which Sikorsky and Boeing Vertol were competing. Details of these contending aircraft are described in this section. In addition, the winner of the advanced attack helicopter (AAH) competition is named and described.

It's characteristic of pilots to brag about the planes they fly, and particularly about their ability to fly them. So, as with the "man bites dog" story, it's news when a pilot admits his particular plane was a flop, even if it has taken more than 30 years for that "news" to surface. The plane was the World War II North American B-25G, a version of the Mitchell bomber that packed in its nose a 75-mm cannon. According to its pilot, he and his crew seldom hit anything.

If we can be forgiven in this Yearbook for looking back more than three decades to the B-25G, perhaps we may be permitted to look ahead to two other developments with broad implications for the future of military aviation.

One is the choice between two unusual designs—the Boeing YC-14 and McDonnell Douglas YC-15—competing for selection as the U.S. Air Force's advanced medium STOL transport, intended to replace the Lockheed C-130 in the early 1980's. No decision is expected for at least another year.

One is the Advanced Tanker Cargo Aircraft (ATCA) which the Air Force hopes to buy to succeed the KC-135 and enhance its existing airlift fleet by reducing dependence on en route staging bases.

The other is the growing role of women in the military academies, as potential astronauts, and as pilots and navigators in military aircraft. If we put the ladies second in this context, it's because they are confined, for the present, to noncombat roles.

But only a die-hard male chauvinist would bet that won't change.

Allan R. Scholin
December 1976

WAR AND PEACE– THE SOVIET VIEW

Reprinted by Permission
AIR FORCE MAGAZINE
October 1976

by Colin Gray

In this article, the author charges that many Western leaders are suffering from intellectual myopia in regard to "stability"—maintaining the STATUS QUO between the Soviet Union and NATO. Indeed, he observes, because of the cast of Communist dogma, Soviet officials do not, and CANNOT, endorse Western motions of world order. There is clear evidence, which he believes is largely ignored in the West, that the Soviets are building for the long term and intend to have the means to win in any future crisis—be it diplomatic or military.

There is nothing mysterious about the Soviet concept of war. Shelf-loads of authoritative, and strongly indicative, Soviet statements concerning it are there for the reading. If Western commentators are uncertain about Soviet aims and aspirations, they have no one but themselves to blame: The record is quite clear.

The purpose of this discussion is not to endorse Soviet strategic views, nor to condemn Western doctrinal preferences. Rather it is to specify the principal ways in which Western and Soviet strategic thought diverge. Overall, Soviet theorists have a clear view of the value of military power while Western theorists and officials do not. Looking ahead ten to twenty years, this difference could be of critical importance. The Soviet Union is not devoting eleven to thirteen percent of its GNP to defense only to achieve "rough parity" with the United States. Soviet commentators know that relative power positions do not remain static. Soviet military, and especially strategic nuclear, power has been *the* elevator of Soviet status in international politics. In all other important respects, the Soviet Union is a third-rate power. In Soviet eyes, detente was *inspired* by Western recognition of the rise in relative Soviet military power. Logically, the greater that power the greater the prospects for peace . . . and so forth. This argument should not be at all controversial; Soviet leaders and analysts have used it for years. Nonetheless, it bears little resemblance to the explanations of detente that were (and are) advanced by Western leaders.

Victory and Stability

Many Western officials and commentators see what they want to see. They ignore the clear evidence of Soviet doctrinal divergence from Western models. The Soviet Union clings to the notion of *victory*. Both at theater and intercontinental levels, Soviet officials are seeking freedom of action. They are acquiring and (probably have achieved) the capability to overrun Western Europe in a short and sharp nonnuclear campaign, while they are building toward the capacity to force partial disarmament of the strategic forces of the United States. At all levels of conflict Soviet analysts and officials appear to endorse Gen. Douglas MacArthur's dictum that "there is no substitute for victory." The contrast with much Western thought and practice could hardly be more direct.

In the European theater, NATO hopes to contain a Warsaw Pact offensive, or—at least—to give ground grudgingly and surge back eventually to the starting line. NATO's strategy and tactics are unequivocally defensive. While NATO hopes to end a war speedily with neither side seriously disadvantaged, the Soviet Union plans to wage a war with what has been termed "Darwinian ferocity," with victory as the goal. Both with respect to the European theater and to the strategic balance, the Soviet Union is purchasing military options that might just give it victory: in the successful resolution of potential confrontations that the West will choose *not* to enter; in successful deterrence in actual crises; and in the conduct of war itself.

So defensive is much Western strategic thinking that there is great lack of understanding of what the Soviet Union is about in its massive force modernization programs. If you believe that the concept of victory can have no rational place in the aspirations of nuclear-armed states, then you have to explain away Soviet military programs and Soviet doctrine as not *really* meaning what they say. Most theories of limited war are almost totally inappropriate for states locked in protracted conflict with the Soviet Union. The theorists of limited war have devoted far too much attention to appropriate limits and far too little to the likely realities of war.

The NATO countries are essentially *status quo* powers, and so have generally adopted a mix of strategic and arms control policies that give the initiative to the other side. The West seeks to defend a structure of world order that seems tolerable. The Soviet Union (save for its current "holdings") is committed to changing that order in a direction that it deems benign. In their military programs and their arms control behavior, Soviet officials do not and, indeed, *cannot*, endorse Western notions of stability.

The Soviet Concept

Because of their generally defensive political and military stance, Western countries are profoundly illfitted to understand the alien strategic mind-set of Soviet officials, and to take timely offsetting action. A similar judgment, of course, applies to Soviet officials. Unfortunately, while their ideology misleads them, it misleads in some extremely dangerous ways. They tend to expect Western leaders to recognize the *objective* deadly danger posed by the socialist camp—and hence to be willing to resort to desperate military adventures. But they also, in best dialectical fashion, expect the capitalist-imperalist world to collapse of its own internal contradictions, and to be capable of being misled by astute Soviet officials.

Apparently presuming, with good cause, that Western officials either will not believe what they see, or will search for and find nonmalignant explanations for Soviet strategic behavior, the Soviet Union is proceeding to acquire whatever military options her economic-scientific traffic will bear. Those who believe that the Soviet military posture relating to Europe reflects nothing more ominous than (a) "the Soviet way," (b) the acquisition of bigger and better bargaining chips for negotiations, and (c) an attempt to balance the NATO threat, have no business discussing affairs of state. A parallel judgment applies to those who are not disturbed by the pace and breadth of Soviet strategic programs. To address the issue of whether Soviet military programs are defensive or offensive in orientation is totally fruitless. One can conceive of a Soviet military offensive launched in Europe for what Soviet officials believed to be sound defensive reasons, *e.g.*, to safeguard the accomplishments of socialism, West German revanchism had to be stamped out.

With certain caveats, one need not guess, even in an educated way, at the purposes that underpin the Soviet defense effort. Indeed, Soviet statements are so frank, not to say brutal, that many Western analysts have difficulty crediting what they read. Important caveats include the following: (a) the Soviet Union, by its own definition, cannot wage an unjust war; (b) the Soviet Union, therefore, cannot launch a surprise attack in the political sense, although it can in the technical military sense (this is never admitted directly); (c) all Soviet military writings have a political purpose, and the level of revealed strategic details is low compared with Western exposition; and (d) when Soviet authorities address primarily a Western audience, their views are slanted toward the propagation of beneficial disinformation. To make sense of the character of the long-term Soviet threat, Western officials should keep the following checklist in mind:

First, the Soviet Union is obliged to regard capitalist and semicapitalist countries as *enemies* that eventually will be overcome by the tide of history—probably with considerable Soviet assistance.

Second, the basic conflictual character of East-West relations is nonnegotiable and cannot be appeased or managed away by technology transfer, cultural exchanges, trade, or any other device.

Third, there can be no "normalization" of Soviet-American relations through detente, except to the important degree that the world is made relatively safe for the prosecution of conflict shorn of the acute danger of nuclear confrontation.

Fourth, the Soviet Union has no interest in institutionalizing the parity principle—as one gullible American arms controller claimed in reference to SALT I. On the contrary, Soviet officials see themselves locked into a dynamic contest of global dimension with the United States—wherein the balance of power is inherently unstable.

Fifth, Soviet officials believe—probably sincerely—that deterrence is fragile and could fail. They also believe (and this should be embossed in gold over desks in Washington and elsewhere) that the greater the relative military strength of the Soviet Union, the more likely it is that deterrence will not fail.

Sixth, deterrence, for which the Soviets have no single parallel term, is not seen in Moscow in the predominantly negative policy framework that it is in the West. Behind the deterrent shield, the Soviet Union seeks to further its essentially destabilizing foreign-policy goals.

Seventh, over the long term, and Soviet thinking is nothing if not pragmatic and cautious, all that is negotiable in East-West relations is how the West is to die. The Soviet Union cannot seek to institutionalize peaceful coexistence with Western countries on the basis of a recognition of the legitimate interests of others. That is an ideological impossibility.

Eighth, for reasons both of *real-politik* and ideology, Soviet ambitions are open-ended. However, Soviet expectations, in the short and medium term, are pragmatic and bounded. What they accomplish with their military forces depends largely on what opportunities come along. Over the past decade they have been purchasing options.

Ninth, without assigning precise political intentions to the Soviet Union, it is nonetheless clear that its leaders take the possibility of war, at all levels, far more seriously than do their Western counterparts. The Soviet Union has a very impressive program for the survival of essential industry and services in nuclear war—

the United States does not. Soviet forces in Europe are prepared to wage a war—NATO is not.

The author, Colin Gray, is a staff member of the Hudson Institute. He has written extensively on defense matters for publications both in Europe and North America, including a number of articles expressly for AIR FORCE Magazine. Dr. Gray's book, THE SOVIET-AMERICAN ARMS RACE, was recently published by D. C. Heath.

On the basis of the growing congruence between what they do and what they say, it must be judged that Soviet officials wish to acquire the ability to do as well as possible in wars at all levels. They may not, in the event, do very well at all. But, we assume enormous and unnecessary risks if we choose not to read the writing on the wall. Against a self-professing "deadly enemy" we array what? A NATO, the operational deficiencies of which are so familiar that they are largely accepted as the necessary price of a multinational undertaking. That price, let it be recognized, would probably translate into *defeat* in short order. At the strategic level, the major and possibly catastrophic asymmetry imposed by the Soviet domestic war-survival program is dismissed as being largely on paper (perhaps it is—but what if it is not?) or easily offset, while a functionally parallel program for the United States is deemed politically unfeasible. A serious US civil defense program (and its ramifications for industry and public education) certainly is politically unfeasible—but how long is it since a President put the full weight of his office behind a major program? In the absence of political leadership, virtually every major strategic program is politically unfeasible.

The Western Concept

While the Soviet Union energetically prepares her war-waging options at all levels, what form should Western defense activity and thought take?

First, few people believe that a Soviet-American armed conflict is at all likely. Fewer still believe that either side could emerge from such a conflict with what could fairly be described as victory.

Second, in sharply descending order of interest, Western theorists, commentators, and even officials, address the problems of war deterrence, of intrawar deterrence, and of war termination. How one prepares for, and then conducts, war against an adversary who is determined to *win* is a question that many people choose simply not to pose.

Third, at the theater and intercontinental levels, Western thought and action are focused upon prewar deterrence, and then—if necessary—on conflict control. These are sensible concerns, but they could leave us vulnerable to a dramatically different Soviet style and concept of war. Few would deny that a Soviet offensive in Central Europe would have to be stopped, or slowed down, within the first few days of a war, if it is to be stopped at all. Also, it is widely recognized that the Soviets are postured for a surprise attack and a rapid breakthrough. But NATO is not ready to meet a conventional blitzkrieg with conventional means (without many weeks of warning time), and it is close to a certainty that nuclear firepower would not be released by political authorities in time to do any good. One may object that "war is very unlikely," but it is not healthy for one side to be ready to move forward on short notice, while the defender is not ready to offer a serious defense.

Fourth, American doctrine for the employment of strategic forces has been recast in favor of greater flexibility. But, notwithstanding the logic of the flexibility doctrine, it is not at all clear that the most probable Soviet style in nuclear war-waging has been taken seriously enough. Should deterrence fail, the Soviet Union may well prove less interested in conflict containment than in the effective prosecution of war. Looking to the period after 1980, how should we wage a war with an adversary who has evacuated most of his urban population, has had a long-term industrial dispersal policy, and who pursues major military objectives? Should the United States exercise one or two of its limited nuclear options as the opening bid, it could well find itself facing a Soviet shut-out reply (a Soviet attack on all land-based missile forces, bomber bases, and SSBN facilities–*and* an attack upon those American industries essential for wartime mobilization and postattack recovery). The Soviets might try to control the pace of escalation by playing according to rules recognizable to Americans, *e.g.,* their responding strikes might be large by fashionable American standards, but not so large as to suggest that the Soviet Union had shifted gears into a purely military conflict. Without denying the possibility of Soviet self-restraint, it is prudent to presume that such self-restraint is improbable. In a war for the highest of stakes, involving the most fundamental of values on both sides (and it would be so perceived by Soviet leaders), the Soviet Union would probably reject limited strategic war as inadequate. This is most likely to be so if American missile silos become as vulnerable after 1981 as the "best estimates" of the US intelligence community now predict, and if domestic war-survival programs in the US remain paper exercises.

A Usable Instrument

It is not claimed here that the Soviet Union would welcome war, nor that it will invariably expect to be victorious. All that is claimed is that the Soviet Union is bending every effort to secure the possibility of victory. The Soviets appear to believe that large, capable, and *operationally ready* armed forces could be extremely useful, either to rebuff imperialist military adventures or to exploit situations of opportunity. One need not explain Soviet motives by specific reference to the enduring pressure point of West Berlin, the attraction of intervention in Yugoslavia, or a grand sweep to the English Channel. The Soviet armed forces should do well in any of these eventualities–but those forces also serve Soviet diplomatic ends just by the fact of their existence and forward deployment.

From the outset, too many Western experts choose to place the Soviets in a "no lose" condition. In Europe, we aspire to restore the *status quo ante*, with much of what NATO could do to improve operational readiness being ruled out as "provocative." Meanwhile, with respect to the strategic balance, follow-on US weapon systems that could actually threaten the survivability of Soviet forces are resisted on the grounds that they would be destabilizing. Because our doctrines are so defensive and so attentive to stability, the likelihood that the Soviet Union sees her military forces as a usable, prospectively war-winning instrument of diplomacy is accepted with acute difficulty, if it is accepted at all.

MIG CASE MAY SPUR JAPAN ARMS BUY

Reprinted by Permission
AVIATION WEEK & SPACE TECHNOLOGY
(September 13, 1976)

Defection to Japan by the pilot of an advanced high-altitude Soviet interceptor version of the MiG-25 Mach 3.2 interceptor with his aircraft last week is expected to spur accelerated decisions by the Japanese on planned orders for a U.S. early warning aircraft and a new fighter capable of dealing with the posed threat.

It also probably will result in a decision to improve substantially the capabilities of the Japanese ground-based air defense radar system that lost the MiG-25 as it dipped to low altitude on its final run to the commercial airport of Hakodate, located on Hokkaido, the northern-most of Japan's three major island complexes.

The ground radars lost contact with the MiG-25 as controllers were attempting to vector for a surveillance intercept two Japanese Air Self Defense Force McDonnell Douglas F-4EJ fighters of the Second Air Wing from Chitose air base located near Sapporo, the capital city of Hokkaido.

As a consequence, no intercept was made.

Japan has been debating for some time the need for modernization of its 24-site base air defense ground environment complex, which has a capability comparable to the aging U.S. Sage system, because of systematic reconnaissance flights within the vicinity of the islands by Soviet aircraft.

Despite Soviet protestations that the defecting pilot and aircraft should be returned to Russian control immediately, Japanese officials permitted U.S. technicians to make their first on-site technical evaluation of the MiG-25, including its two 24,250-lb.-thrust Tumansky R-266 turbojets, its avionics system and the metallurgy employed by the Soviets in the mass production of such a high-speed, high-altitude aircraft. The examination should further define the current status of Russian aircraft design and production technology. Of particular importance will be the opportunity to inspect the electronics countermeasures and counter-countermeasures techniques and equipment being deployed on advanced Soviet combat aircraft.

The Russian pilot also was granted his request for asylum in the U.S. He left Tokyo late last week on a regularly scheduled Northwest Airlines flight to Los Angeles.

The Foxbat A version of the MiG-25 that landed in Japan is equipped to carry four 20-mi-range Apex air-to-air missiles, two on pylons under each wing, although none were carried on the defection flight from a Russian air base at Sakharovka, 120 mi. northwest of Vladivostok, on what was to have been a normal training flight.

Vladivostok is approximately 500 mi. from Hakodate.

The aircraft has a normal mission range of 610 naut. mi. but can be extended to 700 naut. mi. by limiting use of the R-266 afterburners.

Following is the chronology, in Tokyo

time, of the defecting MiG-25's approach to and landing at Hakodate from the time it was detected by Japanese base air defense radars on Sept. 6:

- **1:11 p.m.**—Four Japanese air force ground radars detect an unknown target 200 mi. off the west coast of Hokkaido flying to the east at an altitude of 18,700 ft. and at a speed of 440 kt.
- **1:20 p.m.**—The two Japanese air force F-4s are scrambled from Chitose. Simultaneously, the Japanese air force air defense control center is signaling the still unknown target for identification. There is no reply.
- **1:24 p.m.**—MiG-25 crosses into Japanese territory.
- **1:26 p.m.**—The target begins its letdown and disappears from the Japanese radar screens, making it impossible for the ground control center to provide adequate vectors for the pursuing F-4s.
- **1:52 p.m.**—MiG-25 is observed visually by Hakodate airport officials. The pilot circles the airport at an altitude of 1,000 ft., sees that an All Nippon Airways Boeing 727 transport is on the runway prepared for takeoff and veers away.
- **1:57 p.m.**—The aircraft makes an unannounced approach to Hakodate Airport Runway 12 following the 727 takeoff, with fully extended flaps and a deployed drag chute. It touches down approximately halfway along the 6,560-ft. Runway 12, runs over the threshold and rolls another 800 ft. before coming to a stop after striking two small instrument landing system (ILS) antenna poles that prevented its hitting the ILS localizer antenna 20 ft. beyond.

The pilot, Soviet Air Force Lt. Victor Ivanovich Belenko, then jumped from the cockpit, fired his pistol into the air to warn away approaching Japanese construction workers and subsequently demanded that his aircraft be covered by canvas. The MiG-25, after some delay, was covered and towed to a space on the airport parking ramp. It later was moved into an enclosed hangar to avoid viewing by curious Japanese and others interested in what was being done to the aircraft. A 20-ft.-high fence also was erected around the hangar.

The airport was closed to commercial traffic for a 5-hr. period after the MiG-25 landed. Belenko initially told Japanese officials that he had landed because of a fuel starvation problem but later asked for asylum in the U.S. Japan does not permit political asylum. Subsequent examination showed that the aircraft had consumed approximately 95% of its fuel supply.

After the MiG-25 landed at Hakodate, a total of 16 Russian aircraft were spotted in the vicinity of Hokkaido by Japanese ground radars at different periods up until midnight of Sept. 6, resulting in 143 hot scrambles by Japanese air force aircraft.

The Soviets continued to maintain a vigorous watch off the Hokkaido coast for the next two days with military aircraft.

One of the first would-be visitors to travel to Hakodate airport was a Russian citizen, believed by Japanese authorities to have been a political official from a Soviet ship in Hakodate harbor, who arrived 4 hr. after the MiG-25 landing. Like most others seeking access to the aircraft, he was denied entry.

Others asking to view the aircraft included two engineers for Aeroflot, the Soviet state-owned airline, a reporter for Izvestia and the Bristish Air attache in Tokyo.

Japanese police believe that agents of the U.S. Central Intelligence Agency, the Russian KGB and the secret services of other nations also were among those seeking entry.

On Sept. 8, Japanese police formally asked the Japanese Defense Agency to aid in the investigation of the aircraft, ostensibly to determine whether any explosives or dangerous weapons were on board. A group of more than 20 Japanese air force specialists then began a minute study of the MiG-25.

In reply to repeated Soviet demands that the aircraft be returned promptly, Japanese officials said it is evidence to support charges against Belenko that he entered Japan illegally and that it must remain in Japan until the investigation is complete.

Japanese Foreign Ministry maintains that the status of the MiG-25 currently is in the crime investigation stage because of the charges against Belenko and not yet a matter for diplomatic negotiation. The aircraft represents solid evidence that the Russian pilot illegally violated Japanese airspace.

The Russians, in pressing demands for the prompt return of the aircraft and an interview with Belenko, are neglecting both etiquette and traditional international procedures, ministry officials say.

The Soviets, they add, continue to push their demands, although they have not apologized for the fact that a Russian aircraft violated Japanese airspace.

Belenko told Japanese officials that he took off from Sakharovka in a tandem formation of three Foxbats, with his aircraft in the rear position. Shortly after takeoff, Belenko said, he broke from the formation and dove to an altitude of 150 ft. to avoid detection by Soviet radars. He later climbed to an altitude of 18,000 ft. after he had flown beyond the range of Russian ground radars. He flew first to the Japanese air force base at Chitose but, after encountering heavy cloud cover there, diverted to Hakodate.

Belenko joined the Foxbat squadron last fall and had 30 hr. flight time on the MiG-25 at the time of his defection. Previously, he had been an instructor pilot on the Sukhoi Su-15 Flagon E Mach 2.5 fighter.

MiG-25 INSPECTION DISCLOSES HIGHER-THAN-EXPECTED WEIGHT

Reprinted by Permission
AVIATION WEEK & SPACE TECHNOLOGY
(October 25, 1976)

Examination of the MiG-25 interceptor flown to Japan by a defecting Soviet pilot has revealed the aircraft has a takeoff gross weight of 35 metric tons (77,162 lb.)—about 5 metric tons (11,032 lb.) more than had been anticipated.

This figure includes a full fuel weight of 14 metric tons (30,865 lb.) and four air-to-air missiles with the North Atlantic Treaty Organization code name of Anab.

The unexpectedly high weight is due to the aircraft's primarily nickel steel construction, with only 3% titanium in such locations as the wing leading edge and engine nozzle. No advanced materials like carbon fiber or honeycomb were found.

Japanese Air Self Defense Force experts at Hyakuri AFB 50 mi. northeast of Tokyo ran the MiG's Tumansky R-266 engines for 1 hr. 20 min. Oct. 3 after the aircraft was reassembled following breakdown and detailed examination.

During the test, the 200 liters (52.8 gal.) of Soviet T-6 fuel that remained when the aircraft landed at Hakodate airport was burned first, then supplemented with Japanese fuel.

Fuel consumption rate was quite high, although the exact figure was not revealed. Ground static thrust was measured at 11 metric tons (24,251 lb.), as expected.

JASDF officials concluded that the fuel consumption rate was so high that the aircraft could not fly from its Siberian base of Sakharovka to Tokyo and return, indicating the aircraft is intended for point defense rather than offensive operation. Presence of the MiG squadron of 12 operational aircraft and two reserves is believed to be aimed at discouraging overflight of the Soviet Union by USAF/Lockheed SR-71 reconnaissance aircraft based on Okinawa.

The MiG is 22.3 meters (73.2 ft.) long, with a 14-meter (45.9 ft.) wingspan. Overall height is 5.6 meters (18.4 ft.) and wing area is 56 sq. meters (602.7 sq. ft.). Empty weight is a little more than 20 metric tons (44,092 lb.). Wing loading is 625 kilograms (1,378 lb.) and thrust-to-weight ratio is 0.63. Range is about 2,000 km. (1,243 mi.) and service ceiling is 22,000 meters (65,617 ft.). Width of fuselage including intake ducts is 3.4 meters (11.2 ft.); excluding ducts it is 1.6 meters (5.2 ft.).

According to the pilot, Lt. Viktor Belenko, the MiG-25 costs $20-23 million.

Other facts that have come to light recently include:

- **Fuel**—Wet wing has integral tank, and three of four additional tanks can be carried behind the cockpit. Four underwing pylons have no fuel piping and thus are intended only for missiles.
- **Engine**—Powerplant is 6 meters (19.7 ft.) long, compared to 5.3 (17.4 ft.) for the McDonnell Douglas F-4's General Electric J79. Compressor is one meter (3.2 ft.) long and has only five stages. Intake width is 0.9 meters (2.9 ft.). There are only three engine instruments compared to 10 in the F-4. Auxiliary power is not needed to start the engine.
- **Radar**—Search and tracking radar, in phased-array configuration; has a range of 90 km. (55.9 mi.) and only a slight look-down capability. Nose radar antenna diameter is 0.85 meters (2.8 ft.).
- **Avionics**—These are not packaged in plug-in boxes as they are in U.S. aircraft, but are arranged in scattered and complex manner.
- **Communication**—The MiG-25 transmits on two frequencies, one for peacetime and one for war.
- **Gun**—The aircraft has no fixed gun, but has a gunsight and could carry a gun under the fuselage.
- **Control surfaces**—The MiG has no air brake or spoiler.

Negotiations were expected to begin last week for return of the aircraft to the Soviet Union. It is expected that the MiG will be hauled by trailer to the nearby port of Hitachi, where it will be loaded on a Soviet freighter.

MiG-25 Foxbat flown to Japan by Soviet defector Belenko. (Wide World photo)

THE KIEV: THE SHIP; THE AIRCRAFT

Reprinted by Permission
FLIGHT INTERNATIONAL
(November 6/August 7, 1976)
Edited for Space

The Soviet Navy's first "aircraft carrier"—a description as unsatisfactory as the Russian "anti-submarine cruiser"—is both larger and more heavily armed than had been expected by the West. The vessel, *Kiev,* is 936 ft long overall and displaces about 55,800 tons at 30 ft draught or 62,000 tons at 32.8 ft draught deep-load. The ship has no direct Western equivalent, but can perform many of the functions of a medium-sized attack carrier. One major difference, however, is that missiles rather than aircraft are assigned the long-range strike role and share fleet air defence with the Yakovlev Vtol fighters.

The Soviet Navy may be using the *Kiev* class of ships—at least two more are under construction—to gain experience in the techniques of naval aviation before moving on to larger vessels more like a carrier of the US Navy. But the capabilities of the present class are impressive enough. They include anti-submarine warfare, fleet air defence, long-range strike with SS-N-12 missiles, commando operations with helicopters in the troop-carrying role, air support of land forces and, not least, showing the flag.

The 4° angled flight deck is 625 ft long and has an area of 70,000 sq ft, making it two-and-a-half times the size of the helicopter deck on the *Moskva*-class vessels. Loss of deck area does not appear to have bothered the designers, since all three lifts are in the aircraft-handling area rather than in the deck-edge position favoured by the US Navy.

Propulsion is widely reported to be by steam turbines, although the powerplant could be six Kuznetsov NK-144 gas turbines. Total output is some 212,000 s.h.p., giving a speed of 33 kt. The ship's wake tends to indicate the use of only two propellers, but that would mean that each shaft had to transmit 50 per cent more power than is normal and it is therefore likely that four propellers are fitted, in two pairs. Also, the use of only two shafts would result in the ship being disabled if one were put out of action during battle.

Primary Sensors

A Top Sail three-dimensional long-range search and air warning radar is installed on the first tower atop the island superstructure. Top Sail provides long-range target information for the fighters and the SA-N-3 medium-range surface-to

One of the V/STOL (Vertical/Short Take-off and Landing) aircraft hovers over the flight deck of the new Soviet warship KIEV in the Atlantic. The picture was taken from the Royal Navy frigate HMS TORQUAY.

A bird's eye view of the new Soviet warship, KIEV, photographed in the Mediterranean by an RAF Nimrod of No. 203 Squadron from Malta.

1. SUW-N-1 System (FRAS-1)
2. Twin 76mm Gun
3. Cruise Missile Launcher, possibly for SS-N-12
4. SA-N-3 Launcher (Goblet)
5. SA-N-4 Magazine
6. Elevator for YAK-36 (Forger)

This view from the stern of the new Soviet warship, KIEV, taken by an RAF Nimrod of No. 203 Squadron, based in Malta, shows on deck two of the new Soviet V/STOL (Vertical/Short Take Off and Landing) fighters and four Hormone helicopters. (RAF Photo)

air missiles. On top of the third tower is the antenna for a Head Net air-search radar. This radar, which also has some height-finding ability, acts as a back-up for the Top Sail out to 60-70 miles. It may additionally have a surface-surveillance role.

A variable-depth sonar can be deployed through a cut-out in the ship's stern, and a hull-mounted sonar may also be installed—probably at the bow. The eight spheres on the side of the superstructure are Side Globe ECM (electronic countermeasures) equipment.

Weapons

In addition to its Vtol aircraft, *Kiev* carries a formidable array of missiles, rockets, guns and torpedoes. Medium-range air defence is the task of SA-N-3 Goblet missiles fired from two twin launchers, one on the fore-deck and the other on the after end of the island. About 40 reload rounds are likely to be carried.

Short-range air defence is shared by SA-N-4 missiles and cannon. The two twin launchers for the missiles are housed under the circular plates let into the deck to port of the rear SS-N-12 launchers on the foredeck and alongside the after 76mm gun mounting. The guns, of which there are eight arranged in a box at the four corners of the flight deck, are thought to be of the six-barrelled 23mm rapid-fire Gatling type rather than the 30mm twin-barrelled automatic weapon which uses a similar mounting.

There are also two twin 76mm anti-aircraft guns, one forward and the other aft, with their associated Owl Screech fire-control radars. Three mortars for dispensing chaff are mounted behind the forward 76mm (one) and at the stern (two).

Long-range strike against surface targets is the responsibility of missiles, thought to be the SS-N-12 which is also destined to equip Echo II submarines. Earlier weapons, such as the SS-N-3 Shaddock, may be embarked as an interim measure. The eight launcher/containers for these missiles are mounted on the fore-deck. They can be elevated slightly for firing, but the ship itself has to turn to bring them on to the general bearing of a target. It is unlikely that reload rounds are carried.

Anti-submarine weapons comprise ten torpedo tubes (one bank of five let into each side of the hull aft of the figures "860"), two MBU 2500A 12-barrelled rocket launchers on the fore-deck, and a twin launcher for the SUW-N-1 system (also on the fore-deck). The last-named launcher is used for both the FRAS-1 rocket, carrying a nuclear warhead or depth charge, and the SS-N-14 missile which delivers a torpedo. The Ka-25 helicopters also have a prime anti-submarine role.

The second ship of the class, *Minsk,* is being built at the same shipyard (the Nikolaev yard on the Black Sea) as was responsible for the *Kiev*, and the third unit is thought to be under construction at Leningrad. Mick Dodgson/Mark Hewish

---

SOVIET V/STOL COMBAT AIRCRAFT

The new Soviet V/Stol combat aircraft aboard the Soviet carrier *Kiev* (officially classified as an anti-submarine cruiser, to allow her passage through the Bosphorus under the terms of the Montreux Convention) are believed to represent a trials unit rather than a fully operational combat wing or regiment. It is suggested now that there may be as few as a dozen of the aircraft aboard the *Kiev*.

Existence of a Soviet V/Stol combat aircraft was reported first in early 1974,

The new Soviet warship, KIEV, photographed in the Mediterranean by an RAF Nimrod of No. 203 Squadron from Malta. (RAF Photo)

when aircraft of this type began sea trials from the aft landing platform of the helicopter carrier *Moskva* in the Black Sea. It was not an unexpected development; the Soviet Navy was already known to be building the first of the *Kiev* class and the angled flight deck suggested that a V/Stol type would be operated.

The new type is regarded by Western experts as primarily an air-to-surface aircraft, with possibly a secondary air-defence role. It carries a mix of air-to-surface and air-to-air weapons, with four wing pylons and a gunpack beneath the fuselage. There is no search or intercept radar, but there may well be a small ranging radar for air-to-surface use. The aircraft photographed aboard the *Kiev* are single-seaters, although a two-seater exists for conversion training.

The basic layout of the V/Stol is significantly different from anything tried out in the West. It is now believed that there is only one lift/cruise vectored-thrust engine, probably unreheated and exhausting through twin rear nozzles located level with the wing trailing edge. This is not enough to hold the aircraft trimmed for hovering flight; there is believed to be a single lift engine located aft of the cockpit. Control in the hover is by reaction nozzles or "puffer pipes," visible beneath the wingtips and around the tail and presumably also installed in the nose.

New VTOL fighter on deck of KIEV.

The small folding wing is mid-set and the main spar must arch over the eight-tonne-thrust lift/cruise engine. The mid-wing layout and the absence of jet nozzles ahead of the centre of gravity have allowed the Soviet designer to fit a classic tricycle undercarriage, with the main members attached below the wing fold and retracting inwards behind the four pylons.

The Soviet aircraft recalls the dormant German VFW VAK 191, with its lift-plus-lift/cruise arrangement and small wing, but comparisons with the other operational V/Stol combat aircraft, the British Hawker Siddeley Harrier, may be a little more illuminating.

The V/Stol is somewhat bigger than the Harrier, being some 52 ft long to the Harrier's 45 ft. It is probably heavier as well: up to 28,000 lb for a Stol take-off from the *Kiev's* angled deck. For Vtol—probably not a standard procedure, despite the trials on the *Moskva*—maximum weight could be as much as 21,000 lb. This suggests that the main engine would give some 18,000 lb thrust, with the lift engine contributing about 5,000 lb. The aircraft balances in the hover because the lift engine is about three-and-a-half times as far from the c.g. as the main nozzles.

The mixed layout must lead to some problems on Stol take-offs. Scorch-marks on the *Kiev's* deck show where aircraft preparing to take off run up their engines with the main nozzles pointing downwards (like the Harrier, the new type probably has more power than can be applied with brakes on and nozzles aft). But with main nozzles swivelled aft for the take-off roll, the lift engine cannot be opened up because the nose would lift and the tail would strike the deck. So unless the lift engine is not opened up until the nozzles are swung down through the 50° to 60° needed for a Stol rotation—which seems a very chancy way of leaving a carrier—there must be some sort of flat-plate or cascade deflector on the lift-jet efflux, to point it about 45° rearwards for the take-off roll.

The layout also rules out the use of vectoring in forward flight, which gives the Harrier its remarkable agility in air combat. The lift engine is in fact very well concealed under a solid rather than louvred cover in cruise flight and is not intended for use outside take-off and landing. Any attempt to use the aft-set nozzles of the main engine in flight would result in an uncontrollable nose-down pitch, rather than a tighter turn.

The V/Stol has a large vertical fin to cope with the destabilising effects of the airflow through two large intakes well ahead of the c.g. The tail-plane is however quite small and appears to have conventional elevators. This suggests that the puffer-pipe system is more effective in pitch than in yaw, especially as the airflow through the lift-engine must bring pitch-stability problems.

Maximum speed of the type probably depends on the bypass ratio of the lift/cruise engine (which could of course be a turbojet if thrust/weight ratio was judged more important than specific fuel consumption). If the engine is a turbojet or of moderate bypass ratio the V/Stol may be capable of marginally supersonic speed in level flight at medium altitudes (say Mach 1.3 at 36,000 ft). If the bypass ratio is high, however, the Soviet aircraft will be limited, as the Harrier is, by thrust decay with altitude. It is not likely that the engine is an adaptation of a power-plant already in service; V/Stol lift/cruise engines need to be specially designed to withstand high bleed rates without overheating. But whatever its engine, the type is almost certainly subsonic at most levels. Although there is a large splitter plate on the intake there is no evidence of variable geometry.

Armament of the new type, apart from the gunpack, is likely to include iron and "smart" bombs, unguided rockets and missiles. So far there is no evidence of the sort of optical and electronic sighting heads fitted to the MiG-23B. It is nevertheless likely that the armament of the V/Stol includes the short-range AS-7 Kerry air-to-surface missile, and the aircraft might carry AA-2 Atolls in the air-to-air role.

ISRAELI COMMANDO C-130 RAID FREES 115

Reprinted by Permission
AVIATION WEEK & SPACE TECHNOLOGY
(July 12, 1976)

A 4,800-mi. airborne commando raid by the Israeli Defense Force freed 115 Air France hijacked passengers and crew, killed all seven Popular Front for the Liberation of Palestine hijackers involved plus 20 Ugandan soldiers and destroyed 11 MiG jet fighters at Entebbe Airport on July 4. Israeli casualties included one officer and three hostages.

The slain officer was American-born Lt. Col. Jonathan Netanyahu, 30, leader of the commandos.

The Israeli raid was conducted by about 150 commandos flying in three Lockheed C-130H turboprop transports from a staging base at Sharm-el-Sheikh at the tip of the Sinai Peninsula.

A fourth Israeli airplane, a Boeing 707 converted for military use by Israel Aircraft Industries, was flown to Nairobi, Kenya. It was equipped as an airborne hospital to receive the expected wounded from the Entebbe raid and accompanied the C-130Hs back to Israel.

Fighter cover from Israeli air force Mirages and Phantoms was provided for as far as possible over the Red Sea on both outbound and inbound missions. Fighters were not refueled in the air because that would have involved the use of lights and extensive radio transmissions that could have unmasked the mission. The C-130Hs flew across Ethiopia to Uganga, maintaining radio silence until they were close to Entebbe. They apparently radioed the Entebbe control tower that they were bringing the Palestinian prisoners from Israeli jails demanded by the hijackers.

They approached Entebbe in bright moonlight strongly reflected off nearby Lake Victoria and did not use the main 12,000-ft. runway built several years ago by an Israeli construction firm. Instead, they landed in quick succession on the old 7,900-ft. runway, which placed them about one-half mile from the old terminal building where the hostages were held by the hijackers surrounded by explosives and Ugandan soldiers. Taxiing to the end of the runway, they quickly disgorged their armed jeeps and raced for the terminal. The hijackers and about 20 Ugandan soldiers were killed in the initial attack and all but two hostages, who were killed in the cross-fire, were extracted and led back to the waiting planes. The old terminal was then blown up by the commandos.

Some Israeli commandos set up road blocks and ambushes to handle any sorties from a large Ugandan army camp several hundred yards from the airport perimeter but none materialized. Eleven Ugandan MiG jet fighters—seven MiG-21s and four MiG-17s—were destroyed during the onslaught. This represented about half the Russian-supplied Ugandan air force. Other MiGs were on the field but were too far away to merit further delay in evacuating the hostages. The entire operation from wheels down to takeoff with the rescued hostages took less than 90 min. The Israeli armed jeeps and some mortars were abandoned on the airfield.

The Israeli raid was aided by the extensive knowledge of Uganda in general, and the Entebbe airport specifically, gained during the years up to 1972 when the Israelis were the chief military and economic advisers to this new African nation. Col. Baruch Bar-Lev, who had headed the Israeli military mission in that era, had been at Entebbe for several days as a negotiator with Uganda president Idi Amin.

An Israeli construction firm had built the new terminal and runway, and the Israelis had a detailed knowledge of the field. Immediately after the raid, local communications were temporarily blacked out indicating some type of jamming either from the C-130Hs, which carry ECM equipment, or from ground agents. The Entebbe airport is equipped with a VOR omni-range and a non-directional beacon.

The C-130Hs landed in Nairobi where they were refueled and joined by the Israeli air force 707 for the return flight to Israel. There was no evidence of pursuit by the remaining Ugandan MiG fighters.

Badly wounded hostages and Israeli soldiers were off-loaded at Nairobi into local hospitals where a third hostage died. A sick Israeli passenger who was put into an Entebbe hospital earlier in the week was still there with negotiations for her release being conducted by British diplomats.

Israeli Air Force has 21 C-130H aircraft with three more still to be delivered by Lockheed under a U.S. government foreign military sales contract.

The hijacking began June 27 when an Air France A-300B en route from Tel Aviv to Paris was taken over at pistol point shortly after takeoff from Athens, a scheduled intermediate stop. The aircraft contained 246 passengers, including two Germans and two Palestinian Arabs, who were the hijackers, plus a crew of 12. The hijackers directed the crew to divert to Benghazi, and on June 28, ordered the aircraft to Entebbe. There, the four hijackers were joined by three more Palestinians on the ground.

The hijackers and hostages abandoned the Air France Airbus for the old terminal building where explosives were planted and Ugandan soldiers formed a perimeter guard. The hijackers demanded the release of 53 Palestinian terrorists held in the jails of five countries including 40 in Israel.

On June 30, they released 47 women and children. The next day, July 1, they released another 100 hostages, retaining 97 Israelis and dual nationality persons plus 12 Air France crewmembers, who elected to stay with the hostages.

Negotiations continued with a deadline of 8 a.m. Sunday July 4 when the hijackers threatened to kill all the hostages with the explosives they had planted around the old terminal building. The Israeli raid began just before midnight July 3.

Late last week, Air France still had no indication when it would be cleared to recover its Airbus from Entebbe. The airline has had a Boeing 707 standing by at Nairobi with demolition experts, mechanics and an Airbus flight crew, ready to go in and reclaim the aircraft whenever Ugandan officials give approval. The Airbus was suspected to have been rigged with explosives before the passengers, crew and hijackers got off the aircraft in Entebbe.

Air France spokesmen were unable to say if the delay in recovering the aircraft was due to the complications of working through diplomatic channels or if it was due to resistance on the part of the Ugandans.

ISRAELI-BUILT JET STOLEN FROM UGANDA

Reprinted by Permission
AIR PROGRESS (December 1976)

To add "insult to injury" following the Entebbe raid, two American pilots reportedly stole a Westwind business jet from Uganda President Idi Amin and were returning it to its original owner, the Israeli government.

The aircraft was said to be a gift to Amin at the time Israel was building Entebbe Airport. The Westwind—a Jet Commander built in Israel—carried Ugandan registration with that country's military insignia. The American pilots were supposed to fly the jet to Europe for repair of damage incurred during the daring July 4 Israeli raid on the airport which freed hostages in the Air France A300 hijacking. Instead, the pilots flew to Cairo and then to Cyprus. The next stop was Israel but the delivery was delayed for a short time because Israeli airspace had been closed as a precautionary measure when a Dutch airliner was hijacked. In that hijacking in September, the passengers and crew of the KLM DC-9 were released unharmed by Palestinian hijackers in Cyprus. The hijackers aboard the DC-9 had demanded the release of eight Palestinians held in Israeli jails. The KLM airliner had circled the Mediterranean near Israel's coast for hours while the hijackers threatened to blow it up unless their demands were met. Israel refused to release the prisoners and they blocked Ben-Gurion Airport with trucks and fire engines to prevent the DC-9 from landing. A pair of Israeli F-4 fighters headed off the KLM jet before it could enter the country's airspace and the airliner finally landed in Cyprus where the passengers were released.

After the hijacking drama was over, Uganda President Idi Amin finally commented on the "removal" of the Westwind from his country. He claimed he sanctioned the return, saying that he did not want to use something that did not belong to him. It was still believed, however, that the Westwind was flown out of the country without his permission.

THE MILITARY AIRCRAFT SCENE

Reprinted by permission INTERAVIA (February 1976)

Whether the technologists like it or not, it is impossible to divorce economics and politics from the development of technology, particularly in the military aerospace field. The implications of this for the West have become very evident during the past few years as a number of important factors have interacted to produce major changes of thinking. Defence budgets have been unbalanced by high inflation rates, operational methods are being re-examined in the light of the 1973 Middle East war and as a consequence of the growing sophistication in the equipment used by the Warsaw Pact countries, while the high cost of developing new equipment has made export sales almost mandatory.

Inflation, coupled with a recession in the Western World, has had a deleterious effect on the defence budgets of all NATO nations. While the United Kingdom may be the only nation to institute wholesale cuts in defence outlays, with more theatened, other members have felt the effect of the decline in purchasing power.

West Germany has cut its planned order for the MRCA from 420 to 322 and other programmes have been cancelled. Italy's defence budget has been so badly unbalanced by rapidly rising personnel costs that a special 10-year supplementary budget has been created to pay for major investment programmes, the Air Force's share including money for the MRCA and the Macchi MB.339 trainer/strike aircraft. France's problems are no less: cost escalation of the Super Mirage programme led to its being abandoned in favour of a cheaper solution, the Mirage 2000, while budgetary restrictions have meant that orders scheduled to be placed during 1971-75 for the Mirage F-1, Alpha Jet, Jaguar and others have been deferred.

The list could be made longer. However, it is worth noticing that corrective measures are being taken on two fronts. The cost-effectiveness of some aircraft is being enhanced by making them suitable for more than one role—the Alpha Jet and Hawk being typical – and of others by increasing their survivability through fitting ECM systems. Secondly, even stronger export efforts than in the past are being made; while the rush to sell to the Middle East area also has the motive of maintaining good relations with the oil-rich states in that region, any export sale improves the amortization of development expenditure.

The Soviet challenge

Financial restrictions are pushing Western aircraft development in one direction, while the East Bloc nations, spearheaded by the Soviet Union are moving in another. The most remarkable point is that the latest Soviet aircraft display a sophistication that was almost unthinkable only a few years ago. During the 1950s and 1960s, the tendency was for all but the most specialized combat aircraft (e.g. reconnaissance and anti-submarine warfare) to be simple and robust, and produced in large numbers.

However, this approach has been significantly changed. While the MiG-21 and the Su-7, another relatively simple design, still constitute a major proportion of the Warsaw Pact inventories, aircraft such as the MiG-23 and -25 and the Sukhoi Su-19 are comparable to the latest American products.

All these new-technology types reflect a successful Soviet drive for quality rather than quantity and money has evidently been no object. Straightforward cost comparisons are meaningless but the differing philosophies of East and West must be causing NATO defence planners serious concern. At a time when European defence budgets are squeezed by social and other needs, and American military expenditure is considered insufficient in some quarters, the Soviety Union is prepared to let the consumer suffer to boost its arms capability.

While there is no reason to believe that NATO lags behind the Warsaw Pact in terms of the capability of its latest military aircraft, in the "numbers game" it certainly does. Some 1,700-2,000 combat machines are being built annually by the Soviet Union, which practically monopolizes high-technology production in the Warsaw Pact area; the cumulative total for the five major manufacturing nations of NATO falls rather short of this figure.

The Western reply

While the lack of standardization among NATO nations is much worse on the ground or at sea than in the air, the heterogeneous collection of aircraft operated by the air forces of Belgium, Canada, Denmark, France (included in the term NATO, despite the official looseness of the association), West Germany, Italy, Netherlands, Norway and the United Kingdom is remarkable. Indeed, the crucial positive aspect of the four-nation F-104G replacement competition was that Belgium, Denmark, Netherlands and Norway agreed to buy the same aircraft.

At this point in time, it is not clear just how willing the West is to accept standardization, since every nation with an aircraft industry of any size has purely domestic interests to consider. The Americans would be happy to see standardization on their equipment, since it reduces Western Europe's competitive position; Britain made no serious effort to arrive at a common trainer/strike aircraft solution with France and West Germany, despite the simultaneity of requirements; and France still wishes to be sole master of its own combat aircraft production, etc.

The viewpoints of the various national factions are comprehensible and defensible. To take the case of France, two recent events show the legitimate concern of the Paris Government. Firstly, it is not widely known that no less than 140 aircraft were prematurely withdrawn from French Air Force use last year – 40 F-100 Super Sabres and 100 MAP-funded Mystere IVs – and put into storage at US bases in Britain. Secondly, the decision to abandon the Super Mirage combat type might have appeared a suitable moment to accept "standardization" on a foreign design, except that the industrial potential for French companies was minimal – hence the decision to go for the Mirage 2000 solution.

Thus, there are three mainstreams of combat aircraft development today. The USA prepares to meet both long- and short-range threats from the Warsaw Pact; Western Europe to meet the short-range threat from the East; and the East Bloc to counter long- and short-range developments in NATO.

FLYING THE B-1

A PILOT'S VIEW

Reprinted by Permission
AIR FORCE MAGAZINE (June 1976)

by Major George W. Larson, Jr., USAF

Edited for Space

No matter what the computers and wind-tunnel studies say, final judgment of an aircraft's performance lies with the pilot. In this exclusive report, a veteran USAF test pilot tells about flying the B-1, and gives his view of how it will perform its assigned mission.

I'm fortunate to have been selected as one of five pilots now flying the B-1 at Edwards AFB, Calif. Representing the Strategic Air Command and the Air Force Test and Evaluation Center (AFTEC), I am primarily concerned with evaluating the B-1's operational utility and effectiveness. This translates to how well the pilot/B-1 combination can perform the intended mission. Based on my experience at the controls, I'm convinced the Air Force has an aircraft with the potential to become a well-suited, state-of-the-art, strategic bomber. Let me elaborate from the pilot's point of view.

Looking at the B-1 from the outside during routine exterior preflights, you almost feel that the Rockwell International aerodynamicists have come up with a sleek, oversized, high-performance fighter. This impression is strengthened on climbing into the cockpit, where a control stick (not a wheel) and left-hand throttle quadrants await each pilot. Strapping in, you immediately notice two things: optimumly located tape flight and engine instruments, and visibility that surpasses any large commercial or military jet in the air today. Further investigation of the crew station reveals a totally automatic fuel and center-of-gravity management system. Without it, a pilot would have to accomplish manually many center-of-gravity and associated fuel transfer changes during a normal mission.

The usual complement of navigational aids and communication equipment is located with crew members in mind. On the third B-1, the primary attitude indicators have been replaced with Vertical Situation Displays (VSD). The VSD is no more than a cathode ray tube, but it permits, among other things, a combined display of attitude, command steering, angle of attack, airspeed, radar altitude, heading, and weapon release timing. This reduces the area coverage required for pilot cross check of cockpit displays. All primary flight controls are either duplicated for each pilot or accessible to both pilots.

Starting the engines doesn't require the associated ground equipment I have grown used to with other bomber aircraft. Two on-board Auxiliary Power Units (APUs) provide all electrical and pneumatic power necessary for simultaneous engine starts. These APUs also take care of electrical, hydraulic, and cooling requirements for normal preflight actions.

Taxiing the aircraft is easy with nose-wheel steering through the rudder pedals. Smooth, positive differential braking is effective in the event of a nose-wheel steering malfunction. The old groaning and screeching and shuddering associated with other large aircraft brake systems are not present.

With the wings at full forward sweep (fifteen degrees), slats extended, and full flaps, the B-1 is configured for takeoff. As the four General Electric F101 engines are placed in full augmentor (the B-1 term is "augmentor," not "afterburner") there is a smooth, rapid acceleration to liftoff speed. Only minimum aft stick displacement is needed at rotation speed, and you find yourself airborne in approximately 3,000 to 4,000 feet. Longer takeoff distances will be required as test-program increases in gross weight are scheduled.

After takeoff, aircraft retrimming is necessary as the flaps are retracted. Since each pilot has a wingsweep control, either can sweep the wings aft to twenty-five degrees, the configuration for medium- or high-altitude subsonic cruise. The throttles are retarded to intermediate power (previously known as military power) in preparation for a climb to cruise altitude.

Maneuvering the aircraft in pitch or roll is a pleasant surprise. Only small control displacements (one and two inches depending upon airspeed) are required. The response to a control stick input is rapid. There are no sluggish or delayed control responses.

Since the bomber tactics do not normally include close formation flying, the first real test of the pilot and the flight control system is air refueling. With earlier large bombers, this could become a tedious and demanding task. Refueling the B-1 is much easier. Only very minimal control inputs are required, thrust response is rapid and effective, and visibility is excellent. Even with a higher-than-normal adrenalin level and some apprehension, I was able to refuel the B-1 to maximum inflight gross weight without a disconnect during my first attempt. Unlike recent bomber aircraft, the refueling receptable is located in front of the pilots. This provides an excellent secondary reference in determining closure rates while refueling.

On the Deck

The second and perhaps the acid test for the B-1 man-machine interface is high-speed, low-level flight. This entails flying the aircraft as low as treetop height over any type of terrain at speeds close to 600 miles an hour. Preliminary flights in this portion of the B-1 operational envelope lead me to believe that the B-1 has excellent potential. The already rapid control responses increase in this high "q" (dynamic pressure) regime. The responsive flight control system, when integrated with the soon-to-be-installed terrain-following system, is designed to be coupled with automatically generated pitch commands. This will allow the aircraft to be flown hands off at low altitude. The pilot will only monitor flight parameters unless a malfunction requires that he fly terrain-following system commands manually.

An equally important factor affecting pilot performance during low-level, high-speed flight is the effect of turbulence on the cockpit area. Since large aircraft are structurally flexible, moderate to severe turbulence occurring at the aircraft's center of gravity can magnify and result in a bone-jarring ride in the cockpit. To reduce this flexing effect, the B-1 has a Structural Mode Control System (SMCS). Through automatic movement of the "canard-like" control surfaces located on the forward fuselage, longitudinal and lateral structural flex are countered and thereby reduced. Preliminary evaluations of this system indicate that it performs its intended function during all phases of flight.

The B-1 is even more pleasant to fly at supersonic than at subsonic speeds. The aircraft appears to become increasingly stable with an increase in supersonic speeds up to Mach 1.6, the maximum currently being tested.

In the traffic pattern, the B-1 is so responsive you can fly an ILS or an overhead pattern with equal ease and precision. Some pilot adaptation is required prior to touchdown on landing. I consistently feel that I'm higher than necessary when the main gear touches down. The reason is that the pilot sits considerably forward of the main landing gear and is flying the aircraft at an angle of attack of approximately seven degrees during the landing flare. While it makes a grease job more demanding, it does not detract from easily landing the aircraft.

Design and state of the art improvements have given the B-1 handling qualities superior to present large bomber aircraft. It is not a fighter aircraft, but its flight characteristics are more representative of a small, responsive aircraft than those of the B-52. During no phase of flight has any difficulty been experienced in controlling the aircraft. In fact, all aspects of flying the B-1 have been extremely pleasurable. The "bus driver" handle given to bomber pilots in the past should certainly change when the B-1 enters the Air Force inventory. It is more akin to a sports car than a bus.

In a high-speed, low-altitude penetration, the B-1 will use terrain masking to avoid radar detection. Its Structural Mode Control System moderates low-altitude turbulence.

Testing System Effectiveness

So far, the B-1 test program has concentrated on those items essential to basic airworthiness. Such milestones as flying qualities, stability and control, flutter, air refueling, envelope expansion, and initial performance testing have been completed. Now major testing emphasis will be on the total weapon system's effectiveness in all phases of its primary operational mission.

The Base Escape Phase will emphasize the B-1's rapid response to early attack warnings. With the onboard APUs, we will demonstrate the weapon system capability to provide quick reaction during alert launches. Since no ground-power units are needed for the B-1 on alert, it will have optimum flexibility for satellite or dispersed basing requirements. We will demonstrate the reaction capability of critical systems such as the offensive avionics complex to provide the immediate navigation data essential for an alert launch.

Once airborne, the Climb, Cruise, and Navigation Phase begins. The optimum climb schedules and cruise altitudes will be determined, and the Automatic Flight Control System (AFCS) performance demonstrated. The AFCS is designed for great flexibility in that it has Flight Path Hold, Altitude Hold, Airspeed Hold, Mach Hold, and Approach modes in pitch, and Roll Attitude Hold, Manual Heading, Automatic Navigation, and Approach modes in roll. The navigation system will be thoroughly exercised to determine its capability to accurately guide the B-1 through a long-range operational mission.

Following the cruise phase of the basic operational mission, the B-1 will be refueled from a KC-135 tanker. Since we already know that the basic aircraft's ability to refuel is excellent, very little other onboard rendezvous capability and optimum formating altitudes need to be shown.

Next comes the meat of the B-1 mission. Dropping off the tanker with a full fuel load, the wings will be swept aft sixty-five degrees and the B-1 will descend to low level for its penetration phase. Much like a giant hawk, the bomber will begin its hedgehopping tactics, utilizing terrain masking when possible, to pass through enemy defenses for a surprise attack. The onboard Terrain Following Radar (TFR) system, coupled with the Automatic Flight Control System (AFCS) and Auto Throttles, will provide this capability.

It is difficult to imagine that any pilot would trust an aircraft to fly itself at high subsonic speeds and treetop altitudes. I'll be the first to admit that system confidence through exposure is mandatory. However, that capability has been demonstrated many times in aircraft like the F-111 series, and the B-1 system is designed to give even better performance in this environment. And during the penetration phase, the pilot will have Forward Looking Infrared (FLIR) to display terrain features ahead of his intended flight path. This FLIR display will be superimposed on his primary flight instrument, the VSD, along with the other essential flight data.

When equipped with its Air Induction Control System (AICS), the B-1 will be capable of penetrating enemy defenses at speeds greater than Mach 2.0 at high altitudes. Without the AICS system installed, the B-1 high-altitude penetration speed is limited to approximately Mach 1.6. Although the initial production models of the B-1 will not have the AICS, the system will be fully tested to optimize inlet geometry with desired supersonic cruise or dash conditions, thus providing the greatest possible flexibility against future enemy defenses.

The accuracy of the navigation system combined with the flexibility of the onboard Stores Management System (SMS) will also be demonstrated and evaluated during the remaining test program. During either the high- or low-altitude penetration phases, the weapon system will be delivering a varying weapons mix on simulated enemy targets. Even though smaller than the B-52, the B-1 has a greater payload.

In this age of modern electronic warfare, the B-1 will have the most advanced Electronic Countermeasures (ECM) system in the Air Force inventory. Because the contract for the defensive or ECM system was awarded after the initial airframe and avionics contracts; a defensive system will not be on board any of the three prototype aircraft currently under test. An initial look at the defensive system will be accomplished on the first preproduction aircraft, the fourth B-1.

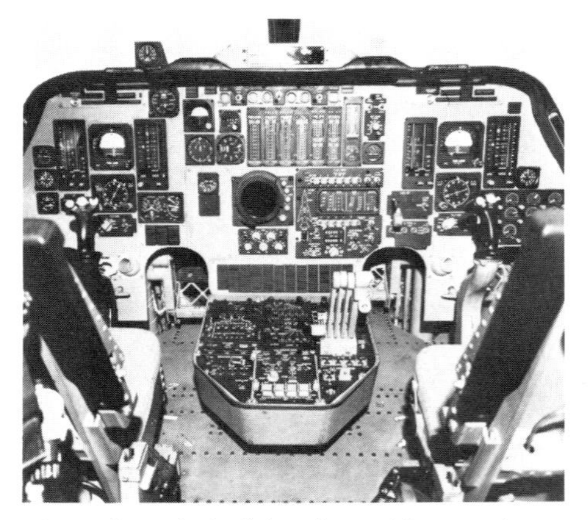

The B-1 cockpit (above) combines optimum instrument layout and exceptional visibility. Unique for an aircraft of this size is the use of a control stick instead of a wheel.

After all the weapons are delivered during the basic B-1 mission, the aircraft will start its Withdrawal and Recovery Phase, continuing terrain following until outside of enemy defenses if desired. The variable geometry wing design of the B-1 will allow reconfiguration to optimize range during this withdrawl phase. Testing will demonstrate capabilities at a lower Mach number, with the wings swept at fifty-five degrees.

On reaching the recovery base, the B-1 systems must demonstrate an Airborne Instrument Landing and Approach (AILA). This is the same as an ILS from the pilot's viewpoint, but with one important difference. The AILA does not depend on ground navigation aids, and is guided totally by the onboard forward-looking attack radar and navigation systems. The flight crew will also assess the B-1's launch and restrike capability.

Even though we have come a long way in demonstrating the basic airframe/engine capability of the B-1, we have even more extensive tests ahead. From what we already know of the systems that are still to be demonstrated, I feel certain that we can achieve the program goal.

It is not my intent to suggest that the B-1 has had no design or systems problems during the test program. It would also be naive to think that no problems will be encountered with the integration of the avionics and airframe on aircraft No. 3. A fly-before-buy developmental test program is specifically designed to determine these deficiencies, and when possible to correct them. In my opinion, the B-1 program has been extremely successful in identifying and correcting deficiencies that have surfaced. I have confidence that the program will continue to operate in this manner. The final objective is a formidable, total strategic weapon system.

There is a great deal of concern about the cost of the B-1, and rightly so. Nobody wants his tax dollars spent for defense if the end result is a system of questionable value to national policy. The B-1 is expensive, if price is the only criterion. However, my criteria are much more encompassing. Do we need a manned bomber? I don't see any other means of assuring a credible nuclear deterrent force without the flexibility of the B-1, including its capability for a show of force and national intent.

While our present B-52 force is effective today, it will be considerably less effective in the combat environment that can be foreseen ten years or so ahead. The B-1, on the other hand, is designed not only to operate effectively in the environment we can foresee, but also to have growth potential to accommodate future defensive or offensive avionics and weapons that may become necessary if technological advances drastically alter the threat. In short, with its larger payload, better performance, and growth potential, the B-1 is a cost-effective system that can assure peace through deterrence into the next century. To abandon it now in favor of a different but vaguely defined manned system, as some have suggested, would only result in a more expensive weapon system and a perhaps critical loss of deterrent capability during the decade or more required for defining, developing, and testing an alternative to the B-1.

It is obvious that I am an advocate of the B-1. However, my advocacy is strongly influenced by one important fact. I have flown the B-1 and am intimately aware of its capabilities and potential. In my opinion, if there isn't a B-1 in our nation's defense forces, we will not be able to effectively support our national policy in the future.

The author, Maj. George W. Larson, Jr., is Chief, B-1 Operations Branch, 4200d Test and Evaluation Squadron (SAC), Edwards AFB, Calif. A graduate of the Air Force Academy and of the Air Force Test Pilot School, and a veteran of eighty-five EB-66 missions over North Vietnam, he has logged more than 3,100 hours in thirty-six different aircraft types, including the B-47, B-52, and FB-111. Prior to his selection in 1972 as SAC's representative on the B-1 Initial Operational Test and Evaluation program, he was assigned to the Bomber/Transport Branch of AFSC's Flight Test Operations at Edwards AFB, Calif.

B-1 – FACTS AND FIGURES

Type	Strategic Bomber—heavy
Designer and Manufacturer	Rockwell International Corporation (B-1 Division)
Powerplant	Four General Electric F101 high bypass ratio turbofan engines, each approximately 30,000 pounds of thrust
Avionic Sybsystems and Integration	Boeing Aerospace Company
Defensive Subsystems	AIL Division, Cutler-Hammer, Inc.
Length	151 feet 2 inches
Height	33 feet 7 inches
Wingspan (aft sweep)	78 feet 2 inches
Wingspan (fwd sweep)	136 feet 8 inches
Weight	350,000 to 400,000 lbs
Speed	Mach 2.0 plus
First Flight	December 23, 1975
Crew-Primary	Two pilots (Aircraft Commander & Copilot) One Offensive Systems Operator (Navigator) One Defensive Systems Operator (Electronic Warfare Officer)
Crew-Additional	One instructor pilot One instructor systems operator
Weapons	75,000 lbs internal—3 bays, 40,000 lbs external (all current and proposed strategic nuclear weapons plus a varied conventional mix)
Avionics	Terrain Following Radar System 2 inertial navigation systems Doppler Radar Forward Looking Infrared Stores Management System Tacan ILS (Instrument Landing System) Defensive System (installed on A/C No. #4) CITS (Central Integrated Test System)

SAY HELLO TO THE WARTHOG

by Major Jack Stitzel
TAC/DRFG

Reprinted by Permission
TAC ATTACK (April 1976)

No, gentlepersons, it hasn't been officially named yet. But it's a safe bet that ultimately, the jocks that fly her will not be guided by whatever name is arrived at by the office types! The A-10, a product of Fairchild-Republic—the same folks who brought us the F-105, F-84, and P-47—is heir to the famed "hog" series. Hog, Super Hog, Ultra Hog . . . and now . . . Wart Hog! Yes, the A-10 is unofficially known as the "Wart Hog."Affectionately conferred, but deserving of a closer look. To the discerning fighter pilot's eye, it's not pretty . . . yea, it borders on terminal ugliness! Aft of the wing, the fuselage is alive with protrusions . . . thousands of rivets . . . which have been left exposed to the breeze. The straight wing certainly isn't stylish in these days of negative dihedral, anhedral swing, and severely bent-wing designs. The bulbous canopy sits high atop the regular-shaped fuselage (no coke-bottle waist or pointy-nose aggressiveness). Pylon-mounted engines remind one of a passenger/utility aircraft and the landing gear remains exposed to the elements. WHEN FULLY UP!! The main gear extend forward into wing sponsons, leaving half of the tires exposed—somewhat reminiscent of the Goony Bird. Of course, we all know the designed simplicity and durability of the Goon, the Wart Hog seems to have that same utilitarian personality. More about that later. All of this adds up to a machine devoid of those visual characteristics we've grown to love and expect from a tactical fighter . . . air-to-air hassler, interceptor, deep striker—the Phantoms, Tomcats, Eagles, ad infinitum.

Let's see, have I covered everything? It's not faster than a speeding bullet, more powerful than a locomotive, nor can it leap buildings in a single bound—and yet, in its own unique way it's super. On the positive side of the ledger are features and capabilities we sorely need and will utilize to the point of pure affection.

There are four labels to describe the A-10's potential. These are: lethality, responsiveness, survivability, simplicity. To break these down to manageable terms, let me expand. On the lethality side, we point first to the gun—the GAU-8A, 30 millimeter cannon. It is nothing short of sensational! To the pilots who have strafed ground targets with 20 millimeter M-61 and been enthralled with its results, you'll want to write home about GAU-8 after using it! I'm talking about double the effective range of the 20MM and seven times its punch. And if you like to stick around the target awhile, think about an ammo drum that will provide you with ten or eleven 2-second bursts.

With 11 pylons on the aircraft, the frag shop is going to have to requisition more storage cabinets for catalogs on A-10 ordnance options. The airplane is being certified to carry the Maverick air-to-surface missile as well as laser bombs to complement the spectrum of conventional ordnance. And to manage it, an armament panel designed by pilots, for pilots. No more leafing through the notes on your knee board to jog your memory on what's on which station. Just read the store off the panel. The type of weapon is handily displayed on the appropriate station, dialed in during load operations.

Turning to responsiveness, the A-10 has long-loiter capacity, measured in hours instead of tens of minutes. Its on-station time will warm the heart of the infantryman. Forward basing, close to the action, is another prerogative. Its high turn rates and good sustained "g" at low airspeed are a natural for that low overcast day with a couple of miles of visibility. In the A-10, after acquiring a target, you can keep it in sight while maneuvering to attack it.

On the survivability issue, the A-10 has less vulnerable area than any other aircraft. It is heavily armored, and the redundant hydraulic flight control systems are a definite plus. Airspeed alone is not a defense against a determined threat. Yes, Warp-3 has a calming effect on the throttle-bender, but a bullet could really care less. The A-10's maneuvering again comes into play, as it must.

As for simplicity, it is a maintenance man's dream. Everything is accessible, no contorted arms required to reach around two corners to get to a connector! I've talked to the bluesuiters who are maintaining it now, and have received enthusiastic endorsement.

In conclusion, on the eve of the A-10 introduction to the tactical inventory, I'd like to register a vote of confidence and a bright future for this specialized close air support aircraft. All who fly it, fix it, and maintain it will come to fondly respect the Wart Hog.

Fairchild A-10A Specification
(Estimated data, production version)

Power Plant: Two 8,985 lb st (4,075 kgp) General Electric TF34-GE-100 turbofans. Internal fuel capacity, 10,650 lb (4,830 kg).

Performance: Max speed, 449 mph (723 km/h) at sea level (clean) and 443 mph (713 km/h) at 5,000 ft (1,525 m) with six Mk-82 bombs; cruising speed, up to 357 mph (575 km/h); stabilized dive speed, at 35,125-lb (15,965-kg) weight at 45-deg angle below 8,000 ft (2,438 m), 300 mph (483 km/h); initial rate of climb, 6,000 ft/min (30,9 m/sec) at basic design weight; take-off roll at max take-off weight, 3,660 ft (1,116 m) and at forward airstrip weight, 1,050 ft (320 m); landing roll at max weight. 2,600 ft (792 m); combat radius, close-air support mission, 288 mi (463 km) with 2-hr loiter; combat radius, deep strike mission, 620 mi (998 km); escort mission radius, 297 mi (478 km) at 173 mph (278 km/h); reconnaissance mission radius, 473 mi (761 km) at 230 mph (370 km/h); ferry range, 2,650 mis (4,265 km).

Weights: Empty, 18,783 lb (8,520 kg) basic weight, clean, 25,470 lb (11,533 kg); basic design weight* 28,650 lb (12,995 kg); forward air-strip weight (with four MK-82 bombs), 29,237 lb (13,262 kg); max take-off weight, †44,547 lb (20,205 kg).

Dimensions: Span, 55 ft 0 in (16,76 m); length, 52 ft 7½ in (16,03 m); height, 14 ft 8½ in (4,48 m); wheelbase, 18 ft 4 in (5,59 m); undercarriage track, 17 ft 6 in (5,34 m); wing area, 488 sq ft (45,34 m^2); aspect ratio, 6.2:1; incidence, -1 deg; dihedral (outer panels only), 7 deg.

Armament: One built-in 30-mm GAU-8/A multi-barrel gun with 1,350 rounds; eleven external pylons for maximum external load of 16,000 lb (7,257 kg).

*The basic design weight includes six 500-lb (227-kg) MK-82 bombs, 759 rounds of 30-mm ammunition and sufficient fuel to fly 345 mi (555 km) plus 20 min reserve.

†Prototype gross weight, 45,202 lb (20, 510 kg).

UTTAS : BIRTH OF A HUEY REPLACEMENT

by the Technical Editor

Reprinted by Permission
FLIGHT INTERNATIONAL
(August 21, 1976)

NOTE: In December, the Army selected Sikorsky's YUH-60A as winner of the UTTAS trials. —JSAY

Two contenders, Boeing Vertol's YUH-61A and Sikorsky's YUH-60A, are now engaged in the Army's Uttas (Utility Tactical Transport Aircraft System) competition. The trials began with the acceptance of three prototypes from each company on March 20. The winner will become the Army's first purpose-built squad-carrying helicopter and, with a potential 1,100 aircraft just for home consumption, can hardly fail to become one of the world's top-selling helicopters, as the UH-1 did before it.

The Huey was built in response to a 1955 specification for a casualty-evacuation and training helicopter, and only later was it found to be suitable for a large number of other jobs. Over the years it was continuously and successfully updated to cope with new demands, but the Army now says that the time has come to seek a replacement.

The Huey is an old design, and technical advances since the mid-1950s now make it more economical to introduce a new aircraft than to continue improving an old one. The Huey's all-up weight has gone up from 6,600 lb to 9,500 lb and, says the Army, it lacks the performance that a new helicopter could give. For example, the UH-1's lifting ability is badly affected by the altitude and temperature combinations characterising the tropical areas of the world, in which conflicts seem increasingly likely to arise.

New combat techniques devised for increased efficiency or improved safety have brought further penalties in their train. In Vietnam the Army was obliged to weigh down its helicopters with quantities of armour to safeguard crew, troops and critical equipment, together with air-gunners to provide suppressive fire. Terrain-following, or map-of-the-Earth flying as the Americans call it, forcibly reminded the Army of the helicopter's traditional stability shortcomings, particularly during low-g manoeuvres.

Again, the alarmingly high helicopter losses due to anti-aircraft fire in Vietnam (even when they were provided with a modicum of self-defence) obliged the Army to look much more deeply at vulnerability. As a result, Uttas and the AAH (Armed Attack Helicopter) are armoured to withstand relatively heavy battle damage and still get home. The Fairchild A-10 fixed-wing close-support bomber will spend most of its time near the ground in a similar abrasive environment, and also comes in for this "hardening" process.

Cost of ownership, too, is a factor which was hardly recognised when the Huey began its career. Labour costs are now the single most expensive item in the Army's budget, and there are heavy pressures to reduce the cost of keeping helicopters in the air. They are intrinsically expensive devices, and the service has about 10,000 of them.

Modern structural techniques take account both of vulnerability and cost of ownership. Airframes are now designed to be fail safe, rather than for the specific life-time envisaged by the earlier practice of calling a finite safe life. Improved diagnostic methods enable the health of airframe, engines and equipment to be rapidly checked by relatively unskilled people. One of the biggest policy changes is the widescale stipulation of two engines.

A list of what the US Army considers to be its principal requirements shows clearly what the priorities are: Uttas must carry a fully equipped infantry squad of 11 men, plus gunner, pilot and co-pilot, for 2-3 hr at 4,000 ft and 95°F; cruise at a minimum speed of 145 kt; climb vertically at 450 ft/min-550 ft/min at maximum weight; show a mean time between failure of not less than four flying hours and a mean time between removal of dynamic components (engines, transmissions and rotors) of not less than 1,500 hr; require a periodic inspection level of not less than 300 hr; be able to complete a mission after being hit anywhere on the airframe by a 7.62mm bullet, or to survive a hit by a 32mm shell on certain critical areas; and show a substantial improvement in crashworthiness compared with earlier helicopters. Rapid air-transportability is of course essential.

The Army investigated a Huey follow-on during the late 1960s. In January 1971 it issued RFPs (requests for proposals) to GE and P&W for the powerplant, following six months later with competitive invitations to the airframe industry to produce a helicopter for introduction in 1977-1978. A market for 3,000-4,000 helicopters was foreseen.

On August 30, 1972, the Army asked Boeing-Vertol and Sikorsky to build three competing prototypes each, plus ground-test specimens. The manufacturers handed over their helicopters to the Army earlier this year for development and operational testing.

The results of the tests will be discussed by that service's Operational Test and Evaluation Agency and the winner will be decided by the Army next January. The successful company will then be authorised to begin low-rate production in readiness for Phase III, a 1,100 hr flight programme scheduled for 1978 and involving development and operational testing on five aircraft. These tests will confirm the previous evaluation and check any modifications that may have been introduced. Such a protracted process is unusual, but it indicates the degree of importance attached to the achievement of a full service-readiness. There is no rush to introduce Uttas, since the US Army still has several thousand Hueys on strength.

Uttas is one of the first US programmes to mature under the design-to-cost philosophy. During the design stage, the various teams within the two companies were given cost targets by their project managers, as well as the usual weight, space, power and functional requirements. The firms also set up special teams to review individual programme costs. They identified components and equipment for which cost/performance or cost/weight estimates could be improved, and they also revealed the benefits of designing for the smallest number of parts, for simple shapes, and for maximum interchangeability of parts (e.g. identical components for the left and

right stabilisers). On the Boeing aircraft, the adoption of design-to-cost techniques is estimated to have cut the cost of an airframe by $50,000.

Numerically Uttas will be one of the largest US programmes for many years. With 1,100 aircraft at stake both companies are putting everything they have into winning the contract. The programme is worth 3,000 jobs over ten years, and a revenue of some $2½ billion. Vertol would probably be hardest hit by loss of the contract, because its work is now mainly refurbishment of CH-47 Chinooks. A potentially large project being prototyped by Boeing for the Army, the HLH (Heavy Lift Helicopter) was shelved last year.

Uttas Comparison

US Army constraints and guidelines laid down for its new utility and squad-carrying helicopter resulted in two very similar-looking aircraft. Both have relatively highly loaded rotors by comparison with that of the Huey, and need proportionately more power to drive them. The UH-1 has a 48 ft-diameter rotor and weighs 9,000 lb, while the Boeing Uttas has a 49 ft rotor and grosses at 18,700 lb; the Sikorsky design is slightly larger, with a 53 ft rotor, and weighs a maximum of 22,000 lb. The commercial versions of both aircraft are temporarily on back-burners while the companies devote their energies to the all-important military fly-off.

Boeing Vertol YUH-61A

The Philadelphia company concluded from its analysis of the Army's requirements that the UH-1 follow-on should be as compact as possible to facilitate air-lifting, operation from restricted spaces and to make it more difficult to hit. In addition, it was obvious that substantially better flying qualities were needed for the contour-following type of operations, involving low- or even negative-g manoeuvres essential for survival in what the Americans call a "mid-intensity" war.

The glass-fibre hingeless rotor, available to Boeing Vertol as a result of its

involvement with MBB's B0105 helicopter, was known to be reliable and to confer good manoeuvrability. Because it was stiffer than a conventional articulated rotor, movement in the flapping plane was smaller and the rotor head could be brought nearer the fuselage. This in turn allowed a desirable increase in cabin height (overall height was a limiting factor if extensive disassembly of the rotor system for transport by C-130 or C-141 was to be avoided).

The principal technical innovations are embodied in the hingeless rotor and its glass-fibre blades, the flight-control system, and arrangements to "harden" the aircraft and its equipment against anti-aircraft fire.

Vertol's decision to use glass-fibre for the blades was based on what it considered the fundamental drawbacks of metal blades: the ever-present possibility of corrosion affecting spars and aluminium honeycomb, sensitivity to small defects, rapid failure propagation, and the difficulty of incorporating twist and changes of aerofoil. By contrast, the firm says that glass-fibre does not corrode and that deterioration due to climatic effects is insignificant. Furthermore, defects are far less important, failures propagate slowly, with the change of blade stiffness acting as a timely warning to the crew, and the material lends itself to the production of optimum blade profiles.

The 10 ft tail rotor is also of glass-fibre and is sized to provide a 15°/sec yaw rate while flying sideways at 35 kt at the maximum power and altitude combinations. The high-aspect-ratio vertical fin is cambered to provide side-thrust against rotor torque so that directional stability is retained if the tail rotor is severed from the aircraft, provided that the aircraft has some forward speed.

The flight-control system transmits crew demands or signals from the stability control augmentation system to the rotor via dual-redundant hydro-mechanical links. The low-authority dual-channel electrical-augmentation system controls the horizontal stabiliser over a range of 45° in the hover to neutral in forward flight to minimise bending moments in the shaft, to make the helicopter feel "stiffer" during transition, and to align the fueselage at the most efficient attitude during flight.

The arrangements for increased survivability are apparent throughout the aircraft and take two main forms: wide separation of such dual-redundant equipment as the engine, fuel and control circuits, and increased protection for single critical items of equipment and for regions in control circuits which have to be run together, as in the flight-deck.

Sikorsky YUH-60A

Sikorsky's analysis of the requirements closely paralleled Vertol's conclusions, helped along by the constraints on size imposed by the need for air-portability. The company chose to sacrifice some endurance in order to secure a top speed of 165 kt, near the top of the 145 kt-175 kt band specified by the Army. But the YUH-60A can carry sufficient fuel to meet the required 2-3 hr endurance at top weight by reducing the payload and taking advantage of overload concessions. Sikorsky recognises that this is a compromise, but says that it has emphasised the performance aspect that contributes most to the helicopter's effectiveness.

Among methods to improve reliability and maintenance, the company has eliminated many of the earlier trouble-makers, such as some seals, short-life blades and lubrication requirements, and has simplified the remainder and extended-inspection intervals.

Sikorsky also chose glass-fibre for its blades, but uses this material as a skin over corrosion-proof titanium spars. The blades have swept tips to minimise compressibility effects, are twisted to improve the lift distribution, and have a cambered aerofoil. They incorporate the company's patented BIM (blade inspection method) system for checking the absence of cracks. In contrast with the YUH-61A, Sikorsky's Huey successor has a fully articulated, through aerodynamically clean, main rotor hub to which the blades are attached by means of elastomeric bearings. The bearings are entirely solid, and pitch changes are made by twisting the material. The company says that this technique cuts maintenance time by up to 60 per cent.

The YUH-60A also has an all-flying horizontal stabiliser to improve manoeuvrability in the pitching plane. The tail rotor has 87 per cent fewer parts than earlier designs and is inclined from the vertical to augment lift. The fin provides asymmetric side force (as it does on the Vertol design) so that landings without the tail rotor can be made at speeds down to 40 kt. The stability augmentation system is based on research done by the US Army and employs fluid sensors to pick up disturbances. There are no moving parts, and an 8,000 hr mean time between removals is guaranteed.

The ability to withstand crashes better than previous helicopters was demonstrated convincingly when a YUH-60A spun into wooded country in Long Island after a tail-rotor drive failure (caused, incidentally, when the ground crew forgot to remove a small ladder in the tail boom). Although the aircraft came down through young oak trees, the two pilots were not hurt and the main transmission, virtually undamaged, was used in the official Army 200 hr tests.

U.S. ARMY'S NEW GUNSHIP IS 20 YEARS ABORNIN'

by YEARBOOK Staff

The U. S. Army announced on December 10 that it had chosen the Hughes AH-64 as its new advanced attack helicopter (AAH) to join its operational forces about seven years from now.

The Army's search for its ideal attack helicopter has taken so long that a person who joined the Army when the search began could retire on half pay by the time the AH-64s begin arriving in operational units.

It was in 1963 that the Army first sought proposals for a helicopter gunship. Eventually that competition was won by Lockheed with the AH-56 Cheyenne. But in 1972 technical problems and escalating costs prompted the Army to cancel the Cheyenne program and start all over.

Using money saved by the Cheyenne cancellation, it immediately invited bids on a new gunship to cost not more than $1.6 million in production quantities in terms of 1972 dollars. In June 1973 the Army chose Bell and Hughes to build two flying prototypes of each of their designs, the YAH-63 and YAH-64.

Last spring the prototypes were turned over to the Army for a five-month flyoff competition, which Hughes won.

There are still more checkpoints to be surmounted before Hughes wins a production contract for the two-place AH-64. The December 10 award calls only for building three more prototypes to continue engineering development.

So far, the competing designs were tested for their airworthiness under conditions the Army envisions for a future war. Now the Hughes prototypes will be fitted with all the systems that make up a modern attack aircraft—weapons fire control, all-weather navigation, night vision devices, countermeasures gear, and communications.

This will take 50 months. If the Army—and the Congress—decide the helicopter is all it should be, Hughes can expect a production contract in 1980 or '81.

It will be another two years before the first production model comes off the assembly line. Six months after that the first combat versions should begin arriving in operational units.

The Army plans to buy a total of 536, with production extending late into the 1980's.

Things ran a little faster in World War II. North American, for instance, rolled out the first P-51 Mustang only 117 days after receiving a British contract for a fighter plane, and build almost 15,000 Mustangs from 1940 to 1945.

True, the Army is in no great hurry to buy its new line of attack choppers. To meet the demands of the Vietnam war it had called on Bell to come up with a gunship version of the UH-1 Huey assault helicopter.

The result of that quick and dirty project was the very effective AH-1G Hueycobra, followed by the twin-engined AH-1J for the Marine Corps and the AH-1S equipped with TOW wire-guided antitank missiles. By the time the Hueycobra production line shuts down in 1979, Bell will have produced about 700 gunships.

It must have come as a surprise, then, to the people of Fort Worth, Texas, where Bell has been turning out various models of the Huey for 20 years, that Hughes and not Bell should have won the prize gunship contract.

It was perhaps less surprising to Army pilots who may have to fly gunships in a European-style conflict. They have a high regard for Hughes' first venture into military helicopters, the OH-6A Cayuse light observation helicopter, because of its maneuverability.

The Army found through costly experience in Vietnam that if helicopters are to survive in combat they must fly the "nap of the earth," at or below treetop level, taking advantage of terrain features to screen themselves as long as possible from enemy guns.

These techniques will be even more essential in a European war, where air crews will face enemy forces equipped with heat-seeking missiles that proved devastatingly effective against Israeli helicopters in the 1973 war.

To save weight, armor is used only to protect the two-man crew. Rotor blades, engines, fuel tanks, and control systems are designed to take hits without crippling damage.

The AH-64 also incorporates features to limit crash damage, such as breakaway fuel cells and fuel lines that automatically seal to avoid fuel leakage and post-crash fires. Landing skids are designed to crush in a hard landing.

Besides being able to survive, an attack helicopter must be able to inflict punishment on the enemy, primarily tanks and other armored vehicles, and the AH-64 will carry a particularly lethal sting.

It employs three types of weapons—a 30 mm "chain" gun capable of firing at the rate of 700 rounds per minute, a battery of 76 2.75-inch rockets, and a new laser-guided Hellfire missile, still in development, that will replace the wire-guided TOW missile in the Hueycobra.

In combat configuration the AH-56 will weigh 17,000 pounds. Tandem seating, with the pilot behind and slightly above the copilot-gunner, gives both crew members maximum visibility to either side.

Diameter of the four-bladed rotor is 48 feet, and overall length is 57 feet. Its dimensions are limited by a requirement that two AH-64s can be airlifted in an Air Force C-141 transport.

Stubby wings carry its weapons and augment lift in high-speed flight. Flaps on the wing trailing edge also help support high-G turns.

Powered by a pair of General Electric T700 turboshaft engines producing 1,500 shaft horsepower each, the AH-64 will have a top speed of 190 miles per hour, maximum combat range of 300 miles, or flight endurance of just under three hours.

YC-14/YC-15

Flight testing of the Boeing YC-14 has begun. The Boeing entry in the Advanced Medium STOL Transport (AMST) competition uses upper-surface-blowing to achieve the ability to fly in and out of extremely short fields.

Upper surface blowing is based on the Coanda effect, employing the concept that high speed air will follow the surface of a wing and its accompanying flap system if the curvature is properly designed. Engine exhaust of the YC-14 creates vertical, powered lift.

Externally blown flaps form a powered-lift system for the McDonnell Douglas entry in the AMST competition, the YC-15.

The large, double-segmented, titanium flaps extend completely through the engine exhaust and deflect engine thrust when deployed. The wide spacing between the segments of the flaps accelerates the flow of the large portion of the jet exhaust that goes through the flap structure. This accelerated air is turned downward by the aerodynamic shapes of the flap segments, and (through the Coanda effect) creates increased lift.

With flaps fully extended, the YC-15 will gain about 55 per cent of its lift from the wing structure, 20 per cent from the deflected exhaust, and 25 per cent from the Coanda flow.

CANADA'S AURORA WILL BE WORLD'S MOST ADVANCED PATROL AIRCRAFT

Reprinted by Permission
CANADIAN WINGS (August-September, 1976)

Canada's 18 Aurora long-range maritime patrol aircraft, ordered by the government from Lockheed-California Company, July 21, 1976, will be the most advanced airplanes ever developed for the role.

R.R. Heppe, the company's vice president and general manager for government programs, said the Aurora includes more than three decades of patrol aircraft expertise developed by the U.S. Navy and Lockheed.

Designated the CP-140 internationally, the Aurora is the "logical extension of the computer integrated systems technology developed for the land-based P-3C Orion and the carrier based S-3A Viking anti-submarine warfare aircraft, both of which are now serving with the U.S. Navy," Heppe said.

Among the advanced features of the Aurora, Heppe noted, are an underwater acoustic data processing system to detect submerged submarines that is superior to any in the world today; a forward looking infrared sensor system that makes night time identification of surface vessels as certain as it is in daylight; a search radar that can detect a periscope in medium to heavy seas; and a compact general purpose digital computer that has the capacity of many computers used in industry.

Another outstanding feature of the Aurora is the multiple navigation systems which will provide the most accurate speed, heading and positional information possible on an airborne platform, Heppe said.

"In addition to the new twin inertial navigation systems and the OMEGA system, the Aurora will carry a sonobuoy reference systsm (SRS) developed for the S-3A and a new Doppler radar system," he explained.

A feature unique to the Aurora, Heppe said, is the full provision for a bomb-bay canister which can be fitted with a mapping camera, a side looking radar or an array of remote sensors that will give the aircraft the capability to conduct a variety of civil missions. This equipment, he explained, will enable Aurora's crews to fly ice surveillance patrol, police the country's fisheries, detect air and water pollution, identify natural resources and monitor their exploitation.

The first Aurora will be delivered to the Canadian Forces in May, 1980, and the 18th in March, 1981.

AAF'S FLYING ARTILLERY THE 75-MM BAKER TWO-FIVE

by Lt. Col. Jim Beavers, USAF (Ret.)
Cartoons by Bob Stevens

Reprinted by Permission
AIR FORCE MAGAZINE
(April 1976)

Early in World War II, someone—certainly not an artilleryman—decided to put a 75-mm cannon in the nose of a B-25. One of the first flying cannoneers describes his experiences, hair-raising and hilarious, as he and his squadron mates stooged their unaimable airborne artillery around the Mediterranean looking for a mission.

The B-25G evoked a variety of expressions—mainly of awe—when it first appeared at Columbia Army Air Base, S.C., in the early spring of 1943. Small wonder: In its funny-shaped nose it carried a 75-mm cannon, surely one of the biggest pieces of armament ever mounted on an airframe.

Most expressions were in the form of slack-jawed questions:

"Who fires that thing?"

"Holy mackerel! Doesn't the airplane almost stop when you fire it?"

"Who loads it?"

"What's it sound like in the airplane when the cannon fires?"

And most relevant of all: *"Can you hit anything with it?"* Somebody always had to get that one in.

In sequence, the answers to the foregoing were: the pilot; no; whoever's in the navigator's compartment; loud; and occasionally.

Those of us who were volunteered to train in the airplane certainly thought at the time we could hit things with the cannon. We flew practice gunnery missions in which we shot an occasional hole in a large, nonhostile wooden target at

As we flew down the valley, we could match a billboard reading to a gunsight setting.

point-blank range while skimming over an uninhabited section of Myrtle Beach. There were annoying times on those flights when everything said that the shell should have gone into the target but didn't. Maybe we had been jolted by a thermal during the gunnery run or distracted at the last minute—something more practice and experience would explain.

After a year of combat in the "G," as it came to be called, I was still asked those same questions since the airplane remained an oddity. With real experience behind me, the question of our ability to hit anything irritated me because it was simultaneously too difficult and too easy to answer, but more because it should have been asked before the model was ever built.

Depending on how I felt from time to time, I may have replied "No," and let it go at that. For all practical purposes, that was an accurate answer. Or I may have said, "Yes, under the right circumstances." However, that was not only evasive but open-ended. It was an invitation to ask what were the right circumstances, and the answer to that was a can of worms. If pressed about it, I had to say, "On the ground, in a secure area with the parking brakes set and the muzzle pressed firmly against the target."

Looking for a Mission

My crew and six others were the first to take the G to combat in May of 1943. Another small contingent left close behind us, and by the time we had flown the South Atlantic and collected ourselves at Souk el Arba in Tunisia, we numbered about a baker's dozen. We were assigned to a somewhat bewildered 47th Bomb Wing (M) that normally stocked conventional B-25s, and were dubbed the "47th Gun Squadron."

We didn't know specifically what it was we were supposed to do. And despite a certain amount of officious bustling around our airplanes, it quickly became clear that the staff of the 47th Bomb Wing didn't know either. Which gave rise to the question: What was the G *for?*

There should have been clues in its configuration. At the outset, the G was really a model C with its nose chopped off, eliminating the bombardier's compartment. The 75-mm cannon was installed in what had been the bombardier's crawlway, and the nose was reconstructed around it and two fixed .50-caliber machine guns.

Losing the nose meant losing part of the bombing system. Bombardier, bombsight, bomb bay door control, and intervalometer were lost to the hacksaw.

Somebody decided that the pilot would absorb what remained of the bombardier's functions. The bomb bay door control was moved into the cockpit, as was the intervalometer. The pilot's control wheel was ringed with buttons—one for bomb release, one for cannon firing, one for machine guns, one for radio and interphone operation, and, in a few cases, one for photography.

The arrangement gave the G pilot pretty much the same chores an A-20 pilot had. Since the latter managed without a copilot, equal justice required removing the copilot's seat from the G. His control wheel and rudder pedals remained, but no seat.

It follows, of course, that we had copilots, and I for one was glad we did, though they weren't much help on the long flight to Africa. We had to improvise seats for them, and the best I was able to rig up was an accumulation of luggage that left my assistant roughly at eye level with the parking brake handle. He rode all the way to Africa with his knees up

... that left my assistant roughly level with the parking brake.

around his ears. When I required relief, he reached up in simian fashion to the control wheel and steered chiefly by instinct. I returned to the cockpit occasionally to find us wandering casually around the South Atlantic.

Other modifications based on combat experience were made to the airplane soon after we joined the 47th. They consisted of dropping useless equipment like the lower turret and adding good things like waist and tail guns and seat armor, and two more .50s in the nose. Oh, and a copilot's seat.

The configuration resulting from these alterations was a gun platform with superficially impressive firepower. However, one critical deficiency was never overcome. With the equipment available at the time, there was no way to estimate range for a cannon moving at better than 200 mph and accelerating, and hence no way to aim it at any distance from the target. So what was the airplane for?

We, of course, had our share of rumors at Columbia. The straight word was that the G was designed for attacks against enemy shipping, that its cannon was intended to suppress antiaircraft fire during low-level skip-bombing runs. There were other straight words, but this one dominated.

Skip-bombing had been done with conventional B-25s, armed with one flexible and sometimes one fixed forward-firing .50, during Rommel's evacuation from Cape Bon. They were reportedly real hair-raisers. It was necessary to fly directly over the target vessel in order to skip-bomb, and since the B-25 was extremely vulnerable in making that transit, relative success depended on whether the ship was being defended by antiaircraft fire. The assumption was that the G's cannon would constitute a great equalizer. So much for rumors.

If at First . . .

My crew drew a bye for the first G combat mission. Unaccountably, the target was a German radar station on Sardinia, a selection that seemed to suggest uncertainty in high places about the airplane's intended purpose. The mission was not exactly a turning point in the war. The Gs drew a shower of small-arms fire, and, except for the language barrier, the station might have provided radar vectors home.

That sort of thing was first among many that the G was apparently not designed to do. A significant precedent was set on that first mission, though. It was flown in a standard four-ship fighter formation that became the norm for us.

A comparative history of the G's use in the ETO and in the Pacific seems to point at that tactic as a basic error on our part, which became pretty much set in concrete as the only way to fly. As a result, we never discovered the available massed firepower of larger formations. Our combat tactics also evolved largely from the fighter formation, and they diluted even the collective firepower of the four-ship flight.

After Sardinia, it was decided that we would fly conventional missions at medium altitude while people gave the G some more thought, and we were distributed among the squadrons of the 321st Bomb Group. Doing routine bombing was a simple matter of presetting the intervalometer and dropping on the lead ship. Then command of the 47th Bomb Wing changed hands, and the new incumbent perceived that we were not exploiting the airplane. The next experiment was low-level operations against shipping.

We had already flown several generally meaningless missions of that kind when we weren't needed for medium-altitude operations—long, tedious drills that covered hundreds of miles of open Mediterranean where, it turned out, enemy shipping was least likely to be found at that stage of the war. But one of those missions had gone right up the Italian coastline, and in a small harbor had encountered more floating armed hardware than it could handle. So that was where the action was.

The wing commander decided to see for himself if the G could be used effectively against shipping. As it happened, he gave us our first opportunity to put to the test an awful lot of theory, some of it running all the way back to the drawing board. It was our first encounter with a surface vessel of any size, alone and—it turned out—unarmed.

We took off in the early morning. The wing commander, a brigadier general, was flying copilot in the lead ship. Other than that, it was a flawless Mediterranean day. Flying the four-ship formation we had adopted, we angled northeast past Sicily, then east to intercept the coast of Italy. We turned north about a mile offshore and began a search for shipping. Within a few minutes, we stumbled onto an old tanker.

The flight commander signaled echelon left, and we complied briskly. After a moment's hesitation that surely included second thoughts about all this, he peeled off and thundered down that long and lonely run that would come to dominate our thinking. Individual attacks seemed a natural outgrowth of the fighter formation, and they too became standard. With the benefit of 20-20 hindsight, I can say with some authority that they were another fundamental mistake.

I was last in line. I rolled out of my turn, flipped up the cannon and machine gun safing switches, set the gunsight at some value, and began firing. After these attacks, the ship was not visibly damaged by anything other than the ravages of time.

If I can reconstruct this accurately, my airplane was moving toward the tanker at about 260 miles per hour, which is about 380 feet per second, which is about 127 yards per second. Which is relevant only because the hand-adjusted gunsight was calibrated in yards. The bad thing was that it was calibrated in thousands of yards, at one click per thousand.

And therein lay the G's fatal flaw as an aimable standoff weapon. Setting the gunsight to the nearest approaching thousand yards was sheer guesswork. There were calibrations intermediate to the clicks—ten subdivisions, I believe—and estimating range to the nearest tenth of a thousand yards was guesswork compounded by an order of magnitude. Since the difference between hitting and missing the tanker was at most a matter of twenty yards in slant range, the tanker was never in serious trouble. Like the other pilots, I fired round after round without coming near it.

The general reached the limit of his forebearance during the four uniformly ineffective runs at the tanker. He turned

"Don't fly west, Sir. Fly 272 degrees."

in wrath to the navigator, an apple-cheeked, imperturbable farm boy from Missouri, and demanded a withdrawal course.

The navigator was stripped to the waist and streaming sweat from loading the cannon. He stood calf-deep in expended casings and clutched a provisional next shell at the ready. Without blinking, he said, "Fly west, General."

This struck the general as flippant, and he snarled something to the effect that when he asked for a course, he wanted one a little more precise than a hot-dang cardinal point on the compass.

The rosy-cheeked lieutenant listened, then turned around and returned the shell to its storage rack. He clattered and clanged through the casings to reach his chart, consulted it briefly, and clanged and clattered back to the edge of the flight deck. He tapped the general on the shoulder. "Don't fly west, Sir," he counseled. "Fly 272 degrees."

It was an eminently forgettable day.

Assaulting the Symptoms

Nobody could really fault the general's peevishness, I guess, because an impressive amount of effort had gone into training us, only to have us go to the plate and come away 0 for 4. Much of it had been instigated by the man he replaced. And there had been those Myrtle Beach outings back in the States. What about those?

Realistically, our Stateside training had taught us only a little more than how to fly the airplane, which, in its cannon-carrying configuration, was heavy and not too stable, judging from the number of sandbags lashed into the tail section. That training had familiarized us with the optics, mechanics, and circuitry of the cannon, but that's about all that could be expected in light of the fact that instructors and instructees laid eyes on the first G simultaneously. If the former were less than aggressive about making students press home attacks on the wooden targets along the beach, that was understandable too. The combination of newly winged pilots, newly configured airplanes, ten feet of altitude at speeds of about 260 mph, and recurring explosions in the navigators compartment as the cannon went off unexpectedly was enough to moisten any IP's armpits.

What little Stateside experience we got with the cannon was limited to use of armor-piercing shells. We didn't know the HE (High-explosive) variety existed until they were handed to us, without enlightening comment, in Africa. Figuring out the difference was an individual problem. The HE shell had a safing pin and a bright aluminum disk in the nose with flat edges on two sides that accommodated a wrench we found with ammunition. Some mysterious little numbers around the disk were intended to tell us something, I'm sure, because after the safing pin was removed we could turn the disk to align an arrow with any of them.

We had one G replacement pilot who reported that he was drawing heavy, accurate flak during every training mission against an old beached hulk we sometimes used as a practice target. It evolved that he was using HE shells and was "winding up the fuze to make it go" before heaving each shell into the cannon. Without realizing it, he was cutting the fuze to its minimum setting. When fired, the shell detonated right before his eyes. Putting two and two together, he concluded that an unseen gun battery was matching him shot for shot.

While I wouldn't categorize that man as your basic Rhodes Scholar, it's only fair to reaffirm that nobody gave us any instructions on the HE shell. That little omission typified our training as artillerymen when we brought the G to combat.

Early in our tour, before anybody fully recognized the enormity of the rangefinding problem, we flew training missions predicated on the assumption that it could be learned. The scruffy mountains south of our Tunisian base contained a little horseshoe valley in which the previous wing commander had built a series of billboards approaching a monster bull's-eye right in the toe. They announced the distance in thousands of yards to the target, and as we flew down the valley we could match a billboard reading to a gunsight setting, fire a round and observe the results. Our collective marksmanship remained poor despite being told when to fire by a passing roadsign.

In frustration, the wing commander finally gave us permission to experiment and innovate. I think I was the one who suggested that, since range was the apparently insurmountable difficulty, we might try to get around it by dive-gunnery. The thought here was that a vertical attack would eliminate the range question, since target and ocean would be essentially the same distance from the airplane at any point (what this proved is a little obscure, in retrospect). Diving straight down on the target would reduce the gunnery problem, it was argued, to a two-dimensional matter of azimuth and elevation. That it had never been anything else was lost in the semantics somewhere.

The next day, several of us flew out to sea to try dive-gunnery, using an uninhabitated rock as a target.

The obvious had already occurred to us: (1) a truly vertical attack was not feasible in a B-25, and (2) anything approaching a vertical attack would have to be conducted with engines idled and landing flaps full down. What should have been obvious was not. Throttling back the engines to idle at 12,000 feet in cold, moist air, dropping like a safe with the door closed to less than 500 feet, then opening the throttles for a fast getaway wasn't feasible either.

The tests showed that landing flaps did not serve as dive brakes. Before I had lost 500 feet I had exceeded the allowable flap-down speed. To hang onto the flaps, I put them up and promptly exceeded the maximum allowable speed in any configuration. This proved to be a blessing in disguise as I eased the airplane out of the dive and found both engines dying from carburetor ice. Trying to get them going again during the long run-out while the airspeed bled off to believable numbers, my copilot and I established new time records for four-handed exercises.

That was it for dive-gunnery.

We also briefly examined formation gunnery, on the theory that in a salvo many errors might average into a hit. What we got was a lot of average errors. To my knowledge, only four of us made a brief stab at formation gunnery and quickly dismissed it. Ironically, it was close to a tactic that proved successful in the Pacific, even if not quite the same.

Meanwhile, in the Pacific

New Gs and crews began pouring in from the States as if the airplane were a godsend. Soon there were so many that it was decided to reequip the 310th Bomb Group with them. To those of us who had brought the originals over, it meant we were now flight commanders, for lack of anybody more experienced. My squadron was detached soon thereafter and sent to the Libyan coast for operations with the RAF against German shipping in the Aegean Sea. The other three squadrons of the 310th remained in Tunisia in a quasi-training status, flying an occasional four-ship combat mission that still doggedly involved aiming the cannon, and without notable success.

On the other side of the world, the Fifth Air Force was taking a much more pragmatic view of the G. It concluded early that aiming the cannon was a waste of time. Depending on target size, it put six, nine or twelve Gs in a line abreast and used them as a covering force for strafing A-20s and other B-25s with forward-firing .50-caliber machine guns that were used as gunships. The G pilots were briefed not to aim at individual targets but to fire as many rounds as fast as possible. It was not unusual for each G to get off eighteen to twenty rounds in a single run. The resulting barrage was intended to do one thing—suppress defenses for the strafers to follow. It worked.

Gen. Richard H. Ellis, now Commander in Chief, US Air Forces in Europe, and then one of the A-20 or B-25 pilots (he flew both) who came in for the kill behind the Gs, recalls that it was very comforting to follow them into a heavily defended complex such as an airfield. The barrage tactic was used successfully against enemy shipping and even to soften up beachheads.

Why didn't we think of that—the barrage tactic? There are several answers. General Ellis suggests one. Targets in the Pacific were different from those in Europe, he points out, and were such that low-level attack was a major Fifth Air Force tactic throughout the war. It was uncommon in the Mediterranean.

There was something else. All our early experimentation with the airplane had the objective of finding a way to *aim* the cannon effectively. It had nothing to do with tactics as such but with technique, and involved a sort of naive, GI faith. The airplane was issued to us with a cannon and a gunsight for it calibrated in thousands of yards. To our uninstructed minds, it followed that it was possible to hit a target thousands of yards away. Since that reasoning precluded consideration of the inadequacies of a rapidly moving and not always stable airplane as a gun platform, the problem was to discover what we pilots were doing wrong when we missed. One high-ranking officer concluded it was a disciplinary matter. He proposed that pilots be required to sign statements of charges for shells that went astray. That, too, may have been a major difference between ourselves and the Fifth Air Force.

In early 1944, my squadron was recalled from Libya to a new base on Corsica, where we were to finish out our tours doing conventional bombing. It had been a full year since we first began flying the G, and few of us ever flew it again.

The B-25 remained in the Air Force inventory for years after World War II for administrative and pilot proficiency uses. It was a solid, stable, dependable old bird that could be trusted in fair weather and foul. Two versions of the airplane that were junked immediately at war's end, however, were the G and its successor, the J. It seemed nobody could find a

When fired, the shell detonated right before his eyes!

peacetime application for an airborne 75-mm cannon. I don't find that surprising. Nobody in my theater of operations could figure out what to do with one in wartime.

Undaunted by his combat tour in the B-25G, the author, Jim Beavers, decided to go for twenty. After earning a degree in physics from the University of North Carolina in 1948, he spent most of his Air Force career in R&D work, specializing in nuclear weapons applications. At the time of his retirement in 1963, he was serving on the Air Staff in War Plans. He is now president of a small company in Winter Park, Florida.

THE NEW CADETS

Reprinted by Permission
AIRMAN MAGAZINE (December 1976)

by Capt. John B. Taylor
Photos by SSgt. Herman Kokojan and TSgt. Paul J. Harrington, AAVS

All cadets arriving to begin life at the U.S. Air Force Academy pass through a large portal emblazoned with the famous quotation, "Bring Me Men." Last June, the class of 1980 marched through that gateway and brought with it some women.

There were some doubts, chiefly among sophomores and juniors, the Academy's third- and second-classmen. But, overall, the tradition of the all-male military academy as one of the last bastions of male dominance fell with hardly a whimper.

That came as no surprise to Academy officials who, since 1972, had been quietly preparing for the inevitable admission of women. When the first female cadets arrived at the gates of the Colorado Springs facility, they found no obstacles waiting, only a welcome mat. The age-old barriers had already been brought down, brick by brick, from the inside.

"We were ready," explained Col. James P. McCarthy, Vice Commandant of Cadets. "The same day President Ford signed the act [admitting women] into law, we sent contracts for women's uniforms to the manufacturers and began our recruiting campaign. There was absolutely no foot-dragging.

"Our superintendent, Lt. Gen. James R. Allen, set the tone: we were going to have the best women cadets we could possibly get by planning carefully, and well in advance."

To learn from the experiences of others, a planning group formed by Colonel McCarthy visited Air Force "coed" training facilities such as the Air Force Military Training Center and Officer Training School at Lackland AFB, Tex., as well as the Merchant Marine Academy, which has had women cadets for two years. Group members also toured civilian institutions that had recently made the transition from all-male to coeducational student bodies.

"The study convinced us that women would be able to do the things already expected of a cadet," said McCarthy. "So we adopted the philosophy that men and women would go through a common training program together. That's what you see here today. Not two separate academies; not even two separate programs."

Thus, women in the class of 1980 were integrated directly into existing cadet squadrons. They went through basic cadet training with the men, ran with them, dug ditches with them. Now, as freshmen, or "doolies," they are taking part in all squadron activities, including intramural athletics.

Academy confidence that such a complete integration would work grew from a training program conducted last winter for women destined to become Air Training Officers, or ATOs. The ATOs gave the program a trial run.

BRING ME MEN

The ATO concept was borrowed from the Academy's beginnings in the mid-Fifties. Since there were no upperclassmen to train that first class, junior officers were brought in for the role. Because it worked well then, Academy officials decided to bring aboard about a dozen women lieutenants to serve as role models for the first women cadets.

To prepare them for the role, the lieutenants were put through a condensed version of the summer-long Basic Cadet Training Program, or BCT. Their instructors were primarily cadets.

"Once they got over the fact we were officers and girls, too," recalled Lt. Terry Walter, "they realized all they were supposed to do was train us as cadets. And they did it. They yelled at us, braced us against the walls, and made us 'know knowledge' [recite information by rote], just like we were basic cadets."

The ATOs even went up to the Jacks Valley camp for the tough physical conditioning program that is a part of every freshman's BCT summer. It included such he-man stuff as a combat assault course and training with pugil sticks, which are long poles with padded ends.

"I had never been hit by anyone in my entire life," said Lieutenant Walter. "It's quite a blow the first time another woman clouts you with a pugil stick."

Meanwhile, the cadet wing—still all-male—had been getting lots of mental preparation for the arrival of the women cadets. "We had lectures on the role of women in the Air Force and that sort of thing," said Cadet First Class Grady Booch, one of the ATO's instructors, "but the cadet attitude was pretty much negative until the ATOs came along. We could then actually see women doing the types of things men did, and that they were quite capable and did a good job. That helped a lot."

Though Booch may not have realized it, his comment hit upon part of the reason for the ATO program.

"First of all, we wanted the ATOs to get the experience, knowledge, and skills that an upperclassman would have," said Colonel McCarthy. "They would need that to assist in training the women. But we also hoped that their performance would enhance the acceptance of women by the cadet wing. Our objective was not to change the cadets' minds, but rather to get them to adopt an open-minded attitude toward women at the Academy. The ATOs provided us a chance to do this.

"We involved male cadets in all phases of planning and made sure as many as possible worked on ATO training. The men were really surprised at what the women could do." While the ATO program proved that women could handle the standard program, it also turned up some "bugs."

"One obstacle on the obstacle course involves swinging across a water hazard on a rope," Colonel McCarthy remembered with a smile. "When the ATOs first did it, it was obvious they had never played Tarzan.

"When we analyzed things we discovered that girls don't have the same cultural conditioning about rope-swinging that boys do. They just don't know how to go about it. We had to stop and teach them the techniques involved. After that about half made it across, which is about average for the men, too.

"By the time we were through with the ATOs, we were very confident that we were ready for the class of 1980, and that there would be no surprises."

The ATOs were ready, too, though perhaps not in the role most people assumed they would perform. Actually, male upperclassmen train all cadets, both male and female. The ATOs serve outside

the cadet chain of command.

"We advise the men and make sure the women cadets don't miss out on any 'good deals,'" Lieutenant Walter explained with a laugh. "We want to make sure that living apart from their squadrons doesn't get them out of anything."

But the most important function of the ATOs is to serve as role models. According to Colonel McCarthy, "We want the women cadets to know the Academy is interested in graduating women officers who are competent, professional, and *feminine*. The ATOs show them that they can go through a program long considered masculine and still maintain their femininity."

By June 28, the cadets, ATOs, and Academy staff were ready. Minor changes to the physical plant had been finished—primarily the rehab of several rest rooms in the academic building and on the one floor of Vandenberg Hall that would be home to the first 150 women. All that remained was to admit the class of 1980 and carry on.

BCT for the men and women doolies of '80 was run this past summer by Lt. Col. Chuck Spruill, whose first impression of the women was a vivid one.

"The new cadets had been instructed to report in between 0730 and 1200," he said. "When I got here at 0630 I looked out at the parking lot and there was Miss Joan Olsen sitting on her suitcase, waiting. It was very clear that she wanted to be the first person, not just the first woman, to report in.

"BCT ran 42 days," he reflected. "16 hours a day, every day, with no free time at all. It is a tough, demanding transition between civilian life and the military life-style. The girls did very well. They did things in BCT that the average girl couldn't have done."

"They went out to Jacks Valley and lived in tents with all the dirt and bugs," explained Lieutenant Walter. "They had to run everywhere carrying M-1 rifles, and do a lot of things girls don't normally do—a lot of John Wayne-in-the-trenches stuff."

"It brought tears to my eyes to watch them gut it out," said Lt. Bonnie Stephan, also an ATO and Lieutenant Walter's roommate. "They were always trying to do things as well as the guys, though they were limited by physiological differences."

Only a few concessions were made for the women's physical limitations, primarily in upper body strength. One obstacle on a course had to be adjusted for shorter people, and in some cases the bolt springs on the women's rifles had to be changed from eight pounds of pressure to five. "You could just see the frustration on a small girl's face when she couldn't get that bolt open," Colonel Spruill remembered. "They showed frustration more often than the men. It seemed like they wanted to excel very badly and got upset when they weren't able to fit *their* perception of what it took to measure up."

And there was also the matter of venting frustrations.

"An upperclassman would ask the ATOs how to handle a crying girl," said Lieutenant Walter. "I'd say, 'Handle it the same way you would with a man; tell her it's not acceptable behavior.' We'd explain that it may be culturally normal for a guy to get rid of his frustrations by punching somebody in the nose; however, a basic can't do that to an upperclassman. A girl's normal outlet of crying isn't acceptable here either."

Many of the girls surprised their male counterparts by beating some of them on the obstacle courses, in swimming, and in other competition. Through the summer they also won the admiration of upperclassmen who trained them. "I think over the next four years these women are going to be tough competitors," predicted Colonel Spruill.

While BCT is behind this year's doolies, their fourth-class year is no picnic. They still march between classes with their chins tucked in, greeting each and every upperclassman and walking only on the narrow marble strips inlaid in the pavement of the cadet area. They are still braced in their hallways by upperclassmen and asked to recite "knowledge." Inspections are still the order of the day on many Saturday mornings.

Whether it's in the men's or women's dorm area, it's business as usual. Stern-faced cadets inspect rifles, check for dust on bookshelves, and scrutinize the alignment of clothes in closets and drawers. The only thing different on the women's floor, other than the expanded variety of clothing to be checked, is the sight of somewhat incongruous bottles of Charlie cologne set neatly in medicine cabinets that up till now held only shaving gear and a few bottles of Brut.

It is quite a different matter for the "inspectees." The look on their faces is pure doolie—that is, overwhelmed and a bit scared. One might wonder why a young woman would give up the life of a campus coed to take on the demanding existence of a cadet. The reasons are somewhat surprising.

"I want to fly," said C4C Andrea Michelle Bopp. "It's just something I decided I would do." Bopp figured that the Air Force Academy would be the best place to start an aviation career. She admitted with a smile that she didn't know the Air Force had no women pilots until she applied for admission. But that doesn't phase her. She hadn't known either that there were no women in the Academy. Like most of her women classmates, she has confidence that barriers to flight training will have fallen by the time she graduates.

C4C Chris Ann Reasner came for the challenge. "It sounded really exciting," she said. "I've always been one to do things differently. The Academy has an image of being only for the best. I wanted to know if I could do it."

BCT may have been a turning point for many of the women. "At the beginning it was just a game," Reasner said seriously. "They throw so much at you, you just try to see if you can outdo them by doing things right. You know, beat them at their own game. But now it's different. Now it's more like a way of life."

The women seem aware that not all their fellow cadets are exactly thrilled to have them around, but they accept it philosophically.

"It's more of an adjustment for the men than us," said C4C Gail Benjamin. "We knew what to expect when we came here. But this has always been a male school. Many of them came here because it was a male school. I think I can understand how they feel now.

"Vassar was always a women's school, and when it changed a lot of girls there resented the school going coed," continued Cadet Benjamin, who attended

Vassar for two years before coming to Colorado Springs. "I just think it's up to us to change the men's minds."

Cadet Reasner thinks the jury is still out. "Some of the men who were telling us that women didn't belong seem to be changing their minds a bit. As we get more involved in activities, I think they will be able to accept us." Reasner is doing her part. She volunteered to play intramural lacrosse, a very rough sport, with the men. "I'm going to give it a good try," she said.

Most of the women don't seem too concerned about acceptance by the men. Part of that stems from the fact they live together in a feminine "island," the top-floor dorm area known to the men as "heaven" or "the penthouse."

"It's not so hard to be a woman in a man's environment this way," explained Andrea Bopp, because the women tend to help and support each other. But Chris Reasner may have hit on another answer when she said, "Whatever anybody thinks, we know the *Academy* wants us here."

The young women accept the fact that they are giving up some traditional college activities by attending a military academy. "I just decided that the fraternity parties and all that wasn't as important as what I could get out of a school," said Cadet Reasner. Many doolies gave up the chance to go to school with special friends, too. "When we get mail from home the upperclassmen are always asking, 'Well, did you get kissed off yet?' " laughed Cadet Benjamin.

"I left some special people at home," said Reasner. "I keep getting those letters, but when I left, I left. That was high school, *this* is my life now. Besides, there seems to be quite a selection out here. I'm sure I'll manage."

That last thought points out a problem in consistency for some of the men. "It's fun to watch some of these guys," said Capt. Dave Allen of the Academy's soaring program. "I've noticed that some of the ones who are the most vocal about the women during bull sessions in the hangar are the first to want to take a woman up on doolie orientation flights. I think we're in for an interesting couple of years around here."

The man-woman relationship is causing some confusion within the cadet wing. There is, first of all, the long-standing nonfraternization rule that applies between the doolies and everybody else. Fraternization used to mean unnecessary familiarity among the ranks of men. Now it has other meanings.

Cadet squadron commanders like C1C Norm Thompson try to maintain equal treatment of both men and women cadets within their units, but there are problems.

"My guys may act differently toward the women without knowing it," he explained. "Then the male cadets complain. So everybody goes to the opposite extreme and then the women feel put upon and complain. We've even had women complain when they thought we were easing up on them."

Women cadets are required to come to their squadron area twice a day to read the bulletin boards, and the ATOs and squadron commanders try to make sure the women are not without upperclass "guidance" in their own area.

Overall, most Academy officials agree, it has been a good start. The class of '80 seems to have the publicity and much of the controversy behind it, and "business as usual" may at last be the order of the day.

With the heavy academic loads all cadets carry, there is little time left for jealousy or rivalry, particularly among the 1,500 or so doolies who are discovering that a "class brain" in high school has to

work his—or her—head off just to be average at the Academy.

There has been one significant change in training at the Academy this year, although not dictated by the admission of women. Prior to the arrival of the class of 1980, General Allen instituted what he calls "positive leadership training." Essentially, it did away with negative motivational techniques such as push-ups and squat thrusts. Upperclassmen training the new class may now use only the leadership skills, techniques, and incentives that would be available to them as Air Force officers.

Attrition has been lower this year for both men and women. Most observers chalk that up to the new positive leadership programs, though they don't discount the presence of women as playing a part. "I think they are a stabilizing influence on their class," observed C1C Grady Booch. "It makes it more difficult for a guy to say, 'I can't do it,' when he has just watched a girl accomplish the same thing."

For their part, the women seem to be banding together to make sure they make it as a group. Most women AIRMAN talked to said they knew someone who had wanted to quit, but that she had been talked out of it by friends or roommates.

"There's a lot of peer pressure on them," said Lt. Col. Chuck Spruill. "Collectively they want to be successful. They want to be accepted by the wing as equals, not as a novelty. I think that's probably the inner driving force that is going to keep the majority here."

As an institution, the Air Force Academy has just gone through a tremendous change and, for the most part, has come through it unchanged. "I don't thing the admission of women in any way implies a change in our mission any more than the increase in women on active duty has changed the mission of the Air Force," said Commandant of Cadets Brig. Gen. Stanley C. Beck. "What we are doing here is developing professional officers who are women.

"We have had women serving as airmen, NCOs, and officers for many years now, and they have done excellent work. In my judgment it was only right that women should have an equal opportunity to come to the Air Force Academy and develop themselves professionally.

"I believe it was an idea whose time had come."

WOMEN BEGIN TRAINING AS AIR FORCE PILOTS

by Al Scholin

Reprinted by Permission
TAMPA TRIBUNE-TIMES
(October 3, 1976)

Edited for Space

At Williams Air Force Base, Arizona, next week, 10 young women Air Force officers will climb into the cockpits of two-place Cessna T-37 trainer planes, seeking to become the first women jet pilots in the U.S. armed forces.

The event marks the first step by the Air Force to join the Army and Navy in training women as military pilots.

Some consider it ironic that the Air Force, youngest and perhaps most innovative of the services, should be the last to accept women for pilot training. After all, the Army Air Forces in World War II set up the Women's Air Service Pilot (WASP) program, training women to ferry military planes from factories to operational bases around the world, thereby releasing men for combat. Eventually more than a thousand women became WASPs. But although they wore uniforms and service pilot wings, they were civilian employees.

The current cycle in training women as military pilots began with the Navy. In March 1973 four women Navy officers entered pilot training at the Pensacola Naval Air Station, and four more chosen from civilian life joined them after completing officer candidate school.

The Navy is still experimenting with its program. The second class, from which five graduated, was made up of civilian women who completed naval officer candidate school before going on to Pensacola. The third class, just about to get under way, includes eight women commissioned on graduation from college who will attend aviation officer candidate school at Pensacola with their male counterparts before beginning flight training.

The Army started a year after the Navy, enrolling the first women for helicopter pilot training at Fort Rucker, Alabama, in the summer of 1974. Since then it has overtaken the Navy, with 17 graduates now flying Army aircraft in the U.S. and overseas, and 14 more currently in training at Fort Rucker. Only eight of the graduates are in the active Army. Another eight are in the Army Reserve and one in the National Guard.

Now that the Air Force has decided to join in training women for cockpit assignments, it may soon outstrip the other services in numbers. In addition to the 10 now at Williams AFB, another 10 will enter the training program early this year, and six have been chosen to enter navigator school at Mather AFB, California. All the Air Force candidates were chosen from among women officers already on active duty.

While only the Air Force women are undergoing flight training in jets, the Navy may soon allow some of its women pilots to transition to jet aircraft. Most Navy jets are combat planes, and as of now women may not fly combat aircraft or land on carriers. But the Navy does have some jet transports, including the T-39 Sabreliner and the C-9A..

That incident suggests what may lie ahead for women military pilots, particularly if the Equal Rights Amendment becomes law.

ATCA—KEY TO GLOBAL MOBILITY

by Edgar Ulsamer, Senior Editor

Edited for Space

Reprinted by Permission
AIR FORCE MAGAZINE (April 1976)

According to recent statements by Secretary of State Henry A. Kissinger and other Administration and military leaders, it is the consensus of the White House, the State Department, and the Pentagon that, for the time being, Soviet aggressive designs will remain confined to fomenting nonnuclear conflicts on a regional basis and predominantly along the periphery of the Warsaw Pact bloc. Prepositioning US forces in adequate numbers at or near all potential conflict sites obviously is impossible for political, economic, and other reasons. The job of deterring, or, if necessary, fighting, such conflicts requires highly mobile military forces, largely dependent on comprehensive airlift. The latter term takes in strategic airlift as well as tactical airmobility, both of which are, in the main, under Air Force purview.

Expansion and refinement of airmobility are the goals of a program that, according to USAF Chief of Staff Gen. David C. Jones, is one of the Air Force's most pressing and vital priorities: "We are confronted today with the imperative of being able to operate almost any place in the world with little if any reliance on enroute bases and to project our forces quickly over great distances." Early projection of US conventional power, General Jones says, "can stabilize the situation if conflict has not yet erupted, can mount a better defense if it has, and, most important, can offer the President an alternative between surrendering a vital national interest or crossing the nuclear threshold."

The Defense Department and Air Force budget requests for FY '77, and proposed budgets for subsequent years, emphasize strategic, tactical, and helicopter airlift. Proposed acquisition costs for modernizing the mobility forces of the four services through improved airlift amount to almost $500 million, roughly double the current level, and are programmed for an increase to $1 billion in FY'78, according to Defense Secretary Donald H. Rumsfeld. The Air Force's share increases from about $170 million requested for FY '77 to about $720 million proposed for FY '78. Except for prototype development of AMST, the Advanced Medium STOL Transport (which is to function primarily in intratheater missions while performing long-range sorties when necessary), all pertinent major USAF programs come under the broad heading of strategic airlift, a mission area for which the Air Force is designated as DoD's single manager.

There is no shortfall, according to General Jones, in USAF's ability to move troops or normal bulk cargo. The combined resources of the Military Airlift Command and the Civil Reserve Air Fleet (CRAF) are "quite adequate" in this regard, *provided* that enroute bases are available. Rather, the deficiency lies in what Secretary Rumsfeld termed a less than optimum ability "to deploy the military equipment of our land and tactical air units in a balanced manner." The imbalance is caused by shortfalls in "oversize" cargo–rolling stock, artillery, armored personnel carriers, and similar equipment–even though the C-5s could easily meet delivery requirements of "out-size" cargo such as tanks and other heavy items that can't be carried by other military or CRAF transports.

The US active airlift force consists of seventy C-5s, 234 C-141s, and about 235 C-130s. If Air Force Reserve and Air Guard C-130s are included, the total number of airlift aircraft comes to about 800. Augmenting this force are 153 long-range cargo and ninety-one long-range passenger aircraft of CRAF, some or all of which can be called up in case of national emergency. Of these aircraft,

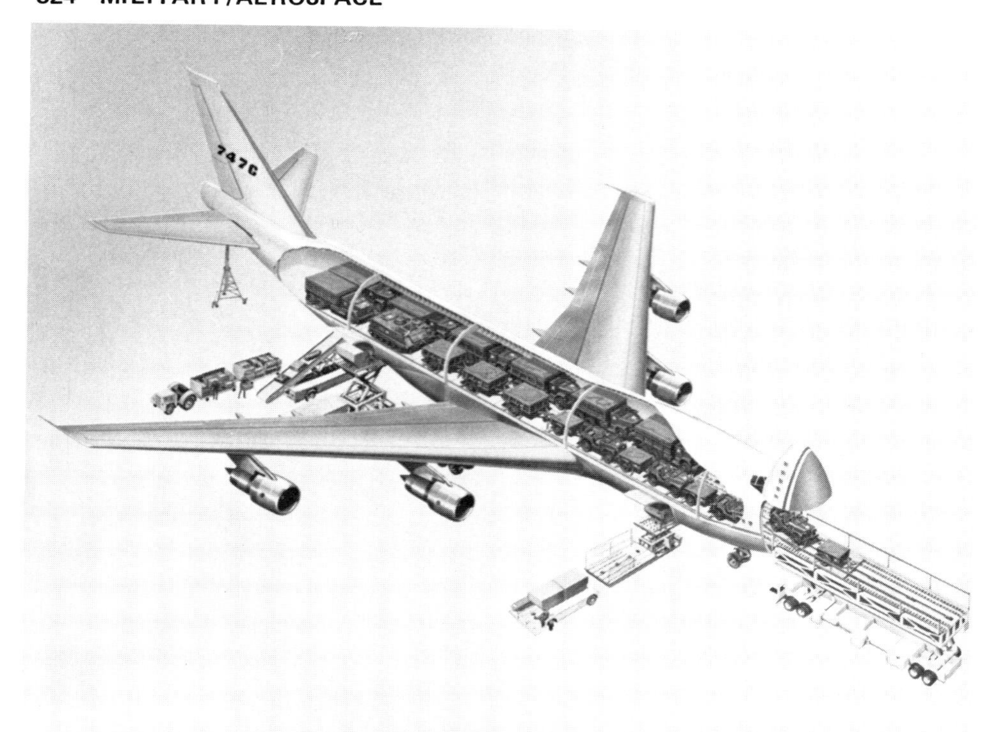

only the C-5s are capable of aerial refueling. The C-141s, which provide about seventy-five percent of the active strategic airlift force, are now dependent on overseas landing authorization and transit services for almost all missions in excess of 3,300 nautical miles.

Air Mobility Enhancement

Planned improvements of the C-141 fleet, currently in early development, include the addition of aerial refueling capacity to give the aircraft global mobility.

The single most cost-effective step to improve strategic airlift would be modification of a number of the wide-body, long-range passenger jets of the US flag carriers. The modification involves adding nose or side cargo doors to the 747 and DC-10 aircraft and reinforcing the flooring to handle heavy military cargo. Plane for plane, this program can produce more of the capacity increase sought by the Airlift Enhancement Plan than any other initiative and at a fraction of the cost of buying and operating organic military aircraft with a like capability.

Advanced Tanker Cargo Aircraft

Fundamental to global mobility of both tactical air and strategic airlift forces is ATCA, the Air Force's proposed new Advanced Tanker Cargo Aircraft. This system is quite different from USAF's 615 range-limited KC-135 aerial tankers. Operating from bases in the continental US, ATCA can provide both aerial refueling and airlift support for tactical combat or airlift forces. According to Secretary Rumsfeld, the system "would dramatically reduce our reliance on foreign bases for support of tactical or cargo aircraft being used in a force projection role." Ultimately, the Air Force plans to buy about forty of these aircraft. The precise number will depend on the aircraft selected and the air refueling mission requirements envisioned for the 1980s and 1990s.

In developing ATCA, the Air Force seeks to satisfy three requirements of "flexible mobility," he said: "Refueling C-5 and C-141 aircraft; refueling the tactical aircraft of the Navy, the Marines, and USAF; and cargo delivery." ATCA might well perform all three missions in one sortie: "It could take off with and top off either C-5 or C-141 aircraft, escort fighters to Europe, and carry some cargo and people so that when the fighters land, they're ready to go into action with the help of ground support personnel and equipment." The Air Force's approach to ATCA is patterned after the airline industry's policy of maximum utilization and multiuse approach.

The case for ATCA, USAF's Chief of Staff points out, principally rests on two factors: "ATCA offers unique capabilities because its range/payload characteristics are unmatched by the KC-135 or any other aircraft, thereby allowing it to go to almost any place in the world *without* enroute bases." Secondly, ATCA is uniquely cost-effective in the broadest sense of that term. Either of the two aircraft under consideration for the ATCA mission–Boeing's 747 and McDonnell Douglas's DC-10 (international long-range version)–represent "sunk cost" systems whose development has been paid for by the manufacturing teams and the airline industry. Cost savings from this alone amount to hundreds of millions of dollars. Further, more than 200 units of each competing model have been sold. Both are of proven reliability and endurance. Unit cost and cost of ownership, therefore, become advantageous to the Air Force, General Jones suggested.

Another intrinsic advantage is derived from the large and almost worldwide support base that the airlines industry has set up for these wide-body, long-range jetliners. Spares and maintenance are available at more than 100 airports scattered all over the world (except for the USSR and the Warsaw Pact countries), thus improving ATCA's operational flexibility and lowering support and operating costs.

An intriguing possibility, General Jones points out, is to purchase ATCA under a total acquisition and maintenance contract, either with the manufacturer alone or in concert with a US flag carrier that has extensive overseas depots. While he emphasized that the military mission and such unique aspects of ATCA as the boom system may preclude a completely commercialized maintenance arrangement, the Air Force plans to exploit all possible economies associated with "piggy-backing" on the commercial aviation system without jeopardizing operating capabilities under crisis or war conditions. A first step in this direction is the Air Force's decision to minimize the initial spares buy for ATCA because the manufacturers and the airlines already maintain a large supply.

On the other hand, General Jones told AIR FORCE Magazine, "we don't expect to place absolute reliance on overseas commercial facilities. The system will have a self-contained deployment capability to give us independence. During peace-time, of course, there is no pressing need for ATCA to operate in such a mode, and we plan to use available maintenance and other logistics support. But during political or military crises, we will shift to operating ATCA in a self-sufficient way."

The benefits of operating hand in glove with the commercial aviation sector far outweigh associated penalities, especially "when there is a need to concentrate on systems with low manpower demands and decreased life-cycle costs, a trait that ATCA epitomizes. We can operate a modest fleet of these aircraft with considerably less organic military manpower than now required," according to General Jones. Both the 747 or the DC-10, he emphasized, have demonstrated a "proven high utilization rate and equally high dispatch reliability, [the latter] approaching ninety-nine percent," he added.

Even when the range deficiencies of the KC-135 are disregarded, a mix of KC-135s (designed more than fifteen years ago) and C-141s cannot compete with ATCA in cost-effectiveness, according to Air Force analyses. The cost-effectiveness of the wide-body jetliners powered by economical, high bypass ratio engines, justified the airline industry's decision in the 1960s to retire, or relegate to standby status, its as yet not amortized fleet of narrow body aircraft—representing an investment of billions of dollars—and shift to the more efficient family of wide-body aircraft that followed in the wake of the C-5 and its TF39 engine.

An added benefit of ATCA's commercial aviation origin is the availability of trained Reserve flight crews during crisis periods, according to General Jones. Reservists who crew 747 or DC-10 airliners can be called into service to operate ATCA without impairing the airlines' ability to operate the wide-body component of CRAF at maximum rate. This is possible because the airlines maintain crew ratios well above what would be required during national emergencies. "Instead of flying at a rate of seventy to eighty hours a month, they can increase flying hours well above that. On the other hand, the airlines are operating the wide-body aircraft at a very high utilization rate, sometimes close to the maximum. As a result, a call-up of some of these crews to fly ATCAs won't affect our ability to operate the CRAF fleet to the maximum rate of utilization the aircraft can sustain," according to General Jones.

A similar high "surge rate" is available for the C-5 fleet which ATCA might have to refuel during emergencies. Each of the Air Force's four active-duty C-5 squadrons is collocated with an Associate Reserve unit that participates in operations and maintenance to permit rapid mobilization. An additional major economic plus associated with a 747/DC-10-based ATCA, General Jones points out, is the availability of comprehensive simulator facilities set up and in operation by the airlines.

Air Force analyses suggest that the eventual introduction of about forty ATCAs into the USAF inventory will capitalize on the available manpower and assure that the C-5s, the CRAF aircraft, and the new Tanker Cargo Aircraft can be utilized at a maximum rate with existing active-duty, Reserve, and airline manpower.

ATCA Decision by Year's End?

Possibly by the end of this calendar year, and certainly not later than the end of FY '77, the Air Force plans to select an aircraft for the ATCA mission, General Jones told this reporter. The decision will not be easy because both the DC-10 and 747 aircraft, for example, offer advantages that are unique in terms of performances, flexibility, and cost-effectiveness. For instance, the fuel off-load capability of the DC-10 over relatively short distances is below that of the heavier four-engine Boeing 747. However, the greater fuel efficiency of the three-engine aircraft

apparently catches up with its competitors on long-distance sorties. This feature gains added importance if, as expected, the cost of the trijet is below that of the four-engine jumbo-jet, he said.

Air Force mission analysis studies indicate that a typical deployment of an F-4 squadron (twenty-four aircraft) from the US to NATO without an en-route stop, or to Korea with one en-route stop in Alaska, would require ten 747-based or fourteen DC-10-based tanker airlift support aircraft. If the mission is confined to refueling only, a 747-based ATCA can take four F-4s and a DC-10 derivative three F-4s to either of these two destinations.

The final selection among candidate aircraft will be a complex process considering such factors as total capability, competitive bids by contractors for production and support arrangements, life-cycle costing, and related conditions, according to General Jones.

The basic requirement for and final configuration of ATCA, General Jones emphasizes, is in "no way affected" by the B-1 strategic bomber program, assertions to the contrary notwithstanding: "The B-1 design mission is based on the characteristics and capabilities of the KC-135 and is in no way dependent on a new tanker. If the KC-135s, which we are reskinning to last into the twenty-first century, are not available for some reason, the B-1 can still perform its job although we might send it against different targets and use less complex tactics. But the B-1 does not need a new tanker with special characteristics such as quicker launch and escape time or other qualities that some people contrive to support their contention that getting the B-1 also means getting a new tanker."

A form of ATCA, General Jones points out, is already approaching operational reality under Iranian aegis: "The Iranians are buying [three] 747s and putting on booms for aerial refueling. Their boom arrangement is not quite the same as we want, but still it is quite similar to ours."

Flight tests by AFSC's Aeronautical Systems Division in 1972 with the 747 showed that aerial refueling from multiple boom stations is feasible and introduces no major vortex (gust) problems for the refueling fighters.

While the Air Force has not yet decided which command is to operate ATCA, General Jones indicated that both MAC and SAC are under consideration.

The Air Force's case for ATCA clearly gains from the system's undeniable cost-effectiveness. Yet, in the last anlysis, no argument in its behalf can be more compelling than the fact that without a system of this type the tactical forces of the Air Force and the other military services, as well as USAF's strategic airlift, might be impaired in, if not prevented from, doing their job in many critical regions.

AMERICA'S SPY PLANE IS FASTEST

by Al Scholin

Reprinted by Permission
TAMPA TRIBUNE-TIMES
(August 8, 1976)

In record-setting flights a few days ago, the U.S. Air Force publicly confirmed what aviation experts on both sides of the Iron Curtain have known for years—that the Lockheed SR-71 Blackbird spy plane is the fastest and highest-flying aircraft in the world.

The SR-71 is the successor to the Air Force U-2 that proved to the world the existence of Soviet missiles in Cuba in 1962. Since that time U.S. intelligence has kept a close watch on everything going on in Fidel Castro's domain, and the SR-71 has undoubtedly contributed a major share of that information through its photography and electronic eavesdropping capabilities.

Though it has been flying since 1964, the Blackbird was kept under tight security wraps until last September when the

Air Force showed it off at Britain's Farnborough air show. On the way over, it flew from New York City to London in one hour and 56 minutes, far eclipsing the former mark of 4 hours and 46 minutes set by a British RAF crew in the F-4 Phantom fighter. For good measure the SR-71 added another astonishing record on its home journey—flying nonstop from London to Los Angeles in 3 hours and 48 minutes.

On July 27 and 28 at Edwards Air Force Base, Calif., the SR-71 laid claim to a new batch of world records, including top speed of 2,194 mph, speed of 2,116 mph over a 1,000-kilometer closed course, and sustained altitude in horizontal flight of 85,126 feet.

Of the records it surpassed, two had been held by USAF's YF-12A interceptor, a plane similar in many respects to the SR-71, and one by the Soviet Union's fastest aircraft, the E266, an experimental version of the MIG-25 Foxbat interceptor. The YF-12A had logged a top speed of 2,080 mph and sustained altitude of 80,258 feet in May 1965, while the E266 in October 1967 had run the closed course at 1,853 mph.

Flown by a two-man crew—pilot and reconnaissance systems operator—the SR-71 is 107 feet long with a wing span of 55 feet 7 inches. Its speed is achieved by a pair of 32,500-pound thrust J58 engines built by Pratt & Whitney at West Palm Beach. The power they produce is equivalent to that of 45 diesel locomotives, and they gulp fuel at the rate of 8,000 gallons an hour. The SR-71's nickname comes from the black paint that covers its skin to help dissipate the tremendous heat generated at high speed.

Despite its exceptional capabilities, only a few SR-71s were built. The actual number has never been made public, but it probably totaled less than 30, and only a dozen are now in service. At least 10 were lost in accidents, though none to hostile action, and the remainder are "mothballed" for replacements.

From their home station at Beale Air Force Base, Calif., and aided by in-flight refueling, the SR-71s can and do roam the skies worldwide, all but invisible to those far below. The U.S. has pledged not to overfly the Soviet Union or China, but this doesn't preclude flights along their perimeter.

NASA TO RECRUIT SPACE SHUTTLE ASTRONAUTS

Reprinted by Permission
THE ATCA BULLETIN (August 1976)

NASA has issued a call for Space Shuttle astronaut candidates. Applications will be accepted until June 30, 1977, and all applicants will be informed of selection by December 1977.

At least 15 pilot candidates and 15 mission specialist candidates will be selected to report to the Lyndon B. Johnson Space Center, Houston, TX, on July 1, 1978, for two years of training and evaluation. Final selection as an astronaut will depend on satisfactory completion of the evaluation period.

Pilot applicants must have a bachelor's degree from an accredited institution in engineering, physical science or mathematics or have completed all requirements for a degree by Dec. 31, 1977. An advanced degree or equivalent experience is desired. They must have at least 1000 hours first pilot time, with 2000 or more desirable. High performance jet aircraft and flight test experience is highly desirable. They must pass a NASA Class I space flight physical. Height between 64 and 76 inches is desirable.

Applicants for mission specialist candidate positions are not required to be pilots. Educational qualifications are the same as for pilot applicants, except that biological science degrees are included. Mission specialist applicants must be able to pass a NASA Class II space flight physical. Height between 60 and 76 inches is desired.

Civilian applicants may obtain a packet of application material from JSC. Requests should be mailed to either Astronaut (Mission Specialist) Candidate Program or Astronaut (Pilot) Candidate Program, Code AHX, NASA Johnson Space Center, Houson, TX 77058.

Military personnel should apply through their respective military departments using procedures which will be disseminated later this year by DOD. Military candidates will be detailed to JSC but will remain in active military status for pay, benefits, leave and other military matters.

ROLLOUT HELD FOR SPACE SHUTTLE ORBITER

by YEARBOOK Staff

Orbiter Vehicle 101, the first reusable Space Shuttle vehicle was rolled out at ceremonies held on September 17. Although Orbiter 101 is the first vehicle off the assembly line, it will not fly into space until the early 1980's. Its first job in 1977 will be as a flight test vehicle.

The public received its first official look at OV-101, NASA's first Space Shuttle Orbiter vehicle, during Palmdale, California, rollout ceremonies September 17, 1976. This first orbiter, named *Enterprise* after television's STAR TREK starship, will first be used to check out flight systems and the vehicle's handling characteristics during approach and landing. When this series of tests is complete, the *Enterprise* will be refurbished and made ready for orbital flight. The second orbital vehicle, presently under construction, will make the first orbital test flight in the spring of 1979. The Orbiter vehicles will form a major part of the National Aeronautics and Space Administration's space transportation system.

The space transportation system is made up of the Orbiter vehicle, two reusable solid-rocket boosters for launch, and an expendable external tank for the liquid oxygen/hydrogen to fuel the three main engines.

The Orbiter vehicle is the main component of this system, which will both reduce the cost and increase the effectiveness of using space for commercial, scientific, and defense needs.

The approximate size of a jet airliner, the Orbiter is 17 meters (57 feet) high, 34 meters (122 feet) long, and has a wing span of 23 meters (78 feet). The gross lift-off weight of the Shuttle system (Orbiter, external fuel tank, and booster rockets) is 1.99 million kilograms (4.4 million pounds) and the reentry weight of the Orbiter is 84,800 kilograms (187,000 pounds).

The payload bay, the "heart" of the Shuttle's reason for being, is 18.3 meters (60 feet) long and 4.6 meters (15 feet) in diameter. It is able to accommodate projects from small satellites to be inserted in orbit to fully-equipped manned scientific laboratories.

The external fuel tank and the solid rocket boosters contain the launch and ascent fuel. The two solid rocket boosters each provide 2.65 million pounds of thrust for launch and ascent. They are jettisoned at an altitude of 43 kilometers (27 miles) after expending their fuel. They descend by parachute to the ocean, where they are recovered and reused in future missions. The external storage tank contains the ascent propellant for the Orbiter's three main engines, each of which provides 470,000 pounds of thrust. After the fuel has been used, the tank is jettisoned and falls into a remote part of the Indian or South Pacific Ocean. The external tank is the only part of the system that is not reused.

The system has been designed for rapid turnaround and the inclusion of nonastronaut trained scientists for payload crew, to increase system effectiveness. For example, crewmembers and passengers will experience a designed maximum gravity load of only 3 G's during launch and less than 1.5 G's during a typical reentry. Scientists and engineers in average physical condition, regardless of sex, can thus be members of the payload crew and passenger load without having special space training. The sea-level atmosphere provided will also create a shirt-sleeve working environment. (The "basic" crew will be three persons, supplemented by up to four additional scientist payload specialists.)

Other capabilities of the system include a nominal two-week (160-hour) turnaround time between missions; a life of 100 to 500 missions for the Orbiter vehicle; a 2,037-kilometer (1,100-nautical mile) cross-range maneuvering ability (cross-range maneuvering is the ability to fly to either side of the ground track after reentry in order to arrive at the landing site); a "safe mission termination" capability throughout the mission; capabilities for missions of three, seven, and thirty days; and a capability for extravehicular activity.

Beginning in early 1977, the *Enterprise* will participate in a series of air-launch tests from the back of a modified NASA 747. This Approach and Landing Test (ALT) series will begin with a series of ground taxi tests, a group of unmanned flights, and finally, beginning in July 1977, a series of free-fall flight tests. The test flights will be held at NASA's Dryden Flight Research Center at Edwards Air Force Base, California.

During these tests, the separation from the 747 will take place at about 23,000 feet. Gear will be deployed at an airspeed of approximately 250 knots, the Orbiter will flare out with a sink rate of less than 10 feet per second, and touchdown will be at 185 knots.

SPACE SHUTTLE DEBUT

Reprinted by Permission
FLIGHT INTERNATIONAL
(September 25, 1976)

Edited for Space

The first Orbiter vehicle, main constituent of America's Space Shuttle, was rolled out at Palmdale in California last week. The Shuttle is a second-generation launch vehicle, and from 1980 will carry into orbit most US and European payloads. **DAVID BAKER** and **MICHAEL WILSON** here present a users' guide to the Shuttle and its growing range of "add-on" equipment.

NASA has coined a new term, the Space Transportation System or STS, to more accurately reflect the capabilities of the composite launch vehicle based on the Space Shuttle as a kind of multi-purpose first stage. It comprises the Shuttle itself, Europe's Spacelab manned orbital laboratory-cum-observatory, and a series of upper stages to boost satellites into higher orbits and start deep-space or interplanetary probes on their journeys. There is also a scheme for an unmanned, heavy-lift vehicle in which the payload, taking the place of the winged Orbiter, is thrust into space by the Shuttle's main engines and boosters. The STS is therefore a building-block launch system, and will replace three of the four expendable rockets—Delta, Atlas-Centaur and Titan-Centaur—which comprise the current fleet of Nasa/DoD launch vehicles. The smallest of the quartet is the Scout, and this is likely to co-exist indefinitely with the Shuttle.

When Nasa's search for a major project to succeed Apollo began to crystallise earlier this decade into a recoverable launch vehicle, the assumption was that it would replace all current US expendable rockets. Indeed, this was how the venture was justified to the US Government. Production rates of the two large rockets, Atlas-Centaur and Titan-Centaur, have now slowed, and their launch costs are likely to be considerably greater than those anticipated for the Shuttle. There will thus be little or no incentive for the commercial or scientific communities to use them when Shuttle operations get under way.

But the smaller Delta remains attractive, and the Shuttle may be economic only if two or more payloads can be carried on one flight. This may be difficult or impossible if the desired orbital inclinations are widely different, a situation which is more likely to affect the scientific-satellite users, with their multitude of orbits, than the commercial customers requiring generally synchronous paths. The former group can often wait for a Shuttle flight going their way, however, while the latter, geared to maintaining a public or commercial service, may be able to "double-up" with another Shuttle customer. So, provided Nasa can organise its route network and schedules suitably, there is likely to be little call even for the Delta after 1980.

In order to keep costs down, the STS will be managed and operated like an airline, with scheduled flights into space leaving once a week from one or other of the two US launch sites. Nasa's chief job now is to market the Shuttle to potential users throughout the world in order to achieve the highest utilisation, particularly in the early years. To this end the agency several years ago prepared a traffic forecast or mission model covering the first 12 years of operation, 1980-1991. This is continuously revised (the latest one covers the period 1979-1992), since the costs of STS operation will be very sensitive to its utilisation. Space budgets in the US have fallen in recent years, and it seems likely that the Shuttle will be initially under-utilised despite Nasa's endeavours.

Based on forecasts of eventual utilisation, Nasa plans to acquire five Orbiters (the delta-winged Orbiter, flown by a crew of four, can be regarded as the key module of the STS). Nasa has ordered one Orbiter, is negotiating for the second, and has signified its intention to buy a third. There is some doubt about who pays for the fourth and fifth Orbiters; the Shuttle is a Nasa programme, but the DoD will fly one-quarter of the missions during the first 14 years.

NASA has selected two, two-man crews for the initial flight tests of the Shuttle Program scheduled to begin in July 1977. The crews are (left to right) Charles G. Fullerton, pilot; Fred W. Haise, Jr., commander; Richard H. Truly, pilot; Joe H. Engle, commander. Haise and Fullerton will make the first flight.

Shuttle and its ancillaries

The basic shuttle comprises the first two stages of a launch vehicle to which may be added one or more third stages to accommodate synchronous-orbit and planetary payloads. The first stage is powered by three liquid-propellant engines, supplemented by two Solid Rocket Boosters (SRBs), which provide the Orbiter with 99 per cent of the impulse required to impart low Earth orbit velocity. Fuel for the Orbiter's liquid-propellant engines is carried in the external tank, mounted under the vehicle's belly. Shortly after the tank has been jettisoned, two Orbital Manoeuvring System (OMS) engines at the rear of the Orbiter ignite briefly to impart the additional impulse needed for a 50 n.m. X 100 n.m. ellipse. This is transformed into a 100 n.m. circular orbit by re-igniting the OMS propulsion unit at first apogee.

The Shuttle's load-carrying capacity reflects a preference for this two-stage configuration, which will be used when the altitude required is less than 650 n.m. for a 10,000lb payload. The maximum lifting capacity of 65,000lb limits altitude to a 220 n.m. circular orbit. If payload/height requirements fall outside these limits a third stage is carried in the cargo bay. The Shuttle then becomes a platform from which a supplementary expendable propulsion system known as the Interim Upper Stage (IUS) and its payload can be separated and put into its own trajectory. The IUS is a DoD-sponsored vehicle to deliver at least 4,500lb to synchronous orbit, and several such stages can be assembled to cater for high-energy planetary or heavy synchronous-orbit payloads which exceed this limit.

In this three-stage configuration the Shuttle promises to accommodate all anticipated Earth-orbit and interplanetary unmanned payloads for the remainder of the century.

Planners consider it likely that by the mid-1980s some customers will be calling for payloads to be retrieved from synchronous orbit for refurbishment and subsequent re-use. With this requirement in mind, Nasa proposes to develop a Space Tug for service in 1985-87. The current basic design is a single-stage liquid oxygen/liquid hydrogen-powered vehicle 30ft long, 15ft in diameter and with a propellant capacity of 50,000lb. Carried to a low Earth orbit by the Shuttle, the Tug would boost its payload to synchronous orbit following separation from the Orbiter cargo bay, just like the IUS. After circularising its orbit and placing the payload on station the rocket would return to a low Earth orbit for retrieval by the Shuttle. The Tug will be lifed for up to 100 flights and will be able to carry 8,000lb to synchronous height, retrieve 3,400lb, or deliver and retrieve 2,400lb. Its synchronous-orbit delivery capability is twice that of the expendable IUS, and retrieval and delivery/retrieval (or so-called round-trip) missions would provide a new level of flexibility for payload planners. Emphasising simplicity and reliability, the Tug will be powered by the RL-10 rocket motor designed for the Centaur upper stage. With a minimum development time of six years, a decision to proceed will have to be taken within the next three years if it is to be introduced by 1986.

Although designed primarily for synchronous-orbit activities, the Tug's generous performance opens up the possibility of larger versions capable of supporting construction of the space station that will supersede Spacelab. A two-stage development of the Tug would deliver 34,500lb to synchronous orbit or perform a round trip with 8,800lb. Another version, the Aeromanoeuvring Tug designed to use Earth's atmosphere for braking, would deliver 12,800lb to synchronous orbit or take 6,700lb on a round-trip flight. It would "skip" through the outer layers of the atmosphere to lower apogee and reduce the propellant required to set up a low orbit after returning from synchronous altitude. Yet another scheme, the two-stage Growth Tug, would require an improved Shuttle capability but promises to deliver 48,700lb to synchronous orbit or send 12,600lb on a round trip. A follow-on Tug, called the Orbital Transport Vehicle (OTV), would be able to deliver 45,000lb to synchronous orbit and, following refuelling from an orbital propellant dump, could return with the same payload mass.

Spacelab is the manned laboratory which, carried aboard the Shuttle Orbiter, will enable experimenters to carry out a wide variety of scientific, technological and biological activities for civil and military customers. Though considerably smaller than America's Skylab, it will—like the Shuttle itself—be reusable and, with specially designed modular experiments and equipment, be capable of rapid turnround on the ground to suit it for different missions.

It comprises a pressurised habitable laboratory, known as a module, in which up to four payload specialists or experiment operators may work, and a number of unpressurised pallets on which are mounted those experiments or payloads calling for direct exposure to space. The configuration can be tailored to the needs of the flight. Some missions may call simply for a module, others for a combination of module and pallets, and yet others will require any number of pallets up to five, but no module. The basic module itself is a cylinder 4.5m in diameter and 2.7m long, but two such units can be joined together, increasing the length to about 5.93m to provide additional space for experiments.

In terms of economy, Spacelab should be one of the most efficient payloads, since it makes use of a relatively large proportion of the facilities offered by the

Approach and landing tests of the Space Shuttle Orbiter will be conducted in early 1977 using a specially modified 747 jet transport. The 34-meter (122-foot) long Orbiter will be carried to a launch altitude of approximately seven kilometers (25,000 feet) by the 747 where it will be released for glide flight to the dry lakebed. Prior to the free flight tests, the Orbiter will be flown mated to the 747 in a series of unmanned tests aimed at examining the aerodynamics of the Orbiter in flight.

Shuttle. Maximum allowable Spacelab weight is 32,000lb, limited by the Orbiter's design landing weight. Between 12,000lb and 20,000lb of this figure, depending on the Spacelab configuration chosen, can be assigned to it.

Spacelab will be the platform for the Western world's manned space activities up to about 1985, when the United States plans to bring along a "Mk 2" Spacelab. This at present appears likely to be an all-American design masterminded by the DoD, in the absence of a more pronounced European commitment in the present Spacelab.

Predicting the market

For the first three years of Shuttle operations (1980-82) payload planners will be cramped by the limited orbital capability from Kennedy. Not until the Vandenberg facility becomes operational will missions be possible with orbital inclinations greater than 55°. Nevertheless, the initial performance envelope will include low Earth orbit, synchronous orbit and escape trajectories. With this restriction in mind, Nasa has prepared a preliminary schedule of missions, or traffic model, on the assumption that US Government, commercial and foreign users will call for a maximum of 60 flights a year by 1985.

A single launch-pad at Kennedy will be available in 1979, followed by a second in June 1982; the first Vandenberg pad will not be ready before early 1983, with the second reaching completion by December 1986. Mission opportunities will be tailored to this schedule and to the Orbiter procurement plan.

On present planning the Department of Defence will fly 27 per cent of all Shuttle flights proposed by the traffic model. Of the 578 flights envisaged in the 1979-1992 period, 70 per cent will be launched from Kennedy, 39 per cent will require seven-day Spacelab operations, and 6 per cent call for 30-day missions with the manned laboratory. About 34 per cent of the flights will involve the use of an IUS or Space Tug for synchronous-orbit delivery or delivery/retrieval missions. Spacelab and IUS/Tug missions account for 79 per cent of all launches tabulated in the 14-year traffic model, leaving 21 per cent for solo Shuttle missions or low-orbit delivery and retrieval work. The new traffic prediction shows a build-up to 60 flights a year from

1985 to 1991, with 40 launches a year from Kennedy and 20 a year from Vandenberg, and replaces the original 1973 traffic model, which envisaged a total of 725 missions.

Cost to the customer of a Shuttle launch is expected to be $20 million at 1976 prices, with Interim Upper Stages charged at 1.8 times the nominal $5 million rate for those users who want a launch at less than three years' notice. It is unlikely, however, that foreign participation will justify specific Shuttle flights, and most STS missions will accommodate a mixture of US and foreign payloads. In this case the cost to the user will be based on the proportional weight and volume requirements of the payload.

Benefits to payload designers

Introduction of the Space Transportation System in 1980 will greatly influence payload planners. To begin with the Shuttle will provide a comparatively comfortable environment for the payload, permitting designers many new freedoms. Payloads of up to 65,000lb can be carried to orbit within a volume of 10,500 cu ft, and satellites weighing up to 32,000lb can be returned.

It is impossible to predict with accuracy the impact of the Shuttle's more favourable acceleration, temperature and vibration environment on payload design, but several technology studies have indicated cost-savings of up to 25 per cent. Nasa has already discovered that costs arising from the use of an expendable rocket with expensive excess lifting ability can be more than offset by a cheaper payload-development phase. The time-consuming process of designing to stringent weight, volume and reliability constraints imposed by expendable rockets can add significantly to the cost of payload development. The Shuttle also provides an "intact abort capability" which permits the Orbiter to return with its payload to a safe landing if the flight is terminated early after a mechanical or "soft-ware" failure. All of these features will serve to reduce space transportation costs beyond the current level.

Organisations providing commercial services such as weather forecasting or communication are at present required to have back-up satellites in orbit ready to take over in the event of the primary spacecraft becoming unserviceable. The Shuttle will remove the need for orbital spares by quickly flying modularised repair packages to ailing or dead satellites in low orbit. A similar service for synchronous satellites will of course have to await the introduction of the Space Tug.

On a typical repair mission, part of a scheduled Orbiter flight would be dedicated to carrying a Tug and an equipment dispenser. The combination would fly under remote control from low Earth orbit to a synchronous satellite. Once on station the dispenser would dock with the satellite, remove the faulty module and replace it with a new unit. The Tug, still attached to the dispenser, would then return to the Shuttle for capture and return to Earth.

Development of a standardised modular satellite would be essential for the effective use of orbital-repair techniques, as would be equipment and experiment packages tailored to specific missions. Low-Earth-orbit satellites could also be built to a common design, with a rotary equipment or experiment dispenser carried within this Orbiter cargo bay. The modularised approach to satellite engineering would, again, reduce payload design costs by permitting a potential user to procure a standard, proven spaceframe and simply install his own experiments, sensors and other equipment.

Modular sensors on an Earth-resources satellite could, for instance, permit adjustments to be made to the choice of spectral surveillance bands from time to time, so increasing the useful life of the satellite and circumventing the need for a costly replacement every few years. This approach would permit the more immediate application of even minor technological improvements. At present, advances in technology have to accumulate over a number of years before they justify a costly new project.

Flight operations

Shuttle launches and recoveries will be confined to the Kennedy Space Centre in Florida and Vandenberg AFB in California. By US regulation, no part of a launch vehicle's ground-track may intersect a major land-mass during the ascent to Earth orbit, and this limitation will be particularly important in the case of the Shuttle. The Eastern seaboard of the United States is so shaped that Shuttle operations from Kennedy will be limited to orbits not exceeding an inclination of 55°. Missions with orbital inclinations greater than this will have to use the USAF launch site at Vandenberg.

Three elements of the Shuttle will return to Earth through the atmosphere during the Orbiter's ascent. The two boosters separate at an altitude of about 25 n.m., with the Shuttle 24 n.m. down-range from the launch site and moving at 4,600ft/sec. The resulting trajectory carries them to a height of 54 n.m. before they fall back towards the sea 210 n.m. from the launch site. Their descent through the atmosphere is slowed by a single drogue and three main parachutes, which lower the 147ft-long units into the sea at 85ft/sec. The external tank, the only expendable part of the Shuttle vehicle, separates at an altitude of 70 n.m. and breaks up in the atmosphere 2,200 n.m. down-range of the launch site.

Each booster is designed for 20 flights and each recovery unit (attachment points, transponders and locator beacon) is expected to survive ten cycles. Shuttle launch costs are assessed on these utilisation rates and on the assumption that each Orbiter can fly at least 100 missions, with the main engines being replaced after 55 launches. Nasa assumes that the thermal-insulation tiles will survive 100 flights, with the reinforced carbon-carbon segments (fitted to 3.5 per cent of the exterior surface area, mainly the wing leading edges) replaced after 60 flights. External tank elements will be shipped from the Martin Marietta plant to Kennedy or Vandenberg as required. The tank costs about $2 million a unit at 1971 prices, or 22 per cent of the total Shuttle launch cost, with the boosters contributing $3.3 million per flight, 31 per cent of each launch.

Shuttle services from the Kennedy Space Centre with Orbiter 102 (the second vehicle) will use the converted Saturn V launch pad (LC-39A) from the first flight in April 1979 until the second pad (LC-39B) becomes available in June 1982. Nasa expects to introduce Orbiter 101, refurbished from its drop-test configuration, in March 1981. It will be followed by Orbiter 103 in March 1982. Operations from Vandenberg will begin in March 1983 with delivery of Orbiter 104, followed a year later by Orbiter 105. Tentative plans envisage the introduction of a second Shuttle launch pad at Vandenberg by December 1986.

Orbiter landings at Vandenberg call for a 7,000ft extension to an existing 8,000ft runway, but emergency landings can be made on the Edwards AFB runway 150 miles east of the launch site. Edwards was ruled out as the prime west-coast landing site because it would require a 747 to return the Orbiters to Vandenberg—with attendant delays in turnround time and expense—and an emergency landing following an aborted launch would necessitate an approach over Los Angeles. Emergency landing sites will be provided at Guam and Hawaii to accommodate east or west-coast flights aborted at an early stage.

Initial flight tests

An important part of the Shuttle flight-qualification programme involves Orbiter 101, the initial vehicle, in a series of air-launched approach and landing tests. It will be carried to 24,000ft atop a modified Boeing 747 and released for a glide return to the Dryden Flight Research Centre at Edwards AFB, California. The tests will also qualify the Shuttle carrier aircraft for ferrying Orbiters between the manufacturer's plant and the launch sites.

Orbiter 101 will fly without main ascent engines, orbital manoeuvring engines or reaction-control equipment. Nor will it carry fuel-cell cryogenic tanks, cargo-bay payload, radiators, star-trackers, or S-band and rendezvous radar. No water, waste-management equipment or food will be carried, and the thermal-protection tiles will be simulated by plastic plates to duplicate the mass characteristics of an operational Orbiter. The leading edges of the wings will be clad with glass-fibre and an instrumentation boom will be mounted on the nose. Fuel-cell reactants will be high-pressure gaseous hydrogen and oxygen (operational flights will use liquid hydrogen and oxygen), and simulated rocket nozzles will ensure realistic airflow characteristics across the boat-tail rear end. Standard aircraft-type ejection seats will be fitted for these drop tests, with blow-out panels in the roof of the flight deck providing an escape path if the vehicle has to be abandoned in the air.

The first of a number of captive flights will take place in February next year, with a gradual progression from simple take-off and landing to long-duration handling trials of the 747/Orbiter combination. The Orbiter will be unmanned for the first 15 flights, but by May 1977 the first of six "captive-active" tests will provide an opportunity for manning, powering up the equipment and testing the flying controls while remaining anchored to the 747. During these tests the crew will be able to check their procedures in preparation for the first of eight manned drop-tests, which Nasa hopes to begin next year.

A typical flight profile begins with a climb to 24,000ft, followed by a turn on to the desired heading relative to the runway selected for the landing. When the checks are completed the combination resumes its climb, to 28,000ft. From this altitude, about 31 n.m. from the runway, the carrier aircraft pitches 9° nose-down and releases the Orbiter at 260kt, 22,000ft. Carried on a single attachment point under the nose and two latches under the rear fuselage, the Orbiter pitches up 6° relative to the 747 and flies off the carrier aircraft. After separation the two vehicles bank in opposite directions to avoid a collision, and the Orbiter glides back to the Dryden Research Centre.

On early flights a tailcone will shroud the simulated main engines, but on later drop-tests the cone will be jettisoned so that the flow across the inert rocket engines of a Shuttle returning from space may be simulated. The tailcone is carried at all times during mated flight to prevent turbulence set up by the aerodynamically dirty boat-tail from impinging on the

747's tail. The Orbiter will normally be released about 55min after take-off and will take about 5min to reach the ground.

Follow-on Shuttle

Future American space-transportation systems will build on Shuttle-derived technology. This includes high-pressure/high-energy rocket engines, re-usable thermal insulation, large-diameter/high-thrust solid-propellant rockets, composite materials for structures, and better performance simulation and environmental-prediction techniques.

Decision on a go-ahead with any one of several proposed Shuttle follow-on configurations obviously depends on the traffic achieved during the early 1980s. If Nasa is correct in its assumptions about initial traffic demands, the present Orbiter will need to be replaced by a second-generation re-usable transporter by the early 1990s.

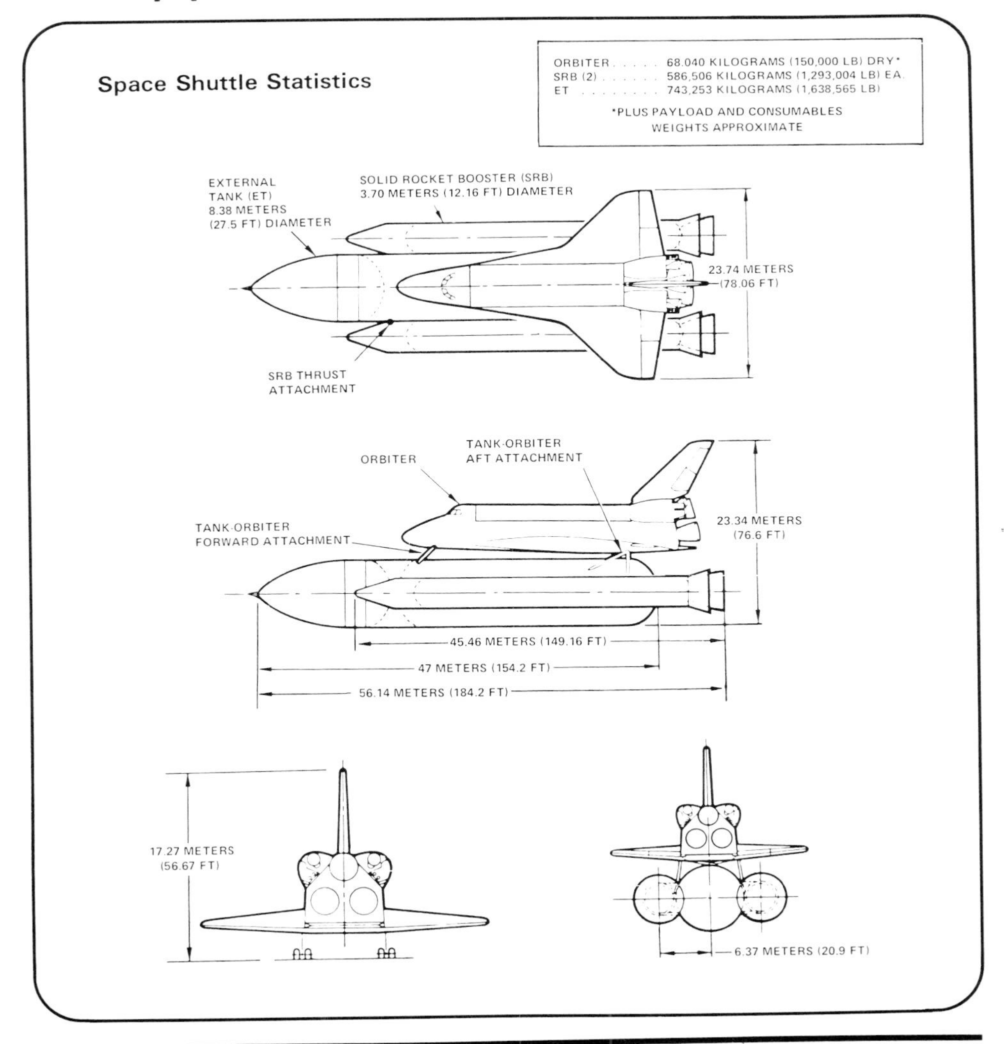

THE QUESTION OF LIFE ON MARS STILL TO BE ANSWERED BY VIKING

by the YEARBOOK Staff

It seems to be a dead planet. All that moves in the chill wind is dust. Ruined towers of long-dead cities do not rear above the rocky horizon. Curious little green men do not peer into the lens of the camera. The surface of Mars seems without life. Whether life exists beneath the surface, scientists are not sure.

Many new facts are available to scientists following the successful landing of two Viking spacecraft on the surface of Mars (Viking I on July 20; Viking II on September 3, 1976). However, these new facts do not clearly answer the question "Is there life on Mars?" Initial experiments have yielded results with both biological and chemical explanations.

Other experiements have yielded more positive answers. The color of the Martian sky, for example, is not black—like that of the moon—but pink. The color is caused by light scattering on dust particles in the thin air. Mars is literally a red planet. The Martian landscape is a reddish rust-brown color, caused by a surface layer of iron oxides.

Experiments have also shown that the principal components of the Martian soil are silicon, iron, calcium, and aluminum. Some scientists believe the soil may be composed of stable hydrates of various elements, because of the large amounts of water released when samples were heated to a high temperature.

Viking I's initial biological experiments generated higher-than-anticipated amounts of oxygen and carbon dioxide from the samples. The amounts produced did not fit previously-observed biological or non-biological experimental results, and are causing more detailed study. Viking II experiments also have produced puzzling readings.

THE FLIGHT TO MARS

The actual flight to Mars began with the launching of two spacecraft in August and September 1975. Carried aloft from the Air Force's Cape Canaveral Launch Station 41 by Titan III/Centaur rockets, the spacecraft were sent forth on a minimum-energy trajectory toward Mars. After 460 million miles and 310 days of travel, the spacecraft caught up with Mars, two-thirds of the way around the Sun from their launch point.

Viking is the most highly automated space mission NASA has ever conducted. The reason for this automation is the time needed to communicate with the spacecraft. Because of the distances involved, it takes more than 37 minutes for a radio message to travel from Mars to Earth and back.

THE VIKING SPACECRAFT

The Viking Spacecraft is actually two vehicles: a planetary orbiter and a lander.

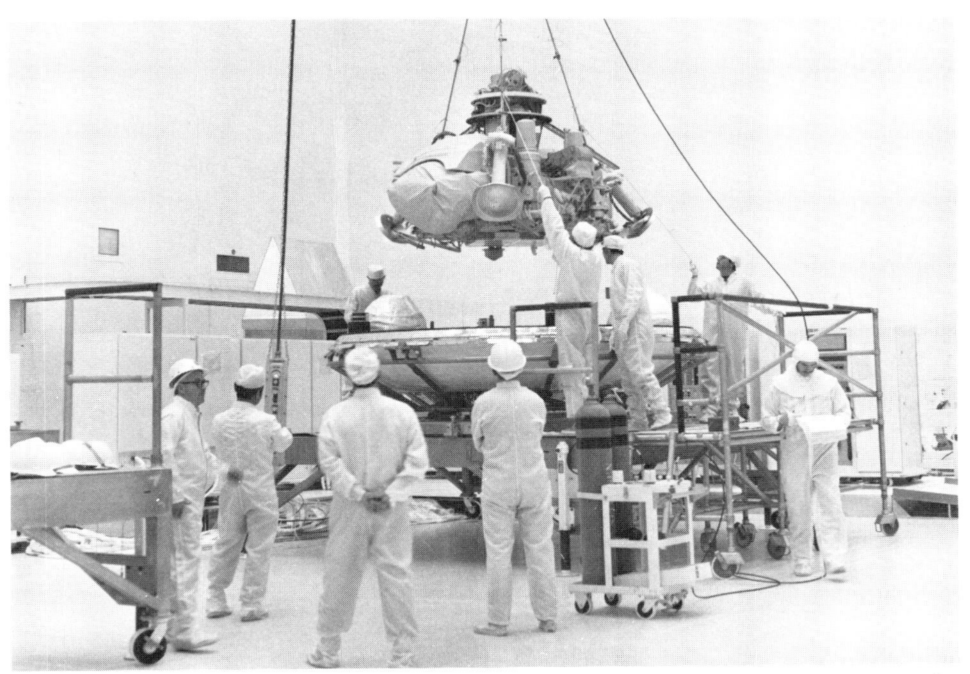

Courtesy of NASA

A Viking lander module is lowered into its capsule, prior to sterilization and launch. Note stowed position of landing legs.

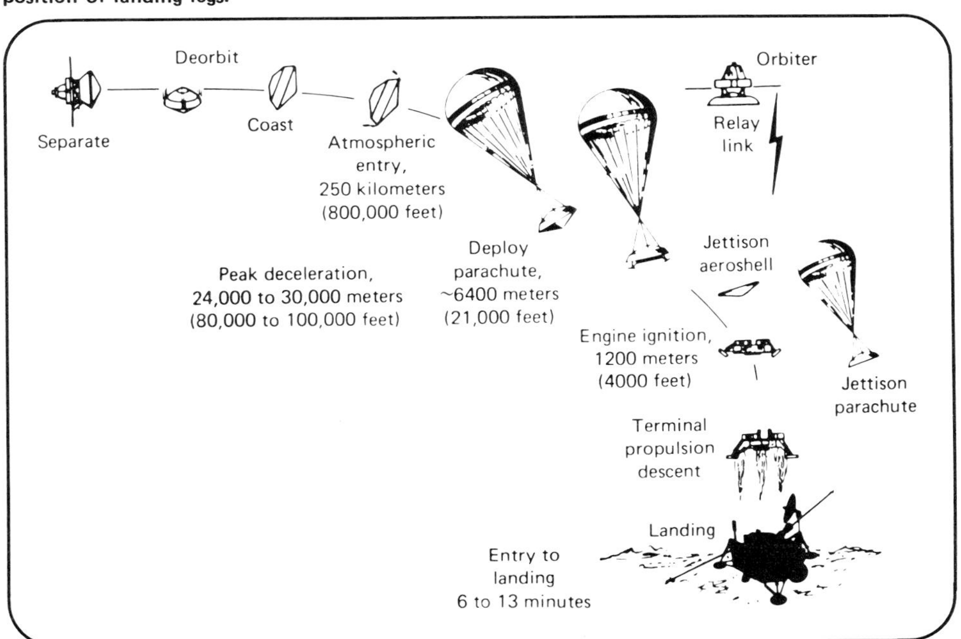

Courtesy of Martin Marietta

Courtesy of NASA

This is the first photograph ever taken on the surface of the planet Mars. It was obtained by Viking I just minutes after the spacecraft landed successfully. The center of the image is about 1.4 meters (five feet) from Viking Lander camera #2. We see both rocks and finely graduated material—sand or dust. Associated with several of the rocks are apparent signs of wind transport of granular material. The large rock in the center is about 10 centimeters (4 inches) across and shows three rough facets. To its lower right is a rock near a smooth portion of the Martian surface probably composed of very fine-grained material. It is possible that the rock was moved during Viking 1 descent maneuvers, revealing the finer-grained basement substratum; or that the fine-grained material has accumulated adjacent to the rock. At right is a portion of footpad #3. Small quantities of fine grained sand and dust are seen at the center of the footpad near the strut and were deposited at landing.

The orbiter was designed to stay in orbit around Mars, serving as a radio relay link with Earth, visually mapping portions of the planet, and mapping atmospheric water vapor and surface temperature variations with infrared instruments.

The lander was designed to descend to and operate on the surface of Mars. During the descent, the lander tested the composition and structure of the Martian atmosphere. Safely landed, the lander tested the physical and magnetic proper-

First panoramic view by Viking 1 from the surface of Mars. The out-of-focus spacecraft component toward left center is the housing for the Viking sample arm, which is not yet deployed. At the horizon to the left is a plateau-like prominence much brighter than the foreground material between the rocks. The horizon features are approximately three kilometers (1.8 miles) away. At left is a collection of fine-grained material reminiscent of sand dunes. The dark sinuous markings in left foreground are of unknown origin. Some unidentified shapes can be perceived on the hilly eminence at the horizon toward the right. A horizontal cloud stratum can be made out halfway from the horizon to the top of the picture.

Courtesy of NASA

ties of the soil, the seismic activity of the planet, photographed the surrounding landscape, took readings of atmospheric pressure, temperature, wind direction, and wind velocity, analyzed soil samples, and conducted biological experiments on the soil.

BIOLOGICAL INVESTIGATIONS

One of the recurring questions Man has asked about Mars has been "Is there—or was there—life on Mars?" Three biological experiments performed by the Viking spacecraft attempted to answer that question: the pyrolytic release, the labled release, and the gas exchange experiments.

The pyrolytic release, or carbon assimilation, experiment searches for signs of photosynthesis by micro-organisms. In the experiment, a small soil sample in a closed container is exposed to light and Martian atmosphere enriched with radioactive carbon-14 compounds. After five days, the radioactive atmosphere is purged from the container, and the soil is heated to 625°C to vaporize any organic material. The vaporized gases are passed through a carbon-14 detector to see if any of the carbon had been taken in by micro-organisms.

Courtesy of NASA

This image shows the trench excavated by Viking surface sampler. Area around the trench has ripple marks produced by Martian wind. The trench which was dug early on the eighth day after landing is about three inches wide, two inches deep and six inches long. Steep dark crater walls show how the grains of the Martian surface material stick together (have adhesion). The doming of the surface at the far end of the trench show that the granular material is dense. The Martian surface material behaves somewhat like moist sand on Earth.

Courtesy of NASA

Courtesy of NASA

Viking 2's first picture on the surface of Mars was taken within minutes after the spacecraft touched down on September 3. The scene reveals a wide variety of rocks littering a surface of fine-grained deposit. Boulders in the 10- to 20-centimeter (4- to 8-inch) size range—some vesicular (containing holes) and some apparently fluted by wind—are common. Many of the pebbles have tabular or platy shapes, suggesting that they may be derived from layered strata. The fluted boulder just above the Lander's footpad displays a dust-covered or scraped surface, suggesting it was overturned or altered by the foot at touchdown. A substantial amount of fine-grained material kicked up by the descent engines has accumulated in the concave interior of the footpad. Center of the image is about 1.4 meters (5 feet) from the camera. Field of view extends 70° from left to right and 20° from top to bottom. Viking 2 landed at a region called Utopia in the northern latitudes about 7500 kilometers (4600 miles) northeast of Viking 1's landing on the Chryse plain 45 days earlier.

The labeled release experiment is based on four assumptions: 1) micro-organisms must be present to recycle organic matter; 2) biochemical reactions are carried on in a liquid (such as water); 3) microorganisms take in compounds and release gas; 4) compounds taken in could include simple carbon molecules. The experiment is simple. A soil sample in a sealed cup is moistened with a carbon-14 rich nutrient solution. For 11 days the sample is incubated, and the atmosphere is sampled for radioactive carbon gases (carbon dioxide, carbon monoxide, methane).

The gas exchange experiment operates on a simple principle: gaseous changes in an atmosphere indicate that life processes are going on. A soil sample is immersed in a nutrient solution, and incubated for 12 days. The atmosphere is analyzed with a gas chromatograph on a regular basis to see if signs of metabolism appear.

These experiments produced puzzling results for the scientists. For example, the gas chromatograph recorded the release of 15 times the anticipated amount of oxygen, plus a much greater amount of carbon dioxide than was expected. Since there were both biological and chemical reasons for such results, the data are being studied very closely by Viking scientists before arriving at any decision.

SPACECRAFT PROBLEMS

Despite the overall success of the Viking mission, there were a few mechanical problems that developed. Considering the distance and length of time involved with the mission, some problems were inevitable.

Close observation of the proposed landing sites resulted in some adjustment in both location and time of landings. The Viking I lander touched down in the Chryse Planita basin; Viking II in the Utopia Planita region.

After landing, the landers performed most of their experiments, including moving rocks with the sampler arm to dig soil unexposed to the ultraviolet radiation

Courtesy of NASA

Shining on the Martian surface near the Viking 2 spacecraft is the aluminum shroud, or cover, which protected the collector head of the surface sampler instrument during Viking's year-long journey from Earth. On September 5, two days after Viking 2 landed, the surface sampler was rotated from its parked position atop the spacecraft and pointed downward about 40 degrees. The shroud was then ejected by a set of eight springs positioned around its base. It struck the porous rock at the bottom of the picture, bounced about 20 inches, hit the surface again and bounced another 20 inches. The scar left by the second bounce is faintly visible halfway between the shroud and the rock it struck. The shroud is 12 inches long and 4½ inches in diameter. The large rock just beyond it is about 2 feet long and about a foot thick. At lower right is the support structure of one of the spacecraft's three landing legs.

bombarding the surface. It was during the performance of these experiments that the problems appeared.

The Viking I sampler arm jammed at first when a locking pin failed to fall free. Additional extension of the arm released the pin, and freed the arm.

The Viking I seismograph failed to unlock and begin functioning. This problem prevents the scientists from pinpointing seismic activity through comparison of readings from both landers.

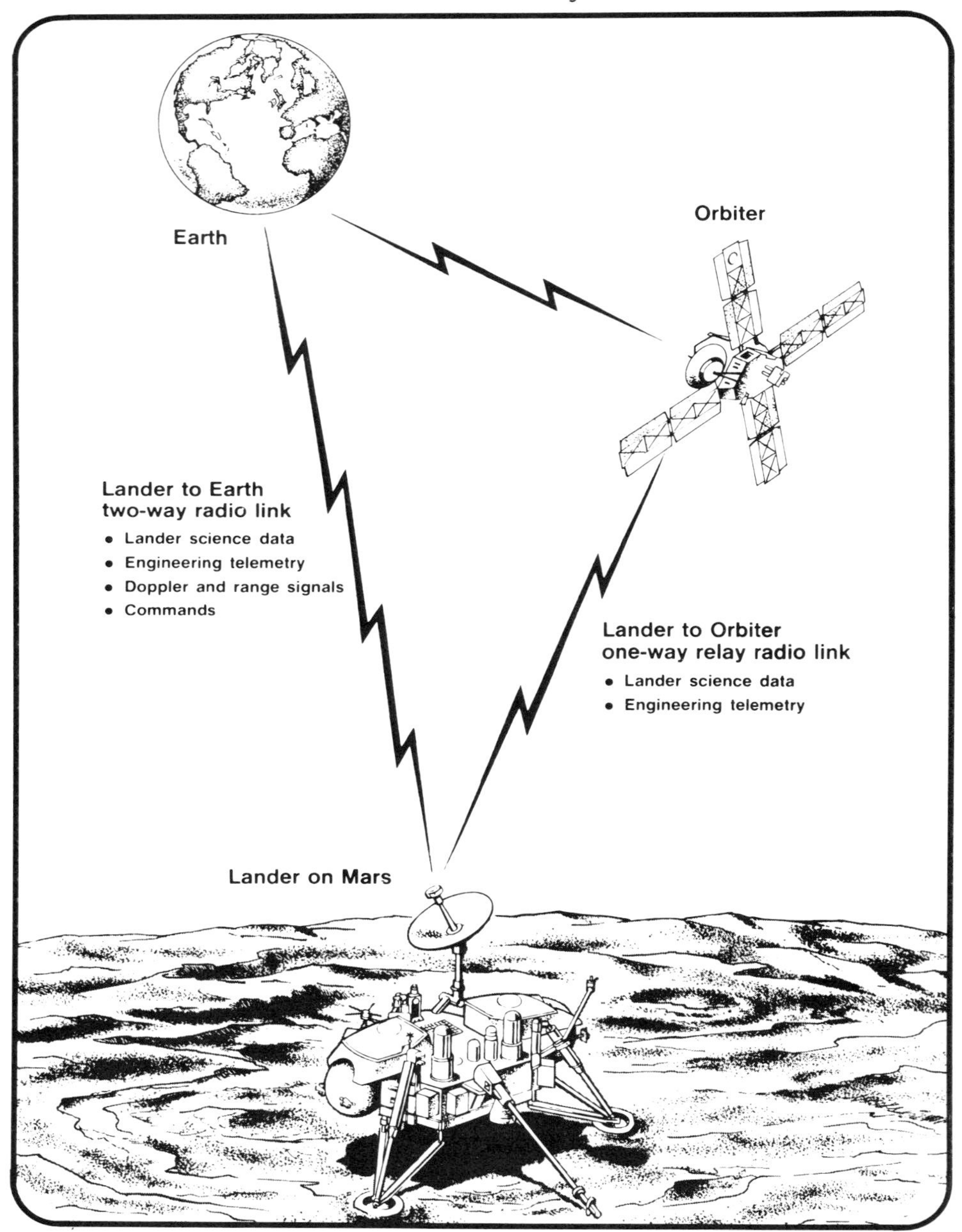

SPORT AVIATION

AVIATION YEARBOOK

INTRODUCTION

In the beginning, all aviation was for sport. Chinese, Japanese, Greeks, and Malays flew gaily decorated kites for amusement and public celebration. By 1790, Frenchmen and Englishmen were floating aloft in wicker baskets slung beneath hot air balloons for sheer diversion and recreation. While Napoleon conquered Europe, one of his subjects made the first successful parachute descent. And years before the 20th century arrived, Germans and Americans were fluttering downhill in the world's first manned gliders.

It didn't take long for all those balloons, aircraft, and parachutes to become virtual captives of military and industrial enterprises. Two great wars and one industrial revolution produced such sophisticated and complex hardware that purely recreational aviation seemed destined for oblivion. And yet, despite that trend, the sheer personal desire for escape, nostalgia, daring, and speed have shaped a totally unique area of flight. The methods and hardware may differ, but when the bonds of gravity are challenged for the pure fun of it, we call that Sport Aviation.

You've heard a lot about astronauts in the past 20 years, even though those popular space travelers have been vastly outnumbered by a very different type of flyer, the aeronaut. Aeronauts know nothing of rocket power and zero "G" orbits as they waft aloft in gaily colored hot air balloons in search of beauty, solitude, and favorable winds.

The non-competitive International Balloon Fiesta took place in Albuquerque, New Mexico, with over 125 of these big floaters wafting for distance, accuracy, and navigation. For six days, more than 200 aeronauts flew, talked, inflated, and cooperated in the unique spirit which has permeated this gentle form of recreational flying.

Capt. Eddie Allen of Indianola, Iowa, made his 3,254th parachute jump this year, saying, "When I get old I'm going to quit jumping." Capt. Eddie is 79. Lots of other youngsters have taken up skydiving, which in 20 years has grown into a sophisticated and challenging sport.

This year's National Collegiate Parachuting Championships were staged at the Jumpwest Parachute Center in Star, Idaho. Despite a capricious pre-winter storm which plummeted temperatures below freezing and dusted the surface with snow, 93 jumpers from 27 schools completed six competitive rounds. When the votes were tallied, the U.S. Air Force Academy walked away with the Gavin Gavel, symbol of the best collegiate parachuting team in the country.

The ultimate level of national skydiving competition was dominated by a 31-year-old high school teacher from Charlotte, North Carolina. Jimmy Davis, the man in white, scored second in style, second in accuracy, and first overall to become the U.S. National Parachute Champion. Davis has almost 4,000 jumps in 13 years of skydiving.

Hang gliding is one of those intensely personal forms of Sport Aviation. Personal because you just glide off a hillside, attached to a cloth kite by a leather harness. Enthusiasts consider it an inexpensive way to enjoy the very essence of flight. Others call it an expensive way to get hurt.

Arguments aside, hang gliding offers some really unique advantages for the would-be flyer. It's truly inexpensive (about $700-$900 for a wing), completely portable, and totally free from regulation and red tape.

There is little doubt that hang gliding has established itself as a growing element in Sport Aviation. Manufacturers and schools are springing up from coast to coast and the lure of simple bird-like soaring will surely attract more and more devotees.

Elsewhere, the more traditional field of sailplane soaring continues to grow in numbers and sophistication. One measure of that sports maturity is the annual Smirnoff Sailplane Derby, a long-distance race from Los Angeles to Washington, D.C. In 16 days of flying, this covey of powerless aircraft raced the elements and each other across three thousand miles of real estate. The race was won by Wally Scott, Sr. in a Schweizer I-35 A.

Along the Allegheny Expressway, Karl Striedick became the first man to fly a sailplane 1,015 miles on an out-and-return flight. He did it for the beauty, the excitement, and because "the eagles do it."

This year saw the demise of a 29-year tradition in Sport Aviation. The all-women Powder Puff Derby was flown for the last time from Sacramento, California, to Wilmington, Delaware. Trina Jarish of Irvine, California, won this year's race in her Beechcraft A-35. There were 176 entrants from 39 states and 4 countries competing in this last running of the Derby, which has succumbed to escalating costs.

One sport airplane made aviation history in 1976. Joe Zinno's Olympian ZB-1 became the first man-powered aircraft to fly from level ground in the Western Hemisphere.

Man-powered aircraft or MPA for short, have received increasing attention since 1959. In that year, Henry Kremer, a British industrialist, offered a prize of 50,000 pounds sterling for the first person to achieve man-powered flight for one mile over a figure-eight course. Since that time, several attempts have been made and no one in the U.S. has come closer than Joe Zinno. With each passing year, the competition becomes more interesting and we will continue to watch it in this section of the YEARBOOK.

While some sporting enthusiasts float aloft in balloons or parachute from perfectly good airplanes, others find recreation in nostalgia. Antique aircraft have become an important part of Sport Aviation and historic aircraft are now eagerly sought by would-be restorers. Elsewhere, aircraft of every description are built or rebuilt in garages and basements and even living rooms by those who seek satisfaction in craftsmanship and personal design.

A discussion of homebuilt aircraft would not be complete without a word about Mr. Jim Bede. Just where are all those little planes of his? Where are all the customers? In fact, where is he? And what about all those homebuilt kits, the ones without engines, without parts. There's news! John Boynton writes about it all in "I, An Actual Customer, Fly An Actual BD-5." Could there really be hope?

These experimental aircraft are a permanent fixture in Sport Aviation with a solid safety record. Unfortunately, builders of one of the most publicized of all homebuilts, the BD-5, have had trouble with crashes in recent months .

Some sport flyers, of course, will always pursue that elusive goal of speed for its own special thrill. At the California National Air Races, at Mojave, souped-up WW II fighters, restored military trainers, and handmade midget Formula Ones thrashed around pylons at speeds which occasionally exceeded 400 mph. Pylon racing has enjoyed a strong comeback in the past several years despite the enormous complexity of outfitting and operating a racing aircraft. Maybe the biggest part of that comeback story is the unbounded enthusiasm of mechanics, handlers, engineers, and pilots who contribute their time and talents for the singular joy of high-speed aerial competition.

In fact, it's simply amazing what people will do for fun and games with an airplane. One guy at Mojave who billed himself as "The Human Fly" rode around the circuit standing atop a DC-8 jetliner. It looks like there is just no end to the possibilities. Sport Aviation is here to stay.

Dan Manningham
December 1976

JARISH WINS LAST POWDER PUFF DERBY

Reprinted by Permission
AVIATION NEWS (August 1976)

Photograph supplied by
THE NINETY-NINES, INC.

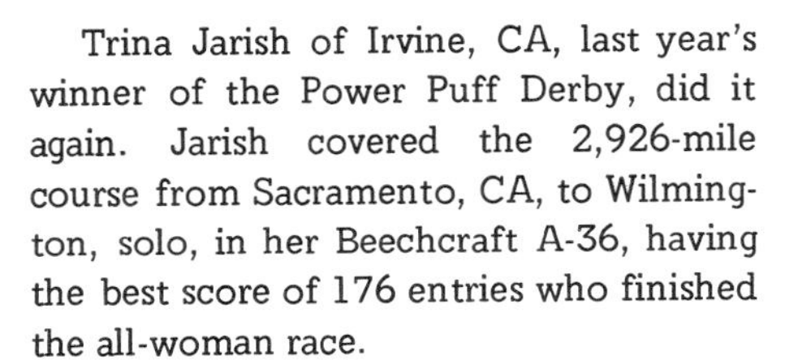

Trina Jarish of Irvine, CA, last year's winner of the Power Puff Derby, did it again. Jarish covered the 2,926-mile course from Sacramento, CA, to Wilmington, solo, in her Beechcraft A-36, having the best score of 176 entries who finished the all-woman race.

Jarish, who was sponsored by Collins Radio, averaged a speed of 209.7–28.7 over her handicap of 181 mph. She won $10,000.

In their third race together, sister team Shirley Cote, Newport Beach, CA, and Joan Paynter, Bakersfield, CA, won $5,000 for second place in a Beechcraft V-35 with a plus score of 27.9.

Just five seconds behind was Helen McGee, Sonora, CA, making the highest speed in the 29-year history of the race, averaging more than 225 mph in a Piper Comanche 400 to take third place.

Veteran racer Marian Jayne and her daughter, Nancy, of Palatine, IL, one of the 15 mother-daughter teams entered, flew a Cessna 210 averaging 27.06 over their handicap to come in fourth.

Honorary Starters for the race were record-holding aviatrix Sheila Scott and Kay Brick, former chairman of the race. At 9:15 a.m., July 9, they dropped the flag at Sacramento's Executive Airport, and racer # 1 was off. The remaining 198 contestants were flagged off in less than two hours by official starter George Griffith of Costa Mesa, CA.

Designated stops for overnight and/or refueling were Fresno, CA, Riverside, CA, Grand Canyon, AZ, Santa Fe, NM, Lubbock, TX, Oklahoma City, OK, Little Rock, AR, Nashville, TN, and Parkersburg, WV. Seventy seven racing planes were still in Riverside at noon the second day waiting for fog to lift.

Delayed by weather, racers were still filtering in Wilmington just before deadline at 8:00 p.m. EDT, July 12. They were from 39 states (including Alaska) and 4 countries, all attempting a perfect cross-country daytime visual flight and their best possible speed. There were five former winners in the race and 60 teams were new racers.

Trina Jarish

Left to right: Bonnie Quenzler, Co-pilot - Pauline Mallary, Pilot - Fifth Place; Marion Jayne, Pilot - Nancy Jayne, Co-pilot - Fourth Place; Helen McGee - Third Place; Joan Paynter, Co-pilot - Shirley Cote, Pilot - Second Place; Trina Jarish - First Place

Pilot-in-command hours covered a wide range. First-time racer, Sharon Fall of Cincinnati, OH, was low-time pilot-in-command in this year's race, having logged only 200 hours. Gini Richardson, Yakima, WA, 1971 winner, was flying her 23rd race with 23,000 hours.

The Powder Puff Derby is the world's longest and largest speed race for light aircraft. Planes must be no more than 12 years old and stock model. At the start planes were inspected for safety and eligibility. Pilots attended educational briefings as part of the pre-race activities.

The perpetual trophy will be displayed in the Smithsonian's new Air and Space Museum in Washington, D.C. The race is being discontinued due to the increased costs of conducting the event.

RESULTS OF FINAL POWDER PUFF DERBY

1-Trina Jarish, Irving, CA, Beechcraft A-36
2-Shirley Cote, Newport Beach, CA; Joan Paynter, Bakersfield, CA, Beechcraft V-35
3-Helen McGee, Sonora, CA, Piper Comanche 400
4-Marion Jayne, Palatine, IL; Nancy Jayne, Palatine, IL, Cessna 210
5-Pauline Mallary, College Park, GA; Bonnie Quenzler, Merritt Island, FL, Beechcraft C23
6-Ginny Wegener, Santa Rose, CA; Lynn Cary, Santa Rose, CA, Cessna P80
7-Pat Jetton, Dallas, TX; Elinor Johnson, Dallas, TX, Beechcraft C33A
8-Pat Forbes, Crystal Lake, IL; Jeanne Rumm, No. Hollywood, CA, Cessna 182
9-Laverne Gudgel, Chowchilla, CA; Marianne McCulloch, Merced, CA, Beechcraft A36
10-Margaret Ringenberg, Grabill, IN, Mooney M20C
11-Evelyn Kropp, Norwich, CN; Jean Batchelder, Laconia, NH, Piper PA-28R
12-Mara Culp, Palm Springs, CA; Alexandra Marshall, Houston, TX, Beechcraft A36
13-Mary Pearson, Center, CA; Harriett Booth, Rancho Santa Fe, CA, Piper PA-28
14-Jeanne McElhatton, San Francisco, CA; Patricia Gladney, Los Altos, CA, Cessna 182
15-Rosemary DeAngelo, Rancho Palos Verdes, CA; Dorothy A. Waltz, Mission Viejo, CA, Cessna 182N
16-Fran Bera, Long Beach, CA, Beechcraft F33A
17-Jeanne Bedinger, Sioux City, IA; Carolyn Rowney, Sioux City, IA, Piper Arrow
18-Sandra Case, Sacramento, CA; Susan West, Sacramento, CA, Beechcraft C23
19-Claudia Beckner, Albuquerque, NM; Roz Kinlen, Albuquerque, NM, Beechcraft A36
20-Norma Futterman, Beverly Hills, CA; Virginia Showers, Los Angeles, CA, Beechcraft A36

21-Dorothy McAllister, Portola Valley, CA; Irene Nealon, San Mateo, CA, Cessna 182
22-Bea Barris, Highland Heights, OH; Lydia Rogers, Broadview Heights, OH, Grumman American AA5B
23-Maybelle Fletcher, Houston, TX; Mary Byers, Pasadena, TX, Cessna 210H
24-Mary Pinkney, Torrence, CA; Kathleen Moskow, Marina del Rey, CA, Cessna 182
25-Jean M. Schiffman, Palos Verdes, CA; Rosetta Wilson, Palos Verdes, CA, Cessna 182
26-Carolyn Pilaar, Greenville, SC; Gary Anne Wheeler, Greenville, SC, Decathlon
27-Jane LaMar, Redding, CA; Mary Grace Sebelius, Redding, CA, Cessna 172
28-Esther Wright, Thomasville, GA; Judy Hall, Macon, GA, Cessna 182
29-Berni Stevenson, No. Hollywood, CA, Mooney M20F
30-Pat Mlady, Wichita, KS; Jackie Luke, Wichita, KS, Cessna 182P
31-Ruth Ruggles, Fort Wayne, ID; Judy Graham, Auburn, ID, Piper PA-28R
32-Jan Gammell, Denver, CO; Ilovene Potter, Federal Way, WA, Piper PA-24B
33-Marjory Robbins, Los Angeles, CA, Piper PA-28
34-Leora Richards, Louisville, KT. Cessna 182P
35-Velda Mapelli, Denver, CO; Stephanie Beuchat, Denver, CO, Piper PA-24-260B
36-Gini Richardson, Yakima, WA, Aero Commander 112
37-Marion Barnick, San Jose, CA; Wilberta Gardiner, San Jose, CA Cessna 172
38-Claudia Jones, Las Vegas, NV; Cathy Jones, Las Vegas, NV, Piper PA-28-151
39-Maisie Stears, Kalamazoo, MI; Esther Bennett, Leonidas, MI, Piper Lance
40-Tina Sturdevant, Valparaiso, IN; Barbara Jennings, Valparaiso, IN, Piper PA32-300
41-Sammy McKay, Grand Blanc, MI; Rene Kirth, Demming, NM, Cessna 210 J
42-Shirley Weinhardt, Williamsport, PA; Catherine Grover, Baltimore, MD, Piper PA-28
43-Ruth Chapman, Mexico, MO; Dorothy Graf, Mexico, MO, Cessna 182N
44-Joyce Odom, Grosse Ile, MI; Alice Gustafson, Orchard Lake, MI, Beechcraft V35
45-Joy Johnson, Houston, TX; Patricia James, Houston, TX, Beechcraft V35B
46-Eva McHenry, San Diego, CA; Diane Stocklin, El Cajon, CA, Piper Cherokee Archer II
47-Marge Bryant, Perkasie, PA; Ginny Terry, Newton, PA, Cessna 182P
48-Loretta Jones, Independence, MO; Helen Hamilton, Blue Springs, MO, Beechcraft A36
49-Tookie Hensley, Riverside, CA; Maurine Wilson, Riverside, CA, Cessna 172
50-Vickie Bruce, Mesa, AZ; Stacy Wachs, Scottsdale, AZ, Cessna 182
51-Ann Lowell, Ft. Sam Houston, TX, Mary Wadington, San Antonio, TX, Piper PA-32
52-Nancy Lynam, Daytona Beach, FL; Eleanor Starkey, Titusville, FL, Beechcraft S35
53-Mary Creason, Grand Haven, MI; Eloise Smith, Kalamazoo, MI, Cessna 182
54-Merle Ann Starer, Philadelphia, PA, Ann Lemmon, Gladwyne, PA, Beechcraft V35
55-Jessie Wimmers, Phoenix, AZ, Wilma Sawatzky, Wichita, KS, Beechcraft V35
56-Susan Adler, Vallejo, CA, Mary Santosuosso, Vista, CA, Cessna 182P
57-Sandra Sullivan, Maple Valley, WA; Bonnie Lou Klein, Bellevue, WA, Cessna 182L
58-Genie O'Kelley, Knoxville, TN; Mickey Childress, Knoxville, TN, Cessna 182P
59-Pamela Gau, Norman, OK; Jill Dacey Norman, OK, Grumman American Tiger
60-Joyce Wells, Larkspur, CA; Alma Hinds, Redding, CA, Cessna 182
61-Esther Grupenhagen, Anaheim, CA, Beechcraft V35B
62-Emily Earlywine, Bedford, IN; Pam Reynolds, Center Point, IA, Cessna 182P
63-Dotty Birdsong, Temple Terrace, FL; Ethel Gibson, St. Petersburg, FL, Beechcraft E33
64-Pam Marley, Phoenix, AZ; Sandy Anson, Phoenix, AZ, Cessna 182
65-Karwyn Blevins, Louisville, KY; Laura Blevings, Louisville, KY, Cessna 172M
66-Belle Hedges, Memphis, TN; Dorothy Wilson, Memphis, TN, Beechcraft V35B
67-Beth Fuhrman, Richmond, CA; Susan Fuhrman, Alameda, CA, Piper Cherokee
68-Valerie Humphreys, Transvaal, So. Africa; Anna Dreyer, Rustenberg, So. Africa, Cessna 172
69-Coralee Tucker, Canoga, CA; Melissa Wreeland, Studio City, CA, Cessna 182P
70-Emma McGuire, Santa Monica, CA; Carole DePue, Henderson, NV, Cessna 182
71-Winnie Duperow, Holt, MI; Amelia Moore, Okemos, MI, Cessna 172
72-Yvette Hortman, Bristol, PA, Grumman American AA-5B
73-Kathleen Wentworth, San Carlos, CA; Rae Gilmore, Belmont, CA, Piper PA-24
74-Barbara Nichols, Burbank, CA; Sally La Forge, Los Angeles, CA, Cessna 182N
75-Joanne Hodges, Amarillo, TX; Barbara Neel, Amarillo, TX, Cessna 172
76-Sandra Simmons, Dallas, TX; Dorothy Warren, Dallas, TX, Grumman American AA-5B
77-Fran Gauger, Vallejo, CA; Lynn Ahrens, San Rafael, CA, Cessna 182
78-Sandra Ledrew, Saratoga, CA; Marjorie Griffin, Los Gatos, CA, Cessna 172

79-Sally Green, Monroe, LA; Leta Drake, Lincoln, NB, Cessna 210
80-Marilyn Cragin, El Paso, TX; Doris Shreve, El Paso, TX, Piper PA-28
81-Lisa Murphy, Columbia, GA; Helen Murphy, Columbia, GA, Beechcraft C23
82-Ali Sharp, Grants Pass, OR, Cessna 177B
83-Shirley Thom, LaCanada, CA; Linda Thom, Englewood, CA, Piper Cherokee
84-Nellie Reynolds, Indianapolis, IN; Lillie Danek, Indianapolis, IN, Cessna 172L
85-Judy Alcombrack, Columbus, OH; Consuelo Huffman, Westerville, OH, Beechcraft V35B
86-Bonnie McSwain, Columbus, OH; Marilynn Miller, Columbus, OH, Grumman American Tiger
87-Sophia Payton, Clearwater, FL; Pat Fairbanks, Cincinnati, OH, Mooney M20C
88-Alberta Nicholson, Salt Lake City, UT; Ruth Kendrick, Ogden, UT, Cessna 182M
89-Barbara Goetz, Fair Oaks, CA; Geraldine Mickelsen, Sacramento, CA, Piper PA24
90-Claudette Parker, Monticello, IN; Linda Leaders, Monticello, IN, Beechcraft V35
91-Francesca Davis, Freeport, Grand Bahamas; Leona Sweeting, Freeport, Grand Bahamas, Cessna 172
92-Julie Ames, Lemore, CA; Diane Mann, Gothenburg, NB, Rockwell Commander 112A
93-Susan Clark, El Cajon, CA; Helen McGee, El Cajon, CA, Piper PA-28
94-Nancy Shaw, Delaware, OH; Helen Stomberg, Delaware, OH, Piper PA-28
95-Kathy Walker, Lansing, IL; Bev Distelhorst, Valpariso, IN, Cessna 172
96-Mary Lowe, Greenville, IL, Betty Board, Bridgeton, MO, Piper PA-28 Arrow
97-Vi Chambers, El Cajon, CA; Dottie Sanders, El Cajon, CA, Cessna 182L
98-Pat Davis, Sunnyside, CA; Jeanine Ceccio, San Jose, CA, Piper PA-28
99-Rachel Bonzon, Santa Monica, CA: Georgia Lambert, Los Angeles, CA, Cessna 182
100-Marie McMillan, Las Vegas, NV; Winnie Howard, Las Vegas, NV, Cessna 172M
101-Jayne Schier, Macomb, IL; Barbara Jenison, Paris, IL, Cessna 172M
102-Esther Gardiner, Reno, NV, Beechcraft F33A
103-Carolyn Zapata, Belmont, CA; Geri Wiecks, San Mateo, CA, Cessna 182
104-Marilyn Jack, Santa Rosa, CA; Gail Bartlett, Monte Rio, CA, Piper PA-28
105-Claire Ellis, Scottsdale, AZ; Cathy Nickolaisen, Mesa, AZ, Cessna 182
106-Jackie Petty, Mountain View, CA; Nancy Rodgers, Los Altos Hills, CA, Piper PA-28
107-Angela Boren, Lubbock, TX; Jana Palmer, Lubbock, TX, Cessna 172M
108-Sherilyn Knight, Windsor, CA; Diane Schirmer, San Francisco, CA, Cessna 182P
109-Sue Swenson, Cincinnati, OH; Pera Beth Swenson, Cincinnati, OH, Piper Arrow
110-Dorothy Dickerhoof, Chanute, KS; Lenore Kensett, Chanute, KS, Beechcraft S35
111-Ruth Bliss, St. Michaels, MY; Caroline Raymond, Poughkeepsie, NY, Piper Archer II
112-Evelyn Snow, Bossier, LA; Betty Heise, Abilene, TX, Piper PA-28
113-Ellinor McElroy, Kent, WA; Elise Smith, Buckley, WA, Cessna 177
114-Shirley Winn, Woodland, CA; Anne Molina, Sacramento, CA, Cessna P206B
115-Beth Howar, Santa Barbara, CA; Joan Steinberger, Goleta, CA, Beechcraft V35B
116-Bonnie Lee Seymour, Carnelian Bay, CA; Linda Claire Seymour, Carnelian Bay, CA, Piper PA-28
117-Micki Thomas, Pompton Lakes, NJ; Diana Caggiano, Pompton Lakes, NJ, Cessna 182
118-Lorraine Newhouse, Tucson, AZ; Terry Robertson, Globe, AZ, Cessna 182P
119-Sally Tanner, Tampa, FL; Linda Tanner, Tampa, FL, Piper PA-32
120-Marilyn Copeland, Wichita, KS, Piper Lance
121-Dorene Christensen, Santa Ana, CA; Janice Christensen, Santa Ana, CA, Piper PA-28R
122-Carolyn Clarke, Salt Lake City, UT; Lila Fielden, Salt Lake City, UT, Bellanca Decathalon
123-Clarissa Quinlan, Anchorage, AK; Betty Rogers, Anchorage, AK, Cessna 172K

124-Mildred Langwell, Grand Terrace, CA; Alberta Brown, Sunnymead, CA, Piper PA-28
125-Annette Fedor, Cleveland, OH; Rosemarie Mintz, Moreland HIlls, OH, Beechcraft A36
126-Janie Allen, Virginia Beach, VA; Linda Hollowell, Virginia Beach, VA, Cessna 182L
127-Lois Felgenbaum, Carbondale, IL; Carol Edwards, New Madrid, MO, Cessna 172M
128-Gene Fitzpatrick, Torrance, CA; Kathleen Woodson, Grissom AFB, IN, Cessna 182P
129-Jan Churchill, Chesapeake City, MY, Dorothy Miller, Graterford, PA, Cessna 182
130-Pam Vander Linden, Fallbrook, CA; Ruth Dilg, Capistrano Beach, CA, Bellanca Viking 17-31A
131-Mary Waite, Pittsburgh, PA; Helen Davison, Gibsonia, PA, Cessna 182
132-Janet Mauritson, Tulsa, OK; Sue McBride, Tulsa, OK, Piper Comanche
133-Joan Kerwin, Wheaton, IL; Caroline Smith, Naperville, IL, Cessna 172M
134-Sally Adams, Hot Springs, AR; Linda Holmes, Albuquerque, NM, Beechcraft E33A
135-Jerry Ann Melton, Dallas, TX; Brenda Strickler, Grapevine, TX, Piper PA-28
136-Alma Hitchings, Howell, NJ; Diane Shaw, Wildwood, NJ, Piper PA-24
137-Mary Ann Hamilton, Fairway, KS; Bobbi Miller, Mission Hills, KS, Piper PA-24C
138-Pauline Geunung, Indianapolis, IN; Lois Hawley, Danville, IN, Cessna 177 RG
139-Marian Piper, Bossier City, IA; Amy Pilkinton, Bossier City, IA, Mooney M20E
140-Sharon Fall, Cincinnati, OH; Sally Berryhill, Fairborn, OH, Cessna 172M
141-Doris Jean Ritchey, San Diego, CA; Catherine Hatch, La Mesa, CA, Cessna 182P
142-Susie Evans, Staton, TX; Cheryl Shaw, Lubbock, TX, Piper Arrow
143-Betty Jane Schermerhorn, Dunrobin, Ont, Canada; Carolyn Thomas, Ottawa, Ont., Canada, Cessna 177 RG
144-Kathleen Snaper, Las Vegas, NV; Lois Erickson, Las Vegas, NV, Beechcraft A-24R
145-Shirley Lehr, Sacramento, CA; Vija Berry, Sacramento, CA, Cessna 210E
146-Lavonna Alter, Coal Valley, IL; Carolyn Pobanz, East Moline, IL, Piper PA-28
147-Norma Vandergriff, Edmond, OK; Gwendolyn Crawford, Enid, OK, Cessna 182P
148-Kathryn Hach, Ames, IA; Jean Ellingson, Northwood, IA, Cessna 182P
149-Jo Dieser, Bradley, CA; Geneva Cranford, Salinas, CA, Cessna 182N
150-Joan Enyeart, Mountain View, CA; Evelyn Lundstrom, Sunnyvale, CA, Cessna 172M
151-Linda Hargraves, El Dorado, AR; Cynthia Sutton, N. Little Rock, AR, Piper Archer II
152-Nanette Gaylord, Denver, CO; Patricia Udall, Boulder, CO, Cessna 177 RG
153-Marjorie Freeman, Chappaqua, NY; Penelope Amabile, Rye, NY, Cessna 172
154-Joyce Williamson, State College, PA; Mary Hull, Montoursville, PA, Piper Comanche PA24
155-Mary Lou Newman, Long Beach, CA; Marilyn Jensen, Huntington Beach, CA, Cessna 172H
156-Peggy Mayo, Bucksport, MA; Carolyn Arnold, Beloit,WI, Cessna 177RG
157-Phyl Vetter, Dover, DL; Martha Howell, Northeast, MY, Piper PA-28R
158-Virginia Rainwater, Reseda, CA, Cessna 182
159-Clarice Bellino, North Caldwell, NJ; Theres a Marais, Cape Province, So. Africa, Piper PA-28
160-Broneta Evans, Manjum, OK; Velma Woodward, Oklahoma City, OK, Piper PA-28R
161-Elsie Wahrer, Barrington, IL; Julia Konger, Hamshire, IL, Beechcraft S35
162-Jane Menzies, Grove City, PA; Susan Simler, Verona, PA, Piper PA-24B
163-Lois Broyles, Leonard, MI; Kathryn Gerhold, Corunna, MI, Cessna 177 RG
164-Jerry Roberts, Haddon Heights, NJ; Kitty Pankow, Haddonfld, NJ, Cessna 206P
165-Margret Bryant, Springfield, OH; Betty Angstadt, Grove City, OH, Piper PA-28
166-Charlene Falkenberg, Hobart, IN; June Basile, Glenview, IL, Mooney M20F
167-Louise White, Asheville, NC; Lucy Merritt, Easley, SC, Mooney M20F
168-Ima Jean Huff, Amarillo, TX; Chris McClain, Amarillo, TX, Mooney M20C
169-Jo McCarrell, Dallas, TX; Patricia Evans, Hurst, TX, Bellanca 17-31A
170-Betty Parthemer, Harrisburg, PA; Ann Turley, New Cumberland, PA, Cessna 182J
171-Irene Brunks, Massapequa, NY; Nina Claremont, Massapequa, NY, Piper PA28R
172-Jana Rae Norrell, Memphis, TN; Barbara Martens, Memphis, TN, Mooney M20C
173-Janis Blackburn, Old Bridge, NJ; Claire Kurica, Howell, NJ, Mooney M20E
174-Betty Elliott, Slingerlands, NY,; Doris Miller, Schenectady, NY, Beechcraft A36
175-Diane Fisher, Ronkonkoma, NY; Heidi Hafner, Bayside, NY, Piper PA-32
176-Tere Lynch, Enid, OK; Pauline Wade, Concord, CA, Cessna 182G

MOTHER-DAUGHTER TEAM WINS ANGEL DERBY

Reprinted by Permission
FLORIDA AVIATION JOURNAL (July 1976)

Photograph supplied by
THE NINETY-NINES, INC.

by Marge Forood

When entries closed on March 15th, 1976, sixty-three aircraft, flown by one-hundred and fourteen women pilots were entered in the twenty-sixth annual "All Women's International Air Race."

For the third time in a decade the fabled "Angel Derby" was sponsored by the city of Fort Lauderdale and was further honored by having the city name the Derby their only special Bicentennial event.

Originating at Quebec City Airport, Canada, the one-thousand seven-hundred and thirty-two mile race followed an historical path along the eastern seaboard of the United States. Refueling cities along the route were Boston, Schenectady, Wilmington, Delaware, Richmond, Wilmington, North Carolina, Savannah, and St. Augustine.

Open to any qualified woman pilot with extensive cross-country experience the "Angel Derby" is the only race of its kind in existence today. Pilots may fly only stock, unsupercharged, fixed-wing aircraft of not less than 145 and not more than 570 horsepower. Each aircraft is assigned a handicap in miles per hour based on minimum speeds obtained from a timed flight test over a measured course. The racer who beats her handicap by the largest margin is the winner.

By impound deadline May 24th, fifty-seven aircraft had checked in at Quebec City Airport for the qualifying inspection. Six entries had been forced to scratch due to technical difficulties. While planes and papers were being carefully scrutinized by the Canadian Ministry of Transportation, the women pilots enjoyed a welcome by the Mayor of Quebec City and three days of warm Canadian hospitality.

At 9:00 A.M. Thursday, May 27, after four days of steady rain, the sun finally made an appearance and the race was on.

Canadian and American dignitaries cheered as the Angels lifted off at one minute intervals and headed for Boston, Massachusetts.

The race deadline was Saturday, May 29, at twelve noon giving the racers two and a half days to complete the course. All flying had to be done under VFR and daylight conditions.

By sunset the first day everyone had come and gone in Boston; four aircraft had stopped for the night in Schenectady, New York; seven had stopped in Wilmington, Delaware; nineteen were RONing in Richmond, Virginia; eighteen were tied down in Wilmington, North Carolina; six were bedded down in Savannah, Georgia, and numbers 2, 12, and 14 had made it to St. Augustine, Florida.

On Friday, May 28th, a cold front moved across the Carolinas, Georgia, and the Florida segments of the race path. The weather deteriorated to the point that eight aircraft were prevented from continuing from Wilmington, North Carolina. Three other racers who sat out the weather in Wilmington and continued on the next day failed to reach Fort

Marion Jayne, Nancy Jayne

Lauderdale by the noon deadline.

By ten Friday morning Angels 2, 12, and 14 had passed the finish line at the Executive Airport tower in Fort Lauderdale. They were greeted with wild cheers from the terminus committee, Virginia Young, Fort Lauderdale's Vice Mayor and the media. By 12:00 noon Angels 1, 4, 11, 14, 17 and 37 had been clocked by the finish line and by Friday evening numbers 5, 8, 16, 20, 22, 24, 27, 30, 33, 34, 41, 45, 46, 49, 53, 61, and 62 were in. The eleven remaining planes raced across the finish line Saturday morning in time to meet the deadline.

For the pilots the race was over, but for the race committee the difficult job of computing Official Times for each pilot had just begun. Only when all the numbers were in and carefully calculated would a winner be determined.

On Monday, May 31st, at the post-race meeting of all the pilots the winners were announced.

Angel "2" placed first.

1–Marion Jayne and her daughter Patricia Keefer from Palatine, Illinois flying an Aero Commander 200D. The Aero Commander handicap was 179.000, Angel One's average speed was 203.581.

Score - 24.581

Angel "1" placed second.

2–Norma Futterman and Virginia Showers from Beverly Hills and Los Angeles, California flying a Beech A-36 Bonanza. Bonanza handicap 176.000. Angel Two average speed was 197.022.

Score - 21.022

Angel "5" placed third

3–Berni Stevenson from North Hollywood, California flying her Mooney M20F Executive solo. Mooney M20F Executive handicap 159.000. Angel Five's average speed 179.400.

Score - 20.400

RED BARON WINS MOJAVE

Idaho Crew Gets It All Together

Reprinted by Permission
AVIATION NEWS (July 1976)

Photos by Permission AIR PROGRESS

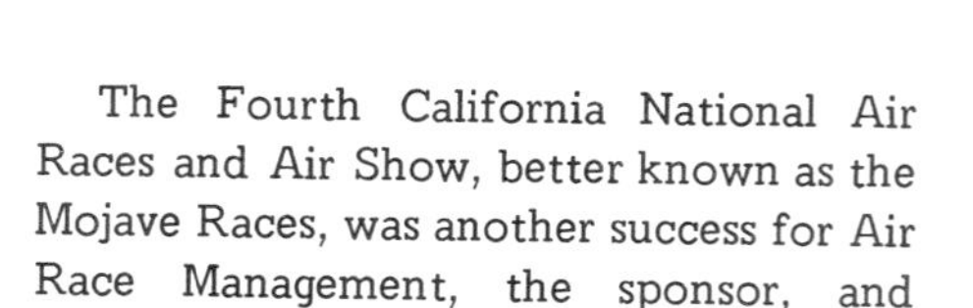

The Fourth California National Air Races and Air Show, better known as the Mojave Races, was another success for Air Race Management, the sponsor, and United States Air Racing Association, the sanctioning organization.

Clay Lacy, president of Air Race Management, and his associates, Dick Sykes, Lyle Shelton, Alice Rand, Dwight Brooks, Gary Thompson, Dave Dwoskin and Richard Runyon, have created an innovative approach to air shows which is proving very popular with the aviation and general public.

For example, the Mojave event pioneered such exciting events as air racing drag races for unlimiteds. Then, this year, they had the "Human Fly," a wing walker perched atop a DC-8 flown by Clay Lacy. This feat attracted press attention all over the country. And, of course they had a good selection of all the air show regulars including Bob Hoover, Duane Cole, Debbie Gary, Mira Slovak, Dick Hunt and others. They also kept the show fast paced, so that spectators were not faced with long periods of inactivity.

The U.S.A.R.A. people held up their end of the bargain, too, providing a first-class racing show for the spectators. All of the usual racing people and aircraft were there, but this year, there were some unexpected changes in the results. Importantly, the Red Baron people got their exciting contra-rotating prop P-51 working right and won handily at an average speed of 406.718 mph.

Complete race results follow:

MOJAVE RACE RESULTS

UNLIMITEDS: 1-No. 5, The Red Baron, Roy McClain, 406.718 mph; 2-No. 81, Precious Medal, Gary Levitz, 376.349; 3-No. 7, Candy Man, Charlie Beck, 349.265; 4-No. 16, Baby Gorilla, Lloyd Hamilton, 348.812; 5-No. 85, Iron Mistress, Clay Klabo (cut pylons No. 1 and No. 4 during lap 2, and pylon No. 6 in lap 1); DNF: No. 69, Miss Candance, Cliff Cummins, engine failure; and No. 86, John Putman, engine failure.

AT-6: 1-No. 44, Sky Prints, Ralph Twombly, 211.320 mph; 2-No. 99, Gotcha, Pat Palmer, 210.671; 3-No. 5, The Red Baron, Roy McClain, 207.757; 4-No. 73, Miss Everything, Ralph Rina, 204.406; 5-No. 9, Cum'n Thru, Marshall Wells, 201.142; 6-No. 3, Two-Five Charlie, 198.577.

The Winning Red Baron RB-51

"The Human Fly" flies.

FORMULA 1: 1-No. 9, Proud Bird, Fred Wofford, 221.888 mph; 2-No. 87, Pogo, Pat Hines, 209.347; 3-No. 41, Shark, Bill Skliar, 204.317; 4-No. 28, Falcon, Bob Downey.

DISQUALIFIED: 1-No. 16, Shoestring, Ray Cote, for engine violation. According to the Violation Review Board, Cote must forfeit all racing awards received at Mojave; he has been suspended from racing at the Reno International Air Races, and has received a two-year probation period. He originally placed first with a speed of 228.546 mph; 2-No. 95, Pegasus, Glen Tuttle, for engine violation. Tuttle must forfeit all racing awards received at Mojave, according to the Violation Review Board. He had placed third in the Gold race with a speed of 211.849 mph.

SPORT BI-PLANE: 1-No. 89, Sorceress, Don Beck, 201.585 mph; 2-No. 1, Sundancer, Pat Hines, 199.949; 3-No. 5, White Knight, Don Fairbanks, 168.610; 4-No. 14, Super Chick, Tom Wrolstad, 164.247; 5-No. 8, Super Mong, Clem Fischer, 157.139; 6-No. 46, Love American Style, Robert Clark, 141.398; 7-No. 90, The Washoe Zephyr, Stan Brown, 140.494.

Under the direction of new U.S.A.R.A. president Don Beck, the sanctioning organization seems to be pursuing an aggressive program to return the sport to the glory which it enjoyed in the 1920s and 30s. Working with effective and innovative promoters like Air Race Management, the U.S.A.R.A. seems destined to achieve an important place for air racing in the motorsports spectrum.

The above sequence, top to bottom, shows the spectacular gear-up landing made by Lyle Shelton's Bearcat "Spirit of '77" after an oil line ruptured.

The "Spirit of '77" after landing on the two lower prop blades and the extended tail wheel.

13 YEARS OF RENO NATIONAL CHAMPIONSHIP RACES

Reprinted by Permission
AVIATION NEWS (October 1976)

Photos by Ronald Miller (Greater Reno Chamber of Commerce)

After 13 years of operation, the Reno National Championship Air Races are still going strong. And with $100,000 in prizes available to the competitors, it's no wonder. Not only does the race provide a bonanza for the competitors, it's a bonanza for the fans, as well.

In addition to 3 days' racing, the fans were treated to one of the best air shows anywhere. For example, there was Frank Tallman, doing aerobatics in his venerable Grumman J2F Duck. Lefty Gardner thrilled everyone flying his P-38 Lightning. The Red Devils, Gene Soucy, Charlie Hillard and Tom Poberezny, and Debbie Gary flew Pitts Specials with style and verve. Bob Hoover provided a feast, flying a T-33, a P-51 and a new Rockwell Commander Shrike. Jim Raymond showed what can be done with a T-6 and Art Scholl flew the Super Chipmunk in his usual spectacular fashion.

The highlight of the show, however, probably was the Confederate Air Force staging their mock attack on Pearl Harbor. Taking the theme from the movie, *Tora, Tora, Tora*, the CAF arrived with a full complement of simulated Japanese Zero fighter who faced opposition from a B-17, a P-39 and a P-40. Planes roared, bombs dropped, people screamed. It was all too real, and very thrilling for the spectators. Executed with superb showmanship, the CAF's show included a simulated forced landing by the B-17. The gigantic old bird flew low across the ramp, getting so low that many spectators feared that it had crashed. Then, with one gear dangling down, it seemed certain to crash on its next pass. The racing got off to a promising start when Don Whittington, Ft. Lauderdale, FL qualified his P-51 at 438.806 mph for a qualifying record. He was followed by John Crocker, San Mateo, CA at 536.633 mph and Roy McClain in the Red Baron RB-51 at 436.094. For the first time in many years, there were no Grumman Bearcats competing in the Unlimited class.

The T-6/SNJ competitors also got off to a good start with Ralph Rina, Huntington Beach, CA, qualifying at 210.116 mph. He was followed by Don Dewalt, Los Angeles, CA, at 209.709 mph and Dick Sykes, Toluca Lake, CA at 208.897 mph. Leader among the Formula 1s was Fred Wofford, Anaheim, CA at 237.362 mph, who was later disqualified for an alleged infraction of the class rules. A surprising second was Judy Wagner, Marina Del Rey, CA at 231.760 mph. Top qualifier in the biplanes was Don Beck, Tahoe Vista, CA at 208.897, another qualifying record.

In spite of his qualifying speed, Don Whittington did not win the Sunday unlimited championship race. He, Waldo (Clay) Klabo and Roy McClain all failed to finish, leaving John Crocker to finish first. His victory was short-lived when he was disqualified for cutting a pylon on lap 3 and for flying over the FAA deadline which separated the racers from the crowd.

The popular winner, then, was Lefty Gardner, Mercedes, TX, flying his beautiful P-51. Darryl Greenamyer, Sun Valley, CA was second. Greenamyer has won more races than any other man, having been victorious in 1965, '66, '67, '68 and '69. The popular Howie Keefe, flying Miss America, was third.

The T-6/SNJ Championship winner was Pat Palmer, Seattle, WA, flying Gotcha at 210.680, a speed higher than his qualifying effort. He was followed by top qualifier, Ralph Rina and John Mosby, Chesterfield, MO.

Vince DeLuca took the winner's laurels in Formula I at an average speed of 228.753. He was followed by veteran competitor, Bob Downey at 222.394 mph and another veteran, Bill Falck, at 221.652 mph. Judy Wagner finished fourth.

Don Beck had an easy time in the Biplane class, winning at 202.153 mph. Logan (Pat) Hines was second at 198.438 mph and Dave Forbes was third at 178.218 mph.

Race Director, Jerry Duty reported that in spite of rain showers on Friday and Saturday, the event attracted an estimated 56,000 spectators and more than 2,500 visiting aircraft to Reno.

Each year, the Reno race seems to grow in importance. Undoubtedly, good planning and a high level of community support must share the credit. Reno has also become an important gathering place for air race fans. They are provided excellent racing, a good supporting air show, the chance to meet and talk with many aviation pioneers and the opportunity to do a little hangar racing themselves. No other city has yet managed to come up with this combination, and so, with 13 years of successful events in a row, Reno's claim to being the air race capital of the world seems secure.

USARA ANNOUNCES POINT CHAMPIONSHIP WINNERS FOR 1976

Reprinted by Permission
UNITED STATES AIR RACING ASSOCIATION (December 1976)

Winners of the 1976 Point Championship Awards of the United States Air Racing Association were announced at the Association's annual Awards Banquet November 20. Award categories and winners are as follows:

PLACE	PILOT	PLANE
Formula One		
1	Bob Downey, Whittier, CA	#28, Falcon
2	Judy Wagner, Torrance, CA	#44, Wagner Solution
3	Vince DeLuca, Rancho Palos Verdes, CA	#71, Little Quickie
T-6/SNJ		
1	Pat Palmer, Seattle, WA	#99, Gotcha
2	Ralph Rina, Huntington Beach, CA	#73, Miss Everything
3	Ralph Twombley, Wellsville, NY	---
Racing Biplane		
1	Don Beck, Tahoe Vista, CA	#89, Sorceress
2	Logan J. "Pat" Hines, Canoga Park, CA	#1, Sundancer
3	Don Fairbanks, Cincinnati, OH	#5, White Knight
Unlimited		
1	Marvin "Lefty" Gardner, Mercedes, TX	---
2	Darryl Greenamyer, Mission Hills, CA	---
3	Roy "Mac" McClain, Eufala, AL	RB-51

THE GAMES RUSSIANS PLAY

What the record books won't tell you about the 1976 World Aerobatic Championships

Reprinted by Permission
THE AOPA PILOT (October 1976)
Photos courtesy Betty Everest

by William Garvey

Detente is a dirty word. Ask the U.S. Aerobatic team.

When the props finally stopped at the close of the Eighth World Aerobatic Championships, held recently in Kiev, Russia, the team prize went to the Russians, the men's individual title was awarded to a Russian and another Russian was named champion of the women's division.

The Americans left the U.S.S.R. with a fourth place team finish, one individual bronze medal and a consummate anger over the machinations and manipulations employed by the host country to guarantee its "victory."

Conrad Kay, an assistant American judge who attended the meet, said, "We learned that the Russians have to win and they don't care how they do it. If they have to lie, cheat, steal or kill to do it, they will."

Well, no one wound up dead and Kay admitted, "there was nothing the Russians did that was not within the confines of the rules—with a couple of exceptions." But he and others involved in the meet provided detail as to how politics, not aerobatic prowess, assured the Russian sweep.

The Russian ploys used in winning were sometimes subtle, sometimes overt, and nearly always successful.

For example, aerobatic pilots must fly within the boundary of an invisible "box" of airspace over the contest area. Points are deducted if the pilot ventures outside the box. At a pre-contest meeting in France, the Russians had promised that an East German-developed framing machine, a device that determines automatically whether a pilot violates the box boundaries, would be used in Kiev. However, when the contestants arrived, there was no machine. Instead, Russian observers were given the responsibility for determining whether or not a pilot went past the boundaries. Their rulings clearly favored the Russian team.

Furthermore, the contest site was, according to one contestant, "similar to looking at a green felt pad." There were no distinguishing landmarks to help orient an aerobatic pilot. On the contrary, Kay said, the Russians had marked an "X" for the landing spot near another "X" marking the center of the box, further adding to the disorientation and causing some pilots to maneuver at the wrong time.

But the most confusing gambit came when the Russian contest officials "moved" the alignment of the box.

Contest rules permit such realignments to adjust for dramatic changes in wind direction and velocity; however, the action was taken reluctantly in the past because of the confusion it would cause among the pilots.

The Russians showed no such restraint. They realigned the box every day from 10 to 30 degrees. Said Kay, "They

Women's 4-minute winners: 1st - Leonova U.S.S.R.; 2nd - Savitskya U.S.S.R.; 3rd - Betty Everest U.S.A.

rotated the box when it was not necessary to do so. The wind was not strong enough to require it and the handicap it placed on the pilots was tremendous.

"The only people who were in a position to handle this was the Russians [pilots]. I assume they practiced there long before the contest began."

Another small, but unnerving factor was the condition of the field itself. Said Kay, "It is the roughest field I've ever seen used and called an airport. All our pilots were in fear of losing their airplanes." Aircraft like Henry Haigh's Pitts and National Champion Leo Loudenslager's much-modified Stephens Akro are the products of thousands of hours of testing, reworking, polishing and dollars. It's tough to maintain a perfect frame of mind in flight when you know that upon landing you could wipe out your baby. The Russians flew state-owned Yaks.

One British competitor did groundloop a Pitts and, recalled Loudenslager, Haigh "just about lost his airplane on landing. It got into the most violent porpoise I've ever seen."

Remember, these guys are not pikers; they're the finest aerobatic pilots in the United States.

The moving box, the rough field and the confusing landmarks were just minor irritants compared to the judging. The contest was lost to the judges, not to the flying competition.

Scoring in international aerobatics is complex but, briefly, here's how it works: judges, each from a different country, cast scores after each performance by every competitor. The two highest and two lowest scores submitted are thrown out and then the balance of scores are averaged for the final score.

National bias on any judge's part is precluded by the 10% rule. That means if a judge's scores are consistently 10% above the average for his country's pilots or 10% below average for pilots from other nations, he may be given a warning. After three such warnings he will be disqualified.

In theory that arrangement works nicely. In Russia that arrangement worked only for the Russians.

No reprimands were issued any of the nine judges at the completion of the "Known Compulsory" flight, the first program. However, midway through the "Unknown Compulsory" on the following day, contest officials issued two reprimands to several judges stating they had exceeded the 10% rule during the previous program.

The American judge, Don Taylor, was one of those who received a double reprimand. He argued logically but without effect that the first reprimand should have been issued before the second day's flying began so the judges affected could take more care in their marking. They couldn't change their index midway through the second program, he said.

Suddenly then, the pressure was on the judges to "go along" with the majority since one more "mistake" could get them kicked off the panel. On all but the first day, five of the nine judges were from Iron Curtain countries.

The reprimands became a decisive issue in the fourth and final program, the "Airshow" competition.

Ten judges had arrived at the meet, but there were positions for only nine. Lots were drawn each day to determine which judge would not vote that day. Well, on the final day's drawing the Russian judge was eliminated and things got crazy.

Meet officials delayed the competition for hours while attempting to have the drawing declared invalid. That maneuver brought tempers to a fever pitch and was ultimately unsuccessful. Still, the Russians got their way in the end.

The day's flying finally got under way but after the first competitor completed his aerial work, the officials stopped the contest. They began tallying the judges' scores in an effort to see if any of them had broken the 10% rule.

The judges were aghast. Prior to this the 10% determination was made after the entire day's flying had been completed, not after each individual's performance.

"The implication was obvious," recalled Kay. Even before the tallies were in, the Russian judge reappeared, sat down and opened his brief case at the judge's table. Shortly thereafter, the officials declared that the British judge, one of those with two reprimands against him, had broken the 10% rule and thus was disqualified. The Russian was immediately appointed to take his place.

There was more, much, much more.

Betty Everest wearing bronze medal; given doll for youngest woman pilot.

For example, the Russian judge scored several Russian flights without ever looking at the aircraft in flight. Also, the Russian judge was allowed to review his fellow judge's scores before he himself scored a flight. It was no wonder he never broke the 10% rule.

"It's difficult to describe how brazen those people are," said Bob Carmichael, the American team manager.

He said the scores that came from the judging panel had little relation to the actual flights. "It was almost like watching one contest and seeing the scores come in from a different contest. It was difficult to believe what we were seeing."

Loudenslager recalls completing his first flight, satisfied that his performance would put him in the top ten. Before stepping out of his plane he was informed that he was in 20th place.

"I sat in the cockpit saying, 'Twentieth!' It was so far from what I should have been, it was impossible. The handwriting was on the wall. I knew that was it, particularly for me."

Loudenslager – homebuilder, airline pilot, aerobat extraordinaire–finished the competition in 23rd place.

"I think if I flew every flight flawlessly, I could have gotten no better than

The United States Aerobatic Team.

15th," he said.

Haigh, at 13th, was the highest American male finisher. Clint McHenry finished 14th. Betty Everest performed well against an excellent Russian women's team. She finished 6th overall and took third place in the "Airshow" competition.

The Americans were satisfied that their performance, had it been fairly judged, would have won their team the first- or second-place prize. The British team, they agreed, put in a fine performance, especially Britisher Neil Williams.

As it was, the British won third place overall behind Czechoslovakia and Williams finished fourth in the individual standings.

"Neil Williams finished fourth," said Loudenslager, "and, in my opinion, Neil Williams should have been champion of the world."

Instead, the title went to Viktor Letsko, a Russian.

"That Russian title was so bad," said Loudenslager, "I think the Russian pilots were embarrassed by it."

The irony of it all, said judge Taylor, was that they need not have been. The Russian pilots "could have almost beaten us fair and square," he said. "I'm not saying they would have, but it would have been a close contest."

Fair or foul, they won. The Nestrov team trophy and the Aresti Cup for the top male pilot were surrendered by the Americans in Kiev. That loss and the methods used to effect it will not soon be forgotten.

The American team members spent months raising the $50,000 needed to haul men and machinery to Kiev and back. They spent hundreds of hours perfecting their routines. They were the best in our nation, sent off amid cheers and waving flags to defend the titles Americans won in the last World Championship held in France. They returned embittered losers, victims of a game that had no rules.

"We don't know how to fight this sort of thing," said Kay.

Added Loudenslager, "We spent all the time over there saying, 'How are we going to tell anybody about this? It's going to sound like sour grapes. No one will believe us.' "

OSHKOSH PUTS IT ALL TOGETHER

Photos by Lee Fray, SPORT AVIATION
Reprinted by Permission PRIVATE PILOT (October 1976)

by Leroy Cook

It happens every year. Like moths being drawn to a flame, or lemmings marching to the sea, around the last of July certain pilots turn their faces toward a small city in Wisconsin and begin to chant, "Oshkosh, Oshkosh, Oskosh!" Not the name of an ancient deity, Oshkosh is the site of the annual Experimental Aircraft Association Fly-In, an aviation event that is the most prized goal of sport aviation.

Oshkosh is, well, just Oshkosh. There is nothing else quite like it, and it's next to impossible to understand the Oshkosh devotee's rationale without experiencing it yourself. Beware: after once attending, you may become infected with Oshkosh fever yourself. Tens of thousands of people attend each year's gathering, with thousands of stock airplanes squeezing onto the airport and hundreds of display aircraft lined up for inspection. It's big, all right.

Oshkosh's uniqueness stems from its ability to be all things to all persons. Originally, it was more or less an amateur plane builder's convention, where ideas were swapped and work was admired. Restoring and building are so closely intertwined that the rebuilders of classic and antique airplanes also took an interest, and it wasn't long until the inevitable trade booths began cropping up. After 24 years, the Oshkosh show now encom-

Aircraft camping area at 1976 Oshkosh Convention.

passes just about anything in the realm of fun flying, from light helicopters and autogyros to antiques and warbirds.

This year's meeting was the first to include two weekends, a total of nine days. Hopefully, this was to spread out the traditional weekend influx of bodies and airplanes which inundated Wittman Field. The logistics of supporting an event of this size are staggering; a small city is created for the week, requiring a post office, phone lines, sanitary facilities and public transit. Only with a quarter-century of experience could the EAA committee cope with the job.

By no means is this just a fly-in. It's a camp-in, drive-in, cycle-in, walk-in event, anyway to get there. Not everybody who gets to Oshkosh is an airplane owner, or even a pilot, but they're all dedicated, irrational, airplane freaks. They arrive in everything from a private Greyhound Scenicruiser to an ancient pickup truck, while flyers-in come by King Airs and Cubs. It doesn't matter how one gets there, as long as you make it. A general feeling of good will always prevails at Oshkosh; people smile a lot, because they're happy to be there. Crime is virtually non-existent, and litter disappears within seconds; mama and the kids mingle with everybody else's mama and kids, as everyone sheds their pretensions to become the Common Man.

By and large, the Oshkosh attendees are working-class folk, who take the time away from their jobs and small businesses to be there; they aren't on an expense account, they're here because they love it. The Oshkosh crowd is not interested in flying-for-hire, they're strictly for the fun of it, in some form or another. Homebuilts, or "custom builts," as they're now called, are much in the spotlight, being the dream of an affordable pair of wings for many low-budget pilots, but cost-cutting isn't the only appealing aspect of rolling your own. Individual desires for something different gives as much impetus as anything else. Hundreds of dreams were lined up in neat ranks, whole fleets of designs like Starduster Toos, Thorp T-18s, and Bushby Mustangs, each unique in its own way.

However, if one's tastes run to classic (1946-1953) factory-built aircraft, some of the finest examples in the nation were to be found in the classic display area: Cessna 140s, 170s, 195s, Swifts, Cubs, Champs, Luscombes, Stinsons, Ercoupes, all gleaming proudly. Or for admirers of the earlier days, the antique area provided even more nostalgia: Wacos, Staggerwings, Ryans, Howards, Rearwins, each one the result of many long hours spent searching for impossible parts and long-lost specifications.

Perhaps your thing is building your own rotorcraft, in which case you would have spent your time at the gyrocopter and helicopter area talking with fellow fling-wing jockeys. At the other end of the mile-long display line, the Big Iron devotees were ogling a priceless collection of warbirds, potent reminders of a world in flames a generation ago. Dozens of T-6s, a small squadron of BT-13s, fighters from Wildcats and a P-40 to the era's last gasp, a Hawker Sea Fury and the Fairey Firefly. The huge bulk of Fifi, the Confederate Air Force's B-29 Superfortress, dominated the warbird area, joined by the CAF B-17, B-25s, B-26, and many more, all requiring buckets of money and elbow grease to keep in the air.

Yet, the aircraft display is only part of Oshkosh. The newest and latest offerings in plans, parts and accessories were of great interest to those looking for just the right item; the exhibit building was always crowded, while much informal bulletin-board and campground trading

Zenith built in 9 days during convention nears completion.

Tom Poberezny congratulates Chris Heinz after the flight of the convention-built Zenith.

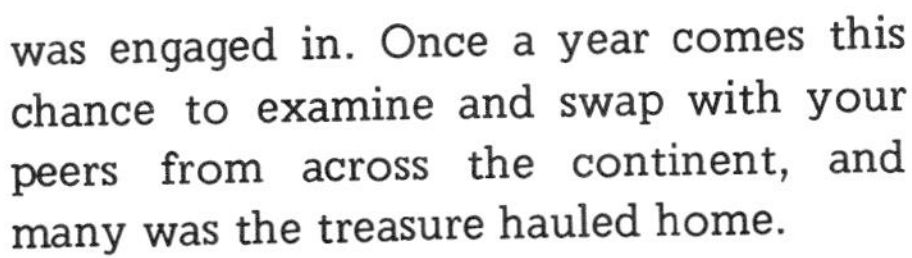

was engaged in. Once a year comes this chance to examine and swap with your peers from across the continent, and many was the treasure hauled home.

In the seminars and displays, Burt Rutan appeared to lead the pack in terms of interest shown, but Molt Taylor and Ken Rand's designs were also strongly scrutinized by crowds of lookers. Jim Bede was his usual ebullient self, hawking the BD-7, of which a prototype/mockup was being displayed. His Xenoah-powered BD-5 was also kept visible, lapping the rest of the fly-by circuit. Two of the surviving owner-built BD-5s were on display, their builders evidently well-pleased with the tiny birds.

Mooney slipped in with a brand-new model 201 down in the factory-built display, which drew considerable interest. The Varga Kachina was also attracting lookers, in the same arena as the Big Three's factory-built offerings.

1976 was the year of the powerplant innovations at Oshkosh, in many ways. If the turboprop Great Lakes biplane was a bit rich for your blood, how about a turbocharged Honda CVCC engine in a J-3 Cub? Jerry Kibler flew the Cub cross-country from California to Oshkosh, waking up little airports along the way with a whining Cub taxiing in. Fred Geschwender showed a complete line of Ford automobile engines converted for aircraft use, and had a 400-cubic inch V-8 installed in an agplane on display. The Polish government's export outlet offered a lineup of the venerable Franklin engines, following its purchase of the factory tooling last year.

For those seeking still more, after a surfeit of displays, conviviality, and gawking at pretty airplanes, there was the late afternoon airshow which never failed to pack the fences. Nightly variations kept it from getting stale throughout the week. The Canadian Snowbirds jet aerobatic team performed superbly the first three days, while the Confederate Air Force's production of the "Tora, Tora, Tora" raid occupied Wednesday evening. Thursday's show, entitled "The History of Flight," was capped by a touch and go from a United Airlines DC-8, with Burt Rutan's Vari-Eze flying chase.

These acts were backed up by the incomparable Bob Hoover, doing his show in a new Shrike Commander and his faithful yellow P-51, and Dwain Trenton, flying the fantastic turboprop Great Lakes. Performing nightly also were Duane Cole, "Bid Ed" Mahler, the EAA

Red Devils, Bob Lyjak, his 1929 Taperwing Waco now refitted with the Wright Whirlwind engine, and the roaring 450 hp Stearmans of Bill Barker, Wayne Pierce and Bob Hueur. Frank Sanders' massive Hawker Sea Fury took the "max smoke" award, with special smoke generators placed under each wingtip to dump their output right into the tip vortexes. Extremely impressive.

Oshkosh 1976 also saw such events as Don Taylor's departure on a second attempt to fly around the world in his homebuilt Thorpe T-18, and Dick Wagner's flight to 20,000 feet in the EAA CUBy J-3 replica, testing automotive gasoline for vapor locking at that height. The C-90 engine suffered no ill effects from the experiment, and the EAA continues to compile data concerning the use of auto fuel in low-compression aircraft engines.

Unscheduled happenings always crop up at Oshkosh. This year, Charlie Hillard's voltage regulator caught on fire in his Pitts, which caused him to drop out of formation suddenly, after the Red Devils' takeoff snaproll Monday afternoon. A B-25 suffered a blown tire on landing after the Warbird fly-by Tuesday, blocking the runway for a time. The show continued despite these minor miscues, as EAA president Paul Poberezny went dashing off in his topless Volkswagen to check on proceedings.

There was, of course, the proliferation of awards which are a timeless Oshkosh attraction. The Grand Champion custom-built airplane was Vern Menzimer's Cavalier 102.5 from Vista, California, a winner at the Corona fly-in earlier in the year also. Best Rotorcraft award went to Ed Kruse of Boca Raton, Florida, while the Grand Champion classic was a Piper Vagabond owned by W. Amundson and R. Oeterson of Stroughton, Wisconsin. The top antique was a 1927 American Eagle, owned by Robert Groff and Claude Gray, from Northridge, California, and the Grand Champion Warbird was Max Hoffman's outstanding P-40 Warhawk. There were plenty of other trophies to go around, to keep runners-up content.

The show numbers followed the declining trend of the past year, with some 1,263 display airplanes attending vs. 1,338 in 1975. Total aircraft movements for the nine days was 64,839, down from 67,314 in 1975's seven days. While the 1975 peak of 16,232 daily movements was not approached, perhaps by design, the first Sunday saw 9,238 movements, while 9,663 crowded in or out on the closing Saturday. By any standards, it was a huge undertaking; dual kudos should go to the EAA Fly-In Committee and the all-volunteer FAA tower personnel.

By giving proper attention to the temporary ATIS and following the notamed arrival paths, there was little trouble in negotiating the traffic even at the height of the influx. Only an occasional black sheep would stray in, displaying seven degrees of frustrated ineptitude and wondering why everybody was upset at him. We liked the tower's handling of the well-equipped chap who reported "ten DME north squawking twelve, for a possible straight-in." He was told "negative, proceed to Omro and report eastbound, behind the aircraft ahead of you." Big or small, we can take you-all, just get in line.

The good humor of the Oshkosh crowd was evident in the atmosphere aboard the jitney wagons and buses that shuttled pilots and passengers to and from parked airplanes. We were riding one such ancient schoolbus when a clutch linkage parted, stranding us in mid-field. Amid jokes and laughter, the riders dis-

Verne Menzimer's all-wood Cavalier: the 1976 Grand Champion Homebuilt.

mounted, someone produced a safety pin while a search party located the fallen part, and a volunteer crawled under to safety pin the clutch back together.

I'll never forget the breathless woman who hailed down the bus to ask if we were going to the other side of the field. "Oh good," she said, "there are four people who just came in with a plane and they need a ride, right over there." At the time, airplanes with four or more people were landing approximately every 15 seconds, each of whom needed a ride. The volunteer driver merely smiled and said, "Yes maam."

Oshkosh remains the cherished gathering spot of fly-for-fun people, and the enjoyment felt by the attendees make it what it is, sport aviation's greatest show.

NOTE: Taylor completed his around-the-world flight on September 30, 1976. —JSAY

A WINNER'S TALE: THE 1976 SMIRNOFF DERBY

Reprinted by Permission SOARING (August 1976)

Edited for Space

by Wally Scott, Sr.

It all started about D-day minus 45. Paul Schweizer called and asked if I would be interested in flying a winner in this year's Smirnoff Sailplane Derby. He wanted to build a 1-35A, a modified version of the 1-35 with a few new ideas off the drawing board, and he needed a pilot to fly it in the Derby. I accepted.

D minus 2-1/2: We arrived in El Mirage by 14:00 on Saturday. My crew was my wife, Boots, and my niece, Lodema James. They were to follow me in a new heavy-duty-everything Chevrolet station wagon.

D minus 1: I was the only one to fly this day.

"WA, this is Smirnoff Ground." Big Ed Butts, the Derby Competition Director, was calling.

"This is WA, Big Ed, but I'm sorry to say that WA is not my call sign this year. My glider competition number is 62."

Ed, remembering 1972 and 1973, said, "Okay, you will be *Mojo.*"

D Day: May 4th dawned with low, low clouds and drizzle. The morning wore on and Big Ed finally called us all together for the briefing, which sounded like a distance day. Our first goal was Carefree Airport, north of Phoenix. A few blue holes opened in the overcast to permit takeoff, but, to get us away from the marine air, the release point had to be Rabbit Dry Lake some miles east of Pearblossom. Things felt pretty stable.

It was time for the race to begin. After the countdown to zero, everyone went for the mountains near Big Bear Lake.

Leaving the mountains, I headed for the first town in the pass and the safety of the airport and arrived there with 1,000 feet. I saw another glider ahead and a little higher. It proved to be Danny Pierson's *Nugget* AU also in deep trouble, but a little bump was found, some water was released, and we were saved. Cu's started forming ahead, and soon what had started as a distance day began looking like a speed dash.

Things went great to Parker, Arizona, where we ran into overdevelopment. Now it was a distance day again. My glide ended atop Moon Mountain, 30 miles

Wally Scott, Sr. Smirnoff

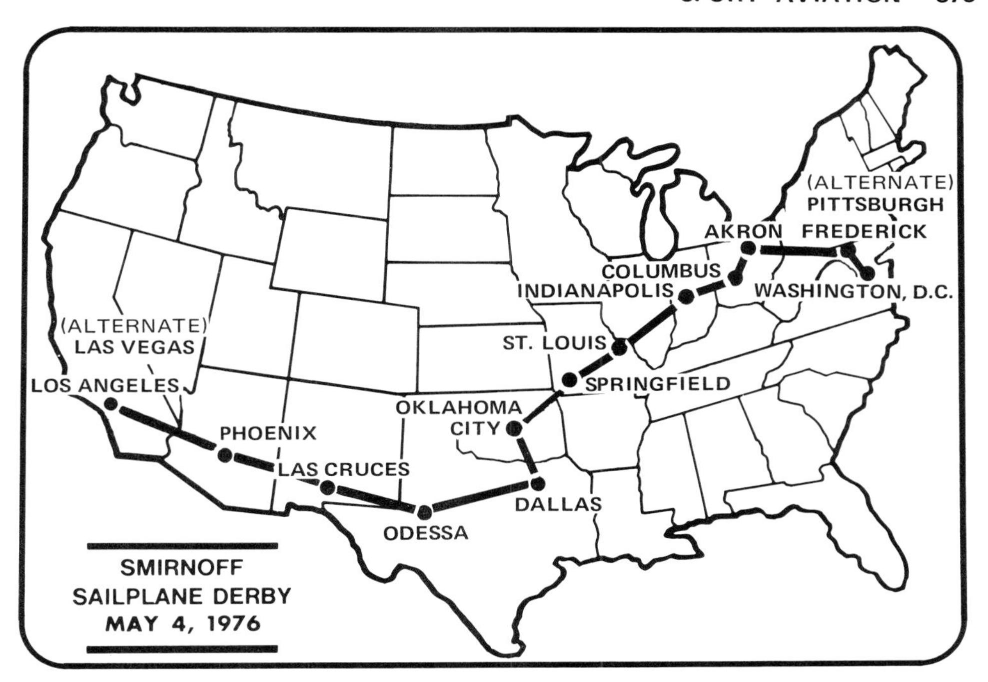

south of Parker. I was 3,000 feet above ground and thought I might just reach the clouds another 10 to 15 miles ahead; but I gradually sank to below ridge level until landing became inevitable.

There were plenty of fields just to the west. As I headed for them, I looked down to find a duster strip directly below. I just had time to lower the gear and land!

My crew had had the first of several hard days. We derigged in the dark, and drove on to Carefree for a 4:30 a.m. check-in. I was in last place for the day and my 150 miles were good for only 430 points.

Day 2: The day's forecast predicted cumulus clouds with 7,500 MSL bases. With much of the terrain near or above this level, I offered to exchange jobs with Director Ed Butts. He declined.

Enroute, Ed called me to find out what flight plan most of the pilots were using; I could tell from their instructions to their crews and from their frightened-sounding voices that we were going toward Globe. The weather turned from fair to worse east of Globe, and we all had a weak 20 or 30 miles to cross under complete overcast. We probably all went places where no glider had any business to be, but we survived.

Things started getting better for a while, until I became enmeshed in strong down from 40-mph westerly winds as I was crossing the big mountain south of Silver City. It was my turn to crawl out of a hole, and though I did so slowly, the heavy wind was carrying me rapidly eastward, raising my hopes for a speed day after all.

About this time Danny started calling for help. When I answered, he said he was in deep trouble and described the terrain as best he could. I told him that I thought I knew his location—that he was probably

Danny Pierson Don Monroe

near Silver City. He checked his map, came back with the relief in his voice, said I was right, and that he could just make the airport—which he did. I relayed to his crew who thanked me, and Danny did, too. I think I may have made a new friend.

Smirnoff Air soon had me in sight as I was climbing. I gradually edged over to the highway between Deming and Las Cruces so that I wouldn't have to worry about a place to land. I worked anything and everything that came along, but slowly sank to within a few hundred feet of the highway. Glancing down, I saw a safe landing place—an old abandoned inspection station, so in I went.

A family in a pickup truck came by, signed my landing card, and said they would call in from Deming. I'd flown farthest—287 miles—and won my first race, though I was third in overall standing.

Day 3: The 267-mile race was to be from Las Cruces to my hometown in Odessa; it began with an ominous look to the weather. A frontal zone and its low pressure system was near.

When finally we did start, I glided straight to the ground without hitting a bump and landed about 15 miles south of Las Cruces in the Rio Grande valley.

Boots and Lodema arrived within minutes and with a passing motorist's help, we derigged and started back for a relight.

As I took off the second time, the clouds took on better form, the sky was recycling, and lift was very good under long streets. I flew as rapidly as possible for about an hour, and managed to cover a lot of ground.

I finally came to the end of the cloudstreet, worked up high, and started edging over to the highway.

"Nan 6," I asked, "how close together are those highway construction markers?" Boots said they looked awfully close together—about 40 feet. Lodema disagreed, and thought them to be about 100 feet apart. This didn't sound promising for a highway landing, but when I edged on over to the highway I could see that they were about 200 yards apart. A highway landing was possible, and would make for easier derigging. I couldn't see my crew on the road.

They were holding about five miles down the road and spotted me as I passed over them with still 400 feet. Even with the strong headwinds, the air was good and I wanted every last bit of contest distance I could get:

"Hook 'em up and come a-running," I radioed, "I'm not throwing away 400 feet." I landed just short of the salt lake, at the west end of the bridge and where I could go no further. Touching down just beyond a barrier marker, I came to rest after only a few feet of ground run.

We had come in 5th for the day, but

just five or six miles short of the 3rd and 4th-place finishers.

Day 4 was for rest and Day 5 was unflyable because we never got enough temperature to burn off the solid overcast. But at least the other pilots and crews learned of the Mojo legend, created and cultivated at Odessa's Permian High School. It stands for the never-say-die spirit, and for accomplishments above and beyond natural abilities. Mojo is known state and nationwide and has become the 7th flag over Texas.

We trailered to Dallas and arrived at about 21:00.

Day 6: Another weather-delayed rest day. Our little black cloud was moving across the country at about the same pace as the Derby.

Day 7: More of the same in the weather department, with very low clouds and a nearly solid overcast. We were committed to fly regardless, because the Red Bird Airport was a busy one and did not want a bunch of gliders cramping its style. Our goal was to be Cimarron Airport west of Oklahoma City.

We all had to make a 30-mile detour to the west in order to stay out of the Dallas/Ft. Worth terminal control area. To the north, conditions improved for 50 miles, but soon we faced a large overdevelopment. It was to be another distance day. None of us had made a designated airport yet; this would be more of the same.

I landed on a farm to market road, next to a farmhouse so I would be near to a telephone. It was deserted.

Day 8: We flew out of Cimarron, Oklahoma, bound for Springfield, Missouri. My course line from Oklahoma City toward Tulsa was quite a bit south of course; even though I was advised to fly to the north, I just couldn't leave the good clouds where I was. The best lift of the

Paul Bikle Don Monroe

day was right over the Tulsa railroad yards next to a refinery.

It had been a day filled with varying conditions, and a very exciting final glide for me. When Bikle had made an airport landing at Mt. Vernon, I had picked out an airport at Aurora, Missouri, that would put me a few miles closer to the goal. I was in doubt of reaching it as I was very low and the airport was hidden behind trees until I crossed over its boundary with just enough altitude to pull up over the top branches, lower the gear, and land. It was nice to be on an airport for easy derigging. But even better, the 249 miles I'd flown gave me first for the day and moved me a lot closer to second place overall.

Day 9: A heavy cirrus overcast was moving in over Springfield from the west, and it was getting heavier by the minute. We finally were instructed to take 3,000-foot tows for timed releases, and, if we had to, we could land at a couple of airports 10 to 15 miles down the road.

No one had ever yet flown over this country on any of the four previous Smirnoff Derbies, much less make it to the Mississippi River and St. Louis. When I got down to 1,000 above ground, a few wispy cu's started forming to the east. I struggled along for another 10 miles, never getting above 2,000 feet until I finally found good lift. What had started out to be a very short distance day was now blossoming into the best day since the beginning of the Derby.

Dropping down out of the hills into the Mississippi basin we saw the end of our good clouds. Before I reached the end of the cloud belt, I had managed to work up as high as possible. It was a fortunate move; the last clouds looked good but contained no lift.

I instructed Nan 6 to leave the interstate and come south to the highway leading to Cedar Hill. Landing near there appeared a certainty. I investigated some cedar breaks and found slight additional lift over brush-covered hills. The convection zone was only 2,000 feet thick, the lift weak, and the wind was a quartering cross headwind. To be so close and yet so far.

I worked my way around the southern edge of St. Louis and was faced with a certain landing near the west bank of the Mississippi. There were not many places to land. A landing place appeared along the west bank of the river. It looked like a cemetery with large parade ground. Then I saw guard towers and tall fencing. I thought it might be a prison. If I landed there, they might not have wanted to let me out. But I knew that I must, and as I swung out over the river to set up my downwind leg I hit a gust of ridge lift from the tall trees along the banks. I zoomed up and pointed the nose of the 1-35 across the river to the east bank, figuring I had just enough altitude with 200 feet. Things went pretty well for a few seconds until I hit heavy down from the lee of the levee on the east bank. I was settling into the west bank when the air calmed and I serenely (?) crossed over the trees, the levee, and landed just short of some high-tension lines.

A Smirnoff pilot had finally flown across the Mississippi! It was a big deal for me. My 190 miles gave me another first and moved me into overall second place.

To add to all this excitement, Nan 6 had trouble near Cedar Hill. Our heavy-duty-everything station wagon had blown an engine.

Day 10: Rest and rain. Our little black cloud had caught up again. The garage was digging into my engine, and it looked bad. Repairs would be impossible by the time the race was to continue, as a new engine was called for.

Day 11: Even heavier rain and tornado alerts. We trailered to Indianapolis, Lodema going with the 1-35, Boots and I bumming a ride with Danny. Our duffle was being transported by several other kind pilots and crews. Here and now let me say this. Not just because of these happenstances, but overall this had to be the best Smirnoff bunch that I have had the pleasure to be associated with.

Day 12: In the early morning hours we had some rain and low clouds. The day was looking more and more like a no-contest day as afternoon approached. Suddenly a big blue hole appeared, cu's started popping, and we immediately staged for timed-release takeoff. My tow pilot was to be none other than A.J. Smith who had come to witness the activities. As Rudy, in the number one spot, was being towed by Smirnoff Air, my crew rolled me into position and hooked me up to A.J.'s 180. We took off to the east, toward Akron, and were to release at cloudbase, but A.J. kept heading east.

"A.J., come back to the airport!" said Smirnoff Ground. We kept going east. "A.J., turn that thing around and come back here this instant."

Well, you don't ignore Big Ed for too long a time. A.J. reluctantly turned back and headed for a cloud over the east end of the airport. We saw Rudy release under the same cloud, a little lower than base. About 10 minutes later, as I was flying on course for Akron, I heard Big Ed say, "A.J., come back to the airport." Again, "A.J., dammit come back to the airport!" A.J. had Paul Bikle on tow and was headed *west* toward St. Louis. I was so amused that I nearly fell out of the air, and we all had a good laugh about all this the next morning at breakfast.

The flying weather on this leg to Akron changed from respectable to good, to weak, and finally to impossible. We all landed about 85 to 95 miles out; Bill Cleary had won the day. I was second,

Ross Briegleb Don Monroe

but Rudy had his hard luck day, making only about 45 miles. The 1-35 and I took over first place overall.

Day 13: Another unlucky day as our low pressure system was there again.

Day 14: Lots of cold wind, but the cold front had passed us by. We would fly. In order not to overtake the front, we would make a stop at Latrobe, Pennsylvania. There were good clouds by noon, but the strong north wind would hinder us somewhat and we would have to go around the Pittsburgh terminal control area. All of us flew to the north except Ross Briegleb who took the southern exposure. It didn't seem to make much difference as nearly everybody landed astern of Pittsburgh except for Danny and me. My landing place was a rocky plowed field nine miles short of Latrobe, but Danny made it in for the first completion of the Derby. He picked up many points on me, and going into the last day's race he was only one point behind me in second place!

Day 15 saw us glued to the airport all day waiting for the weather to break. Ground winds were 45 knots, it was cold with lots of snow reported over the Alleghenies. The alligator would be tough on the Smirnoffs this day and keep us grounded. (But it was kind to the ridge runner. We all wondered if Karl Striedieck might not be bombing down the ridge, and he was.)

Day 16: Victory day. There were still strong westerlies blowing, but clouds started forming during the briefing. I had talked with Paul Schweizer the night before and he had said that I might get some speed points on this day. I would need more than some, I would need them all. Danny had been flying superbly. We were scheduled to take off at 12:30 and release at 3,000 feet over a quarry on the first ridge. It was to be a timed release, not a race-horse start. We were told at the briefing that we could relight and get a new time if necessary. If we landed out we could get back and get a new time also. In this respect I like the timed release better than the race-horse start. If the weather is misjudged you can come in for a new timing on the start. Or if you land out you would not be penalized so much as in the race-horse start.

The timed start also has gamesmanship possibilities. I had briefed my crew as to what to do. They would give all evidence of a fast departure, but go around and hide behind a corner somewhere and await my relight. I thought I was making a clever move, as 12:30 is awfully early in Texas. Our best soaring is toward mid and late afternoon.

I released over the quarry at 3,000 feet and headed north over the ridge. Looking back, I saw that Danny had released and headed south toward better looking clouds. I got out of eyeshot as soon as possible, and told Nan 6 to move out and hold near Latrobe. Watching the airport from afar, when the last glider was on tow I headed back. I worried that my plan might backfire because some heavy high cirrus was rapidly moving in from the west. I would have to make this fast.

I landed with my water load still aboard. I didn't know for sure if I wanted the other pilots to know I was making a relight. I decided to keep my radio quiet, as the advantage of a later start should be all I would need—if I could escape the oncoming cloud cover. But my tow pilot blew my cover, because he headed west and just kept going and I had to go on the air to direct him back to the quarry and then announce my release to Smirnoff Ground. I couldn't find good lift. Had I outsmarted myself? Oh well, this was do or don't, so I streaked out with 3,500 ft.

Basically, the flight across the Alleghe-

Ed Butts Don Monroe

nies was uneventful. There were some anxious moments for all of us at times. Things were pretty choppy, but there was lots of good lift and waves that none of us bothered with. At one point I got to base at 8,500 MSL and had plenty for a fast glide to the Derby finish line at Frederick, not counting the strong tail-wind. The wide valley at Hagerstown was void of clouds, and was in heavy sink. My reserve altitude came and went, but the ridge between Hagerstown and Frederick had a good cloud over it. I had heard Rudy and Bill call the finish gate a little earlier, so knew they had made it. I got to the ridge with enough altitude to make it on in, but decided to insure with another 1,000 feet to take care of any down on the lee side. I had gained 400 feet when I heard Danny calling the gate. That did it. I stopped circling and gunsighted the 1-35's nose toward the airport. The extra 400 feet came in handy (there was some strong down), but I arrived with no problems other than anxiety. On the way, I could tell by Danny's transmissions that he was still some distance out, so I relaxed a little. We had won. *Mojo!*

ONE GRAND FLIGHT

Reprinted by Permission
SOARING (October 1976)

by Karl Striedieck

Like the great soaring birds he admires, studies, and occasionally flies with, Karl Striedieck has a strong sense of territoriality. And his "dominion" has been the out & return record skyways of eastern U.S. mountain ranges. It is true that other soaring pilots in the late 20's and 30's made tentative record-hunting forays in these ridges, but it remained for Karl to prove himself soaring King of the Mountains by successfully capturing and recapturing this record during the past eight years.

He began establishing his dynasty in 1966 when he bought a used, medium-performance Ka-8, hauled it to the top of Bald Eagle Ridge in the Pennsylvania mountains, cleared off enough trees for an airstrip and farm, built a hangar, and lay siege to the out & return record.

His first victory came after four unsuccessful assaults through 1967 and early 1968. On March 3rd, 1968, he astounded the soaring world by completing a 476-mile O&R at a time (winter) and place (the eastern U.S.) and with a ship (his medium performance Ka-8) that should have made such an accomplishment impossible, according to widely-held views of the time.

The feeling that the flight must somehow have been a fluke seemed to be borne out when he lost the record and time rolled by with nothing further being heard from him. But in November of 1971, he again set a new record—this time in an AS-W 15—which exceeded his first flight by more than 100 miles and was posted at 569 miles.

It was clear that Karl's cross-country achievements were no accident, and world recognition came this time with soaring's highest honor—the FAI Lilienthal Medal.

Now others began to move into his territory, but when he lost the record in 1972, he quickly recaptured it with a 639-mile record—only to have it taken away two days later by Jim Smiley who racked up 650 miles! Karl had to wait a year, but as soon as the wintry cold fronts started rolling across the Alleghenies in 1972, he took the record for the fourth time with a 682-mile roundtrip to the South.

In the spring of 1973 Bill Holbrook, a Smirnoff Derby winner, unseated Karl and put his own name on the books with a 783-mile circuit from Lock Haven, Pennsylvania, to Hansonville, Virginia, and back. This was a stunning increase of 101 miles over Karl's best, and it stood until March of this year, when Striedieck broke his own solo tradition and pair-flew with Roy McMaster to win the record jointly with an 807-mile out & return course to Mendota, Virginia.

Now ridges—like cloudstreets—don't go on forever, and in the case of the Allegheny and Appalachian ranges a little town named Lutrell in Tennessee marks the end of the line. It must have occurred to Karl Striedieck that he had been in a continuing race whose increments were increasing each year. Mileages had been going up roughly 100 miles a year—476, 569, 639, 783 (Holbrook), and 807. Karl must also have realized that the first pilot to turn Luttrell would top 900 miles and be the ultimate ruler of the territory. Was there some way to stretch the distance to the magic 1,000-mile figure that had so long excited the imagination of cross-country pilots? That would be a hard act to follow. On the following pages Karl tells how he put it all together on May 19th, 1976.

After Roy McMaster and I made our 807-mile, two-ship, out & return flight along the Allegheny ridges last spring, we

The "Allegheny Expressway" includes long stretches of unlandable forest. Farms and clearings in valley bottoms like the one pictured here are carefully noted by ridge runners as a welcome haven should lift stop unexpectedly.

Striedieck's aerie. Main slope of Bald Eagle Ridge is to the west (right). The forest edge provides a windbreak that permits auto-tow launch even in relatively high winds.

discussed the particulars of what it would take to accomplish a 1,000-mile flight along the same course. We realized it would be necessary to fly off the ridge's southern end in Tennessee and find thermals to fly another fifty miles to make the turnpoint and return to ridge lift.

I've owned the World O&R Record five times in the last eight years. My strategy has been never to declare a turnpoint that was more distant than the minimum necessary to establish a new mark, because until May 19th I never had the weather to go for all the ridge marbles in one flight. In fact, it was only after a what-the-heck check of Penn State U's weather station on the evening of May 18th that I got the idea of giving the 1,000-mile flight a try.

Three things looked unusually good that night: a high was centered far enough west, the pressure gradient was tight, and the winds-aloft forecast was good for the whole route.

In the past my go/no-go decision in the morning has been difficult and fraught with dither because it means getting the observer out of a warm bed at a riduculous hour. I did this more than a dozen times between May 1973 and

March 1976 for flights that got airborne and just as many that were cancelled before takeoff. So to avoid dithering I decided to get up at 4:10 a.m. and, if the wind and weather were good enough, fly locally and get on with it immediately. At 4:15 a.m. I walked to the top of the ridge to listen—and it *was* "go." My brother Walter hurried over at 4:45 a.m. and was soon followed by my observer, Bob McLaughlin. We worked quickly loading water ballast, cameras, barograph, and getting the bird ready for takeoff. Once everything was secure, I slipped into the cockpit and they hooked the towrope from the jeep to the nose of my AS-W 17.

Moments later I was auto-towed into the air. I released, angled over the trees toward the valley, and picked up the slope current. It was 5:35 a.m.—an hour and twenty-five minutes since I had gotten out of bed. That's a snoring-to-soaring record of its own!

Heading north toward Lock Haven where there would be ground observers to verify my start, I had no trouble staying up, although the winds were quartering somewhat from the west. When I arrived at Lock Haven I noticed smoke from a stack was blowing parallel to the ridge. It felt a little strange seeing that. I wondered just how far below the crest I would be able to find lift. I guessed about two feet and took pains to stay high.

I passed through the start gate at 5:55 a.m. and followed the ridge southwest. The lift was not strong. I flew slowly trying to stay above the crests, but at a little gap I got below the top. This is risky early in the morning; I was glad to get up again, and by 6:40 a.m. I passed the farm with little more than an hour having passed since my takeoff. By this point the ridge curved in a more southerly direction. It was safe to put the '17 down in the leaves and smoke along at 110 knots. There was a 3,000-ft. cloud base at the Altoona gap. I coasted across, increased my speed to 130 knots to Bedford, running into ice crystals as I passed Claysburg. The overcast began breaking into cu's. I had been flying straight ahead without circling and I decided to cross the eight-mile Bedford gap the same way, so I tried to pick up as much altitude as possible just before the breach. I could only manage 700 feet above the ridge. Nevertheless, I went straight across and nursed it along to the other side. As a matter of fact, it was Cumberland (160 miles) before I made my first three-sixty. (This reminded me of an earlier try last fall when I flew 190 miles without circling once!)

Sometimes there is a wave at Cumberland. But not this time, unfortunately. I picked my way along the knobs of the Knobbly's until Hopeville where the ridge becomes respectable once more and rises to 3,500-4,000 feet. The lift was strong; I put the nose down and stormed along, but seldom within 200 feet of the crest because of the increased turbulence near the slope. To the west there was a blanket of snow. At 8:45 a.m. I passed Mt. Grove, the turnpoint for my first world out & return distance record. Getting here and back in one day was a max effort a few years ago. On this day I could have made the flight at least twice!

The first really hairy moments occurred when I took a short cut to Covington. The terrain below is unlandable. Just when I was most vulnerable, 1,000-fpm down put me below the ridge I was trying to reach. Working right next to the trees, I clawed back up and headed for Tazewell. It was 10:10 a.m. I had averaged 85 mph, fast enough that I knew I could be at my turnpoint before noon if I could maintain speed. During this time I passed Narrows, my '72 turnpoint. It was

Into the sun. Karl pulls up the AS-W 17 as it slips from beneath the shadow of a cumulus overhead. Note the highways in the background. Like water courses, they frequently mark gaps in the ridge.

9:40 a.m. I was lucky to get there at 1:00 p.m. in the 'old days.'

Four isolated ridges—each about six miles long—are stepping stones to the final link, Clinch Mountain. They are about 4,500 MSL. The wind was booming up there, but the Clinch ridge averages 2,500 feet. I nervously edged down to Clinch, noting the wild waving of tree branches had subsided to an occasional wag and remembering that I was getting farther away from the weather low that was my wind generator. I dropped the speed to 70 knots on the way to Gate City where the ridge turned due west for 10 miles and paralleled the wind flow.

There was no choice but to begin thermaling.

Only about two percent of the last 300 miles is landable, so I had to use 100-fpm lift to survive until I made it to Kermit where the ridge bends southwest again. Ground indications were pessimistic with respect to ridge lift. However, there was enough to allow almost continuous cruising 500 feet above the ridge with occasional circles.

At Luttrell I finally reached the end of the "Allegheny Expressway," a point that for so long has occupied the minds of us ridge runners. What now? It was still 25 miles to my turnpoint, but now, at least, there were landable fields. The thermals were there and at one spot I climbed to 4,500 feet in 300-fpm lift. By 1:30 p.m. I had reached my turnpoint.

I started for home. I had one climb to 4,500 feet, but by the time I was within a couple of miles of the ridge, I was down to 500 feet going around in zero sink. On about the fifth orbit I spied three Black Vultures circling a quarter mile south of me. I quickly moved over to them and effected a save.

Back on the Clinch ridges it was more max L/D flying mixed with occasional thermal circling. Twenty miles from Gate City the cu's reappeared. Passing the town in 500-fpm lift, I dove for Hilton ridge again and found 100-knot cruise lift. At about 2:30 p.m., beyond Mendota, the sky looked just booming—cumulus everywhere and way the hell up there! It was like flying west Texas.

I developed a gap-crossing technique: During the last mile approaching a break, I concentrated on thermal searching. At the first sustained blast I would turn 90 degrees left, climb hovering (or sometimes circling), and ride 500-1,000 fpm to sufficient height to make a crossing. This went on all the way to Tazewell where I punched the clock for what I knew would be a fast 100-mile ride to Covington. It took 40 minutes. That's 150 mph! Pretty darn good for no motor.

At Covington I tried to retrace the same dangerous shortcut from Mt. Grove I had used on the way down. But over Falling Spring (appropriately) I hit 1,000-fpm down. I had to turn downwind and run for the backup ridge, a peak a little north of the town. I got there with 500 feet and figure-eighted to the top of this 4,090-ft. beauty. Here I took a wind sample by pointing straight into the wind and hovering over the trees. The airspeed indicated 50 knots! After covering 20 miles on this backup ridge, I transitioned to the primary route at Mt. Grove and thrashed on.

By this time my hands were aching. I had been continuously squeezing the stick to avoid setting off PIO's—pilot-induced oscillations—in the bumps. As a matter of fact, I flew with both hands on the stick and arms locked against my legs. The geometry of the '17's controls is such that a positive gust forces the stick back, adding yet more *g*'s. I couldn't fly much over 100 knots due to the sharp, chattering, tortured air currents. Each time I tried to sneak up to 120 knots or so, I was rewarded with 3 *g*'s to the head.

Between Mt. Grove and Cumberland cu's were plentiful; no particular sweat, just staying high and plugging along all the way back to Hopeville. I got high in preparation for the low stretches ahead. I was able to figure out the wave effect of a mountain ridge to the left and use it somewhat. There were occasional 2,000-fpm thermals, too, but I didn't bother to fly them because ridge lift is faster.

Thanks to good visibility, I had earlier seen the trailing edge of the front that had gone through the day before. I could see a lot of cirrus up north which I thought was over the coast. However, as I got closer, I got under thicker clouds. They were beginning to cut out the

Helpmate: Although Sue Striedieck missed the takeoff of her husband's 1,015-mile flight, she has auto-towed him with their Jeep to launch most of his flights from Bald Eagle Ridge. She is also an exceptional crew, preferring to handle it alone—even when pulling the trailer across the U.S. on the transcontinental Smirnoff Derby.

heating but I got across the Cumberland Gap without any trouble.

Beyond this point was a quandary. Should I take the safer but longer back route? This meant jumping to the next ridge three miles east and passing Saxton, James Creek, Williamsburg, and then, when I reached Spruce Creek, penetrating back to the main ridge to Altoona.

I decided against it. It would take too long. I committed myself and crossed the Bedford Gap. Once I did that, I couldn't get to the back-up ridge.

As I moved along, the clouds thickened into a solid overcast. Just when I was nearing the end of the ridge, the raindrops started coming and I thought, "Oh, man! Here goes the lift. I'll never make it across the Altoona Gap!"

I had to make two tries. Once I shot out and immediately ran into huge sink. I turned around, got back, and tried for altitude. The second time I was high enough to get out under the edge of what was obviously a wave. I didn't get up in it, but it helped me across to the home-stretch ridge.

There was a drizzle coming down. The clouds were high and there was a strong northwest wind thrashing the trees around. But by the time I reached my farm the rain had reduced the performance of the '17 enough that I was down to the crest. By the reservoir near Howard, the lift got so weak and the wings so wet that I dropped below the ridge. This was the scariest part. I was only five miles from the finish and I thought, "Man, this is sad!" I figured I'd better forget about speed. I dumped the water ballast, managed to claw my way back up the slope, and from there flew as slow as I could. The switch of tactic got me to Lock Haven without any more problems.

I pulled the spoilers and landed at 7:00 p.m., 1,015 miles and thirteen and one-half hours after I took off.

About the only ill effects I noticed (erroneous reports of an ulcer notwithstanding) was vertigo when I began walking—and that cleared up about an hour later. There were temperature problems. It was hard to figure what clothing to wear. Temperature at takeoff had been 36 degrees, but I knew I'd be facing a no-clouds sweat shop at noon in Tennessee. I guess you could say it averaged out: I froze in the north, roasted in Tennessee, and was comfortable at the half-way point!

I might make some comments regarding this flight and ridge flying in general.

As far as I'm concerned, this flight is "The Big One." No straight-out 1,000-miler with a 30-mph tailwind is going to take as much effort as this one did. If someone does 1,500 miles in front of a squall line, takes seven lightning strikes, goes through two tornadoes and a water spout, I will reexamine my position, perhaps. This sort of ridge-running is very demanding physically, exhausting mentally, and is hazardous. Rarely are you more than one minute from landing if your ridge lift quits. Rarely, also, are you above a safe landing field. The majority of Appalachian farmlets are hillside clearings that would "class-26" a glass ship. A good portion of the valleys are forested. Turbulence is a lurking danger that will peel off your wings just a hundred feet above the trees. And then there are those blasted snow showers.

A few words about big ships versus little ships: Having driven light, medium, and heavy gliders amongst the trees for a few years, my conclusions are probably pretty much what one would expect—the AS-W 17 is Boss of the Ridge when it comes to performance and turbulence absorption. The AS-W 17 has about 15 knots over a Standard *Cirrus* at all speeds and pulls about 25% less in *g*'s. When it comes to landing, the smaller the ship the better. A 1-26 is the bird to have then.

The future? World goal and world straight-distance marks are attainable on the right day in April. And the distance triangle is a natural, too. As for out & return, maybe a few more miles could be squeezed out on a super day in a super ship—but I doubt that it will be.

But as long as I can fly, I'll still soar down the Allegheny Expressway because it is beautiful. And exciting. And scary. And sensational. And if the eagles do it, it's got to be right.

FINLAND HAS VERY UNUSUAL WEATHER

Reprinted by Permission
SOARING (September 1976)

Edited for Space

Text and Photos by Tom Page

The coldest June in 100 years threatened the World Championships in Finland so seriously that organizers and pilots alike were holding their collective breath. Even after the minimum requirements were met, it was evident that weather factors had contributed to many misadventures of top-seeded pilots.

Top Pilots

George Lee of Great Britain captured the Open Class title in a Schleicher AS-W 17 by a significant margin over Julian Ziobro and Henryk Muszczynski of Poland who flew *Jantar 2*'s. Ingo Renner of Australia edged out Gunnar Karlsson of Sweden, both flying PIK-20B's, by the tiny margin of eight points for the Standard Class Championship. A larger gap of 124 points separated George Burton of Great Britain, third, also in a carbon-spar PIK-20B, Renner's substantial prize was his very own PIK-20B, a first in the history of the Championships.

World Champion! George Lee, Open Class – Great Britain

The U.S. team had three pilots in the top ten of the combined classes–Richard Butler at fifth in his own Glasflügel 604, Richard Johnson in a borrowed *Jantar 2* at seventh in the Open Class, and Tom Beltz in a borrowed PIK-20B at fifth in the Standard Class. This was an unofficial national standing matched only by Poland. Ross Briegleb placed 29th in Standard with a Standard *Jantar*, loaned, as was Johnson's ship, by the Polish manufacturers through Glider Aero, their U.S. distributors.

Istvan Hahner of Hungary won the World Class Cup in a 19-meter *Jantar 1* and placed 21st in the Open Class rankings. Only three 19-meter sailplanes competed.

The Finns

Imagine carving an international soaring site out of the woods thirty miles

The team assembled around the U.S. flagpole at the opening ceremonies on Rayskala airfield: Top row (crews)—Walker Mathews, Alice Johnson, Ada Gyenes, Charles Gyenes, Sarah Butler, Herb Mozer, Dave Andrews, David Fletcher, and Mahlon Weir. Not shown are Sherry Reed and the three sequential team managers, Jim Herman, Tom Page, and Dick Schreder.

from International Falls, Minnesota, complete with telecommunications and TV monitors on which the scoring computer displayed scoring changes almost on a real-time basis, and you have a fair notion of the energy of the Finnish hosts. Over a two-year period they cleared the disused WWII airfield of a 20-year regrowth of native pine and birch, built a motel, installed three saunas, and added enough temporary facilities to house nearly 1,100 people on site, counting many in caravans and tents. They prepared about 900 meals, ranging from adequate to excellent, at each sitting. Some 200 obliging Finns made it all run.

The Weather

One can be less than overjoyed at the coldest June in 100 years without insulting the Finns. After all, they import all their weather rather than handcrafting it during the long winter months. The weather at the Nummela practice site was wettish; the practice period at Räyskälä was most promising. But only three of the 13 scheduled contest days looked like

When Tom Page offered to cover the 15th World Gliding Championships for SOARING, his offer was accepted with alacrity and gratitude. His writing was a known quality based on his many contributions dating back to 1954, and his outstanding boxes in the last world championship report were still fresh in editorial memory. Tom has already received the Society's highest award, the Eaton Memorial Trophy, and has just now become a member of the Soaring Hall of Fame on Harris Hill.

the travel posters; six were either unflyable for workable tasks or totally washed out; and four permitted only devalued tasks for one or both classes or aborts for Standard. Thus, the Open Class flew seven days—three of them devalued; the Standards attempted the same seven days—two were significantly devalued and two were devalued to zero—net, five.

The weather was, therefore, a significant generator of long-shot gambles by both task setters and pilots. With long shots a lot of people lose who are not normally losers in a game of such high skills.

The contest had been a close one, first of all, with Mother Nature, and then—under those difficult conditions—with an extraordinary group of pilots flying extraordinary sailplanes. Given the weather *and* the limited number of competition days *and* the talent, it is not surprising that the margins of victory are relatively small.

NEW U.S. TWO-PLACE SAILPLANE COMING

Reprinted by Permission
SOARING (September 1976)

Ted Smith, the noted designer of the Aerostar and Aero Commander executive aircraft, last month invited the aviation press to the unveiling of a mockup of his latest project, a two-place, high-performance, all-metal sailplane to be known as the *Soaring Star*. The three-spar, untwisted, 57-ft. wing will use NASA 64_2415 series airfoil and will be skinned with flush-riveted, coutour-rolled skins that are glazed and painted with polyurethane paint to obtain as smooth a surface as possible. Dive brakes on the wing's upper and lower surfaces are designed for use up to 207 mph V_D and 180 V_{Ne}. Maximum positive and negative gust load factors for the 1,150-lb. craft are 8.7 and -6.4 *g*'s, respectively. Differential Frieze-type ailerons (25°/15°) will reduce adverse yaw to a minimum. There is no provision for water ballast. Estimated max L/D is 36.8 at 60 mph.

Smith began his career fifty years ago when he designed a secondary glider, and though he subsequently created only powered aircraft, his interest in gliders remained. By coincidence, he says, he recently learned that 90% of the sailplanes used in the U.S. are imported from European countries. This triggered a decision to return to sailplane design and his company embarked on the *Soaring Star* project. "We spent many hundreds of hours researching what foreign sailplanes had to offer in structural integrity, performance characteristics, speed to V_D, stall, and many other parameters. With the research behind us we sat down to

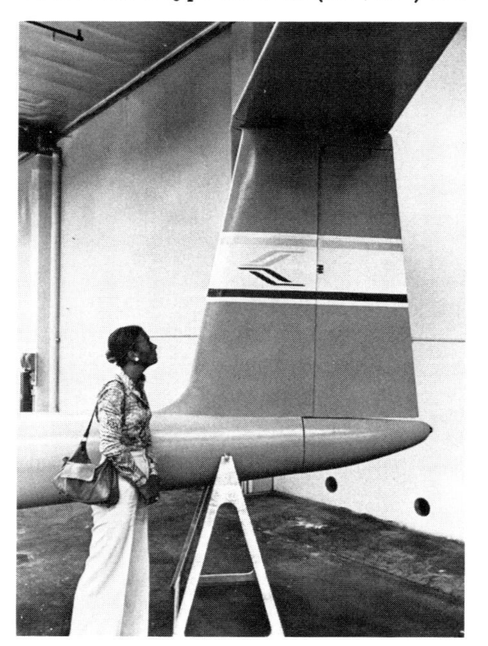

Penny Maines admiring Soaring Star tail logo.

Ted Smith: designer, builder, entrepreneur.

design a two-place sailplane that would be far superior to anything in its category from European imports and also from anything in production in the U.S."

Construction on the prototype has begun in both the Aerostar Van Nuys and Santa Maria plants and the first flights are expected this fall with production to begin next spring. Over sixty have been ordered and will be mainly distributed to Aerostar dealers. A figure of $1,250,000 has been mentioned as Aerostar's production investment in its *Soaring Star*. "It is impossible to quote a firm or estimated price at this time," Ted Smith says, "except to note that when prices are quoted they will be competitive and will include a custom-built trailer."

Soaring hopes to run a full article on the *Soaring Star* when performance figures are made available after flight testing.

NOTE: Ted Smith died December 29, following open heart surgery.

The high performance Soaring Star.

1976 U.S. HANG GLIDING CHAMPIONSHIPS

Dog Mountain, Washington

Reprinted by Permission
GROUND SKIMMER MAGAZINE (August 1976)
Photos by Bettina Gray

by Pork

Dog Mountain, the site of this year's National Championships, is some five miles east of the small logging town of Morton, Washington. The improved launch atop this foothill ridge of the Cascades faces west into the prevailing wind 1,480 feet above Davison Reservoir, and can be flown south, west or north. With the right conditions, it is a truly fine site and sometimes can be soared with as little as 5mph wind coming up the graveled cliff launch. In fact, the air above the valleys on three sides can be soarable at the same time.

One might expect a meet held at such a site to be smooth and pleasant, but with the possible exception of daily freeflying, this was not the case. Fliers were not heard complaining about free-flying, but during the meet itself, conditions proved unbelievably variable, and as a result of frictions between judges, pilots and officials, times and rules changed so fast it was nearly impossible for the judges themselves to keep up with what was happening. When it was all over, though, the USHGA had three winners in each of three classes to send to Europe.

The three classes were standard, open and unlimited. Unlimited class was expected to be dominated by rigid wings, but three SST's entered, flown by Burke Ewing, Dave Vincent and Jim Debauche, whose Icarus V had been run over by a neighbor in his driveway. Their entry in the higher class drew complaints from pilots flying Quicksilvers and Icaruses. The Rogallos didn't win, though they all placed in the final ten.

I regard myself as more of a journalist than a flyer, but I entered my SST in open class, hoping to gain insight into what the pilots were experiencing, and to learn what it feels like to fly in a national competition. As I feared and expected, it was difficult, and bad luck and inferior flying skills removed me from competition at the first cut. Nevertheless, I'm glad I tried. It made me a better and I hope a more cautious pilot. Flying in competition wasn't fun for me, but the feeling I had after making one good flight was worth it.

Dog Mountain is tall and steep—not a good place for the inexperienced flyer. Its slopes are covered with two hundred foot tall Douglas Firs, and infested with a lower growth of alder, madrona and vine

The challenge of human skill against unsettled conditions was a mark of the Dog Mountain competition.

maple, making it a beautiful but deadly place to crash. Half a dozen bald eagles make their home in the area, and several flyers had the thrill of flying tip to tip with our national bird. Don Cohen of San Diego attempted to do a few 360's with an eagle, but the bird could come around so much more quickly and cleanly than his Cal Glider Don could never get out of being "it" in the game of tag.

The first few days of the meet were miserable—likely June is the wrong time for competition here. Rain, mist, sloppy ground and winds shifting around the compass from SE to W to NE made takeoffs nervewracking, and until it was graveled, the launch became so slippery it was like trying to run down a pile of softening butter. Reggie Jones and Tom Peghiny both made mis-stepped, heart-in-the-mouth launches.

Things were made worse by continuing arguments between the top officials. Only about sixty per cent of the qualified pilots showed up, and a number of unqualified pilots including myself wanted to take their places. Some officials were much opposed to the idea, and at first were only going to allow us one flight to qualify, but at last they were persuaded to give us three. That seemed fair enough, but for an idea of how things went, I'll relate a small story of my own experience.

I felt lucky when I stepped up to make my first launch. The wind was reasonably steady at 5-10mph from a westerly direction, and the rain had slackened to a light drizzle. I made a clean takeoff, and though I was feeling mild lift, I cooked along near maximum L/D toward the pylon course a quarter mile away. I reached the course with more than a thousand feet of altitude, worried I wouldn't be able to figure eight properly at that height and bothered a bit by water in my eyes. Everything seemed to be coming together for me, however, and I concentrated on making clean turns and watched for the flags to indicate I was on course. From watching other flights I knew I had a good run going and was getting about twice as many pylons as most of the previous pilots. My last turn was especially pleasing, because I came out of it with good altitude and a line on the target. I made a light touchdown well within the circle and whooped with joy

Steve Patmont launches the Mitchell Wing, which predictably drew much attention from spectators and pilots alike.

Unidentified pilot heads for competition area under the critical eyes of launch area observers.

when Bob Wills told me I had nine pylons. Then I heard my name on the P.A. and was called to the judges' stand to be informed my flight was scoreless because the clock didn't work. I won't tell you what I said to the judges, but I apologized to them later. I never had such a good flight again–no one's fault, really, just the breaks of the game. Poor Reggie Jones had the same thing happen to him twice, consecutively.

I witnessed two crashes the first day of competition. They occurred during a brief sunny period when the wind was coming from a ninety degree southern cross angle at the west launch. All pilots had to fly from the same launch or the whole class had to be run again. The first stall on launch was made by Jim Springer of Tennessee. He fell some thirty feet with his 20-20 Swallowtail into a maze of sticks and logs below the cliff. He was uninjured, though his control bar was pretzeled and he had the wind knocked out of the lower part of his body. "It was my decision," he said, bent over and breathing hard. "I was in a hurry because I missed my class this morning." Crosswind launches continued, and the next stall happened to Pat Hickman, winner of the Oregon Open, who was flying a glider of his own design, a Tradewinds III. Pat hooked his truncated tip first on the launch and then on the debris, but escaped with only a ripped sail, which put him out of competition. I saw no more accidents until the last Saturday, when Scott Price of Seattle lost control of his Quicksilver twice in gusty winds over the spectators as he approached the target, and went in among the concession booths. Luck again prevailed, and no serious damage occurred to pilot or airplane, though the incident was to earn the "Turkey of the Meet" award for Scott, who also took third place in the unlimited class.

On Monday, May 31, the takeoff and setup areas were improved with gravel, and pilots felt better about launching. The road from highway 12 into the spectator area was also graveled and graded, making it much more pleasant for pilots and spectators alike, who had to contend with huge chuckholes created by rain, log trucks and hundreds of autos. The citizens of Morton and Glenoma behaved very well toward this invasion of hang gliders, and except for a couple of instances of late-night police harassment involving young people in vans, there seemed remarkably little friction between pilots and the public.

Flying continued during these days of elimination, often halted by squalls moving up the lake about once an hour. The rains made unhappy times for the many pilots waiting them out on top, and once while the standard class was waiting out a spell of bad weather Chris Price, Terry Sweeney and a few others argued the merits of altitude gains, one-on-one competition and other possible scoring methods to determine the best pilot.

During the week the weather improved slowly. On Wednesday two of the hottest wings got tough breaks. Brian Porter, flying the Easy Riser with which he won the world open made a fine but nearly scoreless flight, because the pylons were so close together he couldn't see them well from the position he occupied in his airplane. Steve Patmont was flying a new Mitchell Wing, and would have had a fine score indeed had he not misjudged wind force over the target (like so many rogallo pilots), and missed the foul line. High scores in open class that day went to:

Glider	Pilot	Score
Cirrus III	Brian Huston	15.628
SST	Dean Tanji	12.653
Albatross	Bruce Waugh	12.500
SST	Bob Wills	12.186
Merlin	Joe Greblo	10.641
Dragonfly MK II	Rick Monteverde	10.257

Shear lift that occurred late Wednesday afternoon caused great flights and difficult landings.

Thursday, June 3, brought blue skies, puffy white cumulus clouds with gray bottoms and retreating snow moving up nearby mountains. Brisk 15-18mph winds built whitecaps on the lake and long wind lines snaking across the water toward the Dog Mountain launch. The lift band extended out nearly half a mile, but had lumps and soft spots near its end, not far from the distance pylon that was part of the day's task. Pilots camping at the south end of the lake put up a great lean-to of fir driftwood and plastic sheeting the size of a small three-room house.

Conditions were so variable Thursday that while the average scores were from two to four points, pilots who got flights when the lift was hot occupied another realm altogether. For instance, one group of six pilots who flew when it was "happening" over the course obtained an average score of around twenty-two points. They were Tom Peghiny, Dean Tanji, Roland Davies, Conrad Agte, Stewart Smith and Trip Mellinger. High scores Thursday:

Glider	Pilot	Score
Kestrel	Conrad Agte	72.080
Phoenix VI-B	Trip Mellinger	24.592
Moyes 120	Roland Davies	9.379

Scores went rapidly downward from that point. They were figured by elaborate calculations from a fast time to the

Scott Price, a member of the Dog Mountain Air Force, placed 3rd Unlimited Class.

speed pylon about half a mile out from the launch, a slow time over the course, the number of pylons maneuvered around, followed by a spot landing. High score in the unlimited class was earned by Tom Vayda, flying a Flegling. He got 6.475 points, and Dan Alban, with a Bobcat II in standard class had 3.146.

Out of the competition at this point, I took over the announcer's chores for a few hours and then went to the top for free-flying. About seventy pilots were already there. The "magic lift" started again, and gliders played king of the hill on the north face. Brian Jensen, Dave Vincent and Rich Grigsby had top dog position at consecutive times. Then a Cessna 120 came along and with an astounding maneuver cut his engine and soared the face below the launch, to the cheers of those on top, after which he made a "dead stick" landing below on a road known as "the drag strip" which ran into the lake. When I left the hill to go into town I took a last look back at Dog from the hill going into Morton. Twenty gliders were soaring high over the mountain and the lake. A Manta Fledgling was 3,000' up and over a mile down the lake, a nice sunset scene. It was Brian Porter's twenty-second birthday.

Good times were enjoyed evenings in The Wheel restaurant and The Spar tavern. One could gossip with Bill Bennett, shoot pool with Dave Cronk, play fussball with Roy Haggard or down an "Oly" with Dave Witt. Brian Porter got more free birthday beer than he could drink.

Davison Lake frames a competitor after take-off from 1500' Dog Mountain.

But Friday morning brought another cut to the flier's ranks, and pilots and birds continued to fly the face of Dog Mountain. A few mild bandtailed pigeons were seen, and a crow harassed an eagle all over the mountain for ten or fifteen minutes, diving on our national symbol with such accuracy and vehemence that the larger bird was forced to seek another hill to soar.

Saturday was the first clear day of the meet. Mounts Ranier and Adams were starkly visible from the launch, and while pilots waited on top for the morning downwind condition to switch, a hot-air balloon did touch and go landings on the flat green valley floor some 1,500' below. The setup area behind the launch contrasted sharply with the green hillsides and snowy peaks, displaying a triangular maze of brightly colored high-performance gliders.

However beautiful Saturday was, variable conditions persisted as did the squabbling between judges and judges and pilots and judges. At one point a number of pilots threatened to walk out, declaring the meet null and void, but they were dissuaded from this action and the meet continued toward its nearing end. There were some two thousand spectators who paid entrance Saturday, and their number seemed to double as good weather returned Sunday, proving the monetary worth of a big-time meet.

On June 6, a pleasant Sunday, the fourth national championships were concluded. Gusty and turbulent winds seemed to haunt the landing area, and many of the best pilots choked under the pressure or were betrayed by the air over the target and fell from high standings. One of them was Bob Wills, who had consistently high scores all week, but hit a gust in his last turn toward the target and was forced down short, a repetition of the ill luck that plagued him at the world meet last April. The three winners in each class won trophies and a trip to Europe to fly for the United States in world competition.

Most of the contestants left Sunday. I stayed around until Monday, because I sold my glider and then bought Burke Ewing's lovely new SST, hand-painted by his wife Rene. At the site Monday morning, I talked with Bruce Waugh as he loaded his small foreign car.

"How do you feel about second place?"

"It was all right," Bruce allowed, as he stooped his lanky frame to stuff his sleeping bag in the trunk. (The day before, he and John McVey had tied for first in the unofficial "Tallest Hang Glider Contest"). "It was like old home week for Keith Nichols and I. We lived together for a couple of months before the world meet. I've got the glider tuned, now."

Before I left this place of excitement, controversy and some of the greatest free-flying I've ever seen, I took one last look around the Dog Mountain site. The lake was rising about an inch an hour. Already it covered half the area available for landing, and in a few days would cover everything but a small place about the size of the competition target. I walked back to my van, past the silent bullseye and the deserted concession stands. A light southerly wind under a thin layer of stratus clouds turned back the leaves of the alders at the hill's base. *Soarable*, I thought, and looked upward toward the peak of Dog Mountain. It was fogged in.

Pilot prepares to lower landing gear during National competition.

OPEN CLASS STANDINGS

Place	Pilot's Name	Glider Flown	Total
1	Keith Nichols	Albatross	4777
2	Bruce Waugh	Albatross	4740
3	Zane Wadley	Dragonfly	3961
4	Rich Grigsby	Cumulus VB	3949
5	Dean Tanji	Wills Wing SST	3815
6	Bob Wills	Wills Wing SST	3767
7	Roland Davies	Moyes	3739
8	Barry Gordon	Cirrus III	3673
9	Conrad Agte	Kestrel	3556
10	Steve Hasterlik	Phoenix VI-B	3472

STANDARD CLASS STANDINGS

Place	Pilot's Name	Glider Flown	Total
1	Robert Reed	Bennett 19/13	5912
2	Dan Alban	Bobcat II	5910
3	John McVey	Windlord III	5593
4	Terry Lorentson	Eipper 19/17	5443
5	Doug Heath	Bobcat II	5377

UNLIMITED CLASS STANDINGS

Place	Pilot's Name	Glider Flown	Total
1	Ken Koklenski	Fledgling	5312
2	Tom Vayda	Fledgling	4495
3	Scott Price	Quicksilver	4353
4	Buzz Bussmeir	Quicksilver	4202
5	Jim Debauche	Wills Wing 100	3640

1976 ALBUQUERQUE INTERNATIONAL BALLOON FIESTA

Written especially for
AVIATION YEARBOOK 1977

by Dick Brown, Editor
BALLOONING JOURNAL

There is a dry dusty field, which was once an irrigated alfalfa field, just on the northern edge of Albuquerque, New Mexico. For nine consecutive days in October, this site blossomed into one of the world's most spectacular events in sport aviation—the Annual Albuquerque International Balloon Fiesta.

Under a flawless crystal-blue sky, thousands of sleepy-eyed spectators watched as the first of 150 hot air balloons set themselves free. It was Saturday, October 9, 1976—the grand opening day of Albuquerque's fifth annual balloon fiesta. With 50 to 75 thousand cubic feet of hot air imprisoned in each of their spectrum-colored nylon bags, aeronauts and lucky riders lifted off in a dramatic display of man at the mercy of the capricious wind.

Albuquerque is ideally suited for giant balloon rallies. There, the terrain, although at an average elevation of 5,300 feet, is basically flat and extends across the entire 30-mile wide Rio Grande valley. On the west, beyond the volcanic cinder cones, the mesa drops off abruptly into another river system. On the east, the Sandia Mountains rise a mile or so above the valley floor. When not under the influence of major pressure disturbances, the winds follow classic mountain-valley patterns. Early morning Albuquerque breezes are notoriously light and variable.

It is a fact that light winds and wide open spaces, combined with a successful series of balloon fiestas, have led directly to Albuquerque's reputation as the Balloon Capital of the World. With 55 resident aerostats and an immaculate safety record, Albuquerque has earned its "balloon capital" distinction.

The people who come to watch the balloons quickly discover that ballooning is a magnetic sport. Year after year, it attracts and captures the imagination of those who have never before experienced this unusual sport of flying without wings. For the first day of the fiesta, Albuquerqueans set aside their Saturday morning chores and come out to the balloon field.

By the time the sun squinted over the Sandia crestline, the field was the scene of sprightly bubbles swaying in a northerly breeze. Seven launch directors, in striped black and white jackets, scurried about the field, signalling lift-off clearance to each individual balloon pilot. The whoosh of propane burners blended into one long tumultous roar. One by one, the balloons detached themselves from all earthly things and soared aloft.

Early morning ascension during the first day of the Fiesta. Photo by Dick Brown

The events for the weekend called for a series of the famed Roadrunner-Coyote races which provided out-of-town flyers with an opportunity to study Albuquerque's unique wind system. Invariably in October, the wind at ground level is from the north and the wind at 1,000 feet is from the south, thus allowing the exciting possibility for non-steerable balloons to shoot landings back at the launch field. Incidentally, the international flavor of the fiesta was maintained through the attendance of Dick Wirth (England's current national balloon champion) and Phil Hutchins from Great Britain and Holbert Masson of Grindavik, Iceland.

The Roadrunner-Coyote event, more commonly known as the Hare and Hound race in any other region of the world but the Southwest, is a test of the pilot's ability to track a lead balloon and land as close as possible to that balloon's final landing spot, much the same as a pack of coyotes track down an elusive roadrunner. The Roadrunner balloon, with an early start, tries to outwit the cunning Coyote balloons, often numbering up to 30 per pack, as they maneuver at various altitudes in search of winds to carry them close to the target balloon.

Unlike the World Championships which used the same balloon field the previous year, the 1976 Fiesta was designed strictly for fun. There were no serious competitions; in fact, the winners were often declared to be those having the most fun. The smiles on their faces were their trophies. But as with all competition, serious or otherwise, there were protests. After flying, these were settled by water balloon fights—all in fun—at a special area on the balloon field set aside for jousting. If a balloonist declined a challenge from another balloonist, the challenger won by default.

On Monday afternoon, Race Director Bob Ruppenthal organized a mountain-climbing party at the base of the Sandia Peak Tramway. Albuquerque aeronaut Darryl Gunter lifted off in his balloon "Sky Chariot" and set the pace for climbing at 650 feet per minute. Darryl

Day of mass ascension, Sandia Mountains in background, spectators and their cars at the launch site. Photo by Ed Langmaid

was quickly followed by Tim Thorsen in "Seaflame," Bob Bensen in "Ziggy Stardust," Tom Gabel in "Rock City," Phil Hutchins in "J&B," Clair Bennett in "Four Corners," Kurt Gottlieb in "Cactus Jack," Wally Book in "Constitutional Motivation," Mike Corlew in "Dust Devil," and Bob Ruppenthal in "Goldfinger." Bob's basket-mates were Albuquerque Mayor Harry Kinney and his wife Carol. The ten balloons passed an occasional hang glider on their arduous trek up the jagged western face of the Sandias. They climbed to 12,500 feet (MSL) to clear the wind curl on the leeward side of the crest and sailed in 20-knot breezes to the next valley.

The next morning, all balloons and crews reassembled at the balloon field for more fun competition. One of the events was the Tumbleweed Drop in which the

From 1,000 feet, aerial view of balloons with city of Albuquerque. Photo by Dick Brown

pilots, aided by local maps, select a launch site at least three miles radial distance from a fixed target on the balloon field. The race requires excellent judgement of wind patterns as the pilot must fly his balloon in breezes that hopefully will carry him near enough to the target for a good score. This contest takes on the exciting atmosphere of the old-time flour-bombing airplane races. However, the bombardier's trajectile is a tumbleweed with an attached identification tag. Its aerodynamic qualities are not exactly conducive to bomb-dropping, thus adding another factor to thwart the pilot's skill.

Blackjack is a similar contest and just as much of a crowd-pleaser. Again, the pilot launches his balloon from three miles out and carries three tagged markers on board. The balloon field is divided into squares, numbered one through eleven on the pilot's maps to indicate their numerical scoring value. Like the card game, the objective is to score 21 or less. The markers are only dropped from altitude between 200 and 1,000 feet.

Another race at the Albuquerque Fiesta was one which tested the agility and inflation skill of both the ground crew and the pilot. It's called Giddy Up—Whoa and is designed to keep the balloons in front of the crowds. The balloons must be inflated quickly, but the first to inflate must be the last to take off. Fifteen minutes after the first balloon launches (it was the last to inflate), all others must land and the balloon closest to a fixed target is the winner. The idea is to make use of the most favorable winds in order to minimize the drifting distance.

Tuesday and Wednesday afternoons of the Fiesta found 50 balloons participating in the filming of a multi-media presentation. The aeronauts were required to

Ascension during one of the many Fiesta fun-flying events. Photo by Ed Langmaid

Balloon inflations at the launch site. Photo by Dick Brown

crouch down in their baskets during lift-off, thus giving the appearance of empty balloons ascending together. With considerable cooperation and plenty of laughs from the balloonists, the movie-makers succeeded in getting the balloons to rise in unison. The film, which will be distributed nationally as a children's story, will be reversed to effectively produce a spectacular scene in which a fleet of pilotless balloons descend into a single field.

The fiesta continued through the week, with each day featuring excellent flying conditions. Most balloonists managed to stay out of the air traffic patterns and off the nearby Indian reservations. A gala balloon banquet, replete with Mexican buffets, Mariachi bands, and awards presentations to overall winners, topped the week's buoyant activities. Balloonists from twenty states and three nations were recognized for their grand aerial performances.

The Great Mass Ascension on the final day (October 17th) of the fiesta climaxed the week of comraderie, sunshine, and fun flying. Mass ascensions at large balloon rallies are the traditional way of saying goodbye to the host city. Like a hundred bobbing ships upon an endless sea, every balloon set sail, heralding the end of a spectacular balloon fiesta. With periodic spurts of heat, the balloons maintained neutral buoyancy and literally floated above the field. The aeronauts could look down to see their launch spot disappear in the crowd—one of the largest numbers of balloon-watchers ever assembled at a balloon rally. Citizens, young and old, stood on the roofs of cars, trying to catch a fleeting glimpse of hang gliders swirling among the very balloons from which they were released.

At the end of it all, as the dust settles at the balloon field, each aeronaut leaves with mental snapshots—living memories which keep flickering in the mind: the "Caution—Balloon Crossing" signs erected near the launch field by the New Mexico Highway Department, a pilot's repeated failure to hit his opponent with water balloons, the tall stories of "controlled crashes," the precision bombing raids over the balloon field, and the splash n' dashes in the muddy Rio Grande.

Three drifting aerostats, low over the balloon field. Photo by Ed Langmaid

Not to be forgotten is the monumental amount of hard work that goes into organizing such a giant balloon get-together. In a nutshell, the 1976 Fiesta was hosted by the Albuquerque Aerostat Ascension Association—the nation's largest balloon club, sanctioned by the Balloon Federation of America—the United States representative in the world ballooning organization, and organized by the Albuquerque International Balloon Fiesta, Inc.—a non-profit corporation created to manage the financial and organizational tasks of the annual event.

While Pasadena has its Tournament of Roses, New Orleans its Mardi Gras, Indianapolis its Indy 500, Albuquerque has its Balloon Fiesta. Likewise, while powered flight has its gadgetry and vibration, balloon flight has its simplicity and serenity—unequalled in any other form of sport aviation.

NEITHER RAIN, NOR SNOW, NOR COLD...

The 1975-76 National Collegiate Parachuting Championships

by Mike Truffer, Meet Director

Reprinted by Permission
PARACHUTIST (February 1976)
File photos courtesy USPA

Skip Giles squinted at the sky and turned to me. "It looks like it's clearing up. Should we go with accuracy or RW? The weather will probably stay good for both."

"Accuracy," I replied. "Since we have a round of RW completed, let's work on accuracy."

Bill Bohringer interrupted. "There's one problem with that, Mike. The pea gravel is frozen solid."

Such went the 1975-76 National Collegiate Parachuting Championships. We were at Jumpwest Parachute Center in Star, Idaho. A freak winter storm had moved into that part of the country the day before the meet was scheduled to begin on Thanksgiving morning. Ninety-three competitors from 27 schools stood huddled around heaters in Jumpwest's loft, wondering what it was going to be like to jump with snow on the ground.

In spite of the unusual and sometimes uncomfortable weather, minimums were completed in every event. For 3 rounds of accuracy, 2 of style, and one of RW, the U.S. Air Force Academy dominated the meet to win the Gavin Gavel and be recognized as the best collegiate parachuting team in the country. Jim Salisbury, a student at Florida Tech, placed first in advanced style and second in advanced accuracy to be named 75-76 National Collegiate Parachuting Champion and be awarded the $500 Istel Memorial Scholarship. Leo Leduc of the U.S. Military Academy won the Intermediate Category, and Dennis Hailey of Louisiana Tech placed first in Novice accuracy. Only one round of RW was completed and the team from the University of Montana, (Jon Andrus, Donald Fisher, Robert Murray and Tara Sayles) won with a time of 12.7 seconds for the star-backloop-star sequential jump.

When the weather cleared, the meet went fairly smoothly. Giles headed up the manifest and aircraft dispatch section to keep the jumpers moving and the Cessnas airborne. The competitors cheerfully cooperated. (Each year the staff tells me that they enjoy working for the collegiates because it is a "good vibe" meet. That's for sure.) The judging and recording section, directed by Bill Bohringer, probably had it the worst: they had to stand in the peas or by their telemeters for hours in the sub-freezing weather.

Although the meet was scheduled for four days, we only had 1 1/2 days of jumpable weather. The jumpers stuck around in the loft while Tim Kataras ran

THE WINNERS . . .

The U.S. Air Force Academy Parachute Team, 75-76 Overall School Champions. Photo by USAFA

Advanced Overall Champions. Left to right: John Bowen, 3rd, USAFA; Andy Probert, 2nd, USAFA; Jim Salisbury, 75-76 National Collegiate Parachuting League Champion, Florida Tech; John Kurtz, 5th, USAFA; John Finley, 4th, Linfield College. Photo by USPA

The medal winners in Novice Accuracy. On the left is Bob Davis (University of Nevada, Reno), standing next to Terry Heck of Ferris State. First place finisher Dennis Hailey of Louisiana Tech is not pictured. Photo by USPA

Relative Work Champions. From left to right: Bob Murray, Tara Sayles, Donald Fisher, and Jon Andrus. All are students of the University of Montana. Photo by USPA

Three of the top five in Intermediate Overall, all from the Air Force Academy: Left to right: Bob Donze, 5th; Ed Miller, 2nd; Kevin Roll, 3rd. Not pictured, from the Military Academy: Leo Leduc, Intermediate Champion; and John Lidh, 4th. Photo by USPA

THE STANDINGS . . .

Advanced Overall

	Name	School	Score
1.	Jim Salisbury	Florida Tech	466.7
2.	Andy Probert	USAFA	527.8
3.	John Bowen	USAFA	534.0
4.	John Finley	Linfield College	547.5
5.	John Kurtz	USAFA	557.5
6.	Dave Williams	USMA	561.3
7.	Randy Mattson	USAFA	574.8
8.	Mike Mythen	Central State College	624.5
9.	James Daron	USMA	656.7
10.	Scott Rogers	Southern Oregon State College	821.3
11.	Tom Sorce	University of Nevada, Reno	830.2
12.	Mike Klinke	Montana State	869.7
13.	Kevin Woody	USMA	891.0
14.	Mike Bouton	Boise State	917.3
15.	Robert Murray	University of Montana	953.8
16.	Terry Wilds	Boise State	987.7
17.	Jon Andrus	University of Montana	1015.3
18.	Ken Gano	Eastern Ill. University	1145.8
19.	Donald Fisher	Univ. of Montana	1167.3
20.	Andy Watson	Southern Oregon College	1243.8
21.	Maurice West	University of Nevada	1399.3

Intermediate Overall

	Name	School	Score
1.	Leo Leduc	USMA	528.5
2.	Ed Miller	USAFA	547.3
3.	John Roll	USAFA	580.5
4.	John Lidh	USMA	584.3
5.	Bob Donze	USAFA	599.7
6.	Terry Smith	USAFA	670.7
7.	John Ayers	USMA	713.8
8.	William Ramsey	USAFA	744.5
9.	Dave Engstrom	USMA	766.0
10.	Phillip Matthews	Louisiana Tech	803.2
11.	Mike Franklin	USMA	822.2
12.	Tom Kern	Montana State University	1017.0
13.	Mike Pilliers	University of Nevada, Reno	1129.0
14.	Mike Melner	University of Nevada, Reno	1297.7
15.	Bob Buckles	Montana State University	1335.0
16.	Rich Leathers	University of Nevada, Reno	1356.0
17.	Tara Sayles	University of Montana	1366.3
18.	Phillip Hall	Montana State University	1466.7
19.	Scott Boiko	Illinois State University	1468.5
20.	Bev Saxerud	University of Minnesota	1540.7
21.	Gary Rafuse	Patterson State College	1554.0

Novice Accuracy

1.	Dennis Hailey	Louisiana State	0.90
2.	Bob Davis	University of Nevada, Reno	1.22
3.	Terry Heck	Ferris State	2.13
4.	Paul Dobos	Ferris State	4.20
5.	Dave Fox	Louisiana Tech	5.39
6.	Bob Klosterman	University of Nevada, Reno	5.61
7.	Mike Lynn	Iowa State	7.73
8.	Gary Lamb	Oregon State	8.79
9.	Dave Floyd	University of Minn.	8.96
10.	Dick Marrow	USMA	9.29
11.	Robert Korbus	University of Illinois	9.99
12.	David Wells	Boise State	10.00
13.	Dave Browne	Boise State	11.38
14.	Gregg Campbell	College of Idaho	11.59
15.	Richard Conway	Ferris State	11.72
16.	Carl Blincoe	University of Nevada, Reno	11.76
17.	James Neiderheide	Ferris State	12.19
18.	Tadeusz Szwars	Iowa State	12.48
19.	Thomas Swan	Pennsylvania State	13.61
20.	Bill Allen	Boise State	15.26
21.	Douglas Hoskins	Oregon State	20.00

School Standings

	School	Quality Points
1.	US Air Force Academy	244
2.	US Military Academy	141
3.	University of Nevada, Reno	39
4.	Southern Oregon State	35
5.	Florida Tech	34
6.	University of Montana	30
7.	Idaho State	27
7.	Montana State	27
8.	Louisiana Tech	25
8.	Linfield College	25
9.	Boise State University	18
10.	Ferris State College	15
11.	Central State College	13
12.	Casper College	6
13.	American River College	5
13.	Oregon State	5
14.	Iowa State	4
15.	University of Minnesota	2

Other Colleges in Attendance:

University of Oklahoma
Patterson State College
Mountain Home Community College
College of Idaho
Eastern Illinois University
Pennsylvania State
University of Utah
Simon Fraser University (Canada)
Vancouver City College

Relative Work

1.	University of Montana
2.	U.S. Air Force Academy
T3.	Montana State University
	University of Nevada, Reno
5.	Canada
T6.	U.S. Military Academy
	Boise State University
T8.	Southern Oregon State University
	Idaho State University

video tapes of the '75 Nationals RW and style events. Giles, Bill Jones, and others led seminars or discussions to fill in the spare time.

The Canadian Sport Parachuting League sent a team of four to the meet. Besides being a friendly group of people, the Canadians were damn fine parachutists who did well in every event they entered. (There was some discussion regarding the possibility that they were somehow responsible for the "typically Canadian" weather.)

The competitors got a chance to let us know what they thought of this meet and the NCPL in general at the NCPL General Membership Meeting on Friday evening. NCPL Committee Chairman King Morton attended the meeting and listened closely to suggestions that the Collegiate Championships be "opened up" a bit to encourage greater participation particularly in the RW event. The question of "Why Star?" came up and was quickly followed by the explanation that the NCPL Committee accepts bids from anyone and hopefully more bids will be submitted in the future, instead of the *one* that was received last year.

The banquet, held the following night, proved to be a pleasant but humdrum affair as the audience, filled with turkey, had to endure the awkward speeches of the staff.

So went the 1975-76 National Collegiate Parachuting Championships. Although it wasn't the resounding success of the previous year's meet in Florida, it definitely wasn't the great disappointment that we endured in Illinois the year before that. Right now the NCPL Committee is considering some significant rule changes that will let more competitors do what competitors everywhere seem to be doing more of—relative work. Let the Committee hear your ideas. Encourage your center or club to host next year's championships.

Do any of you collegiate jumpers in Tahiti hear that?

USPA File Photo

LOCK THE WHEEL!

Reprinted by Permission
PLANE & PILOT (January 1976)

by E.A. Owens

Edited for Space

Young and impressionable I was in those days; awed and dazzled by anything aerial. I would do just about anything to check out in a different airplane, but mostly I just wanted to stay airborne. Landings were more than the culmination of a flight; they were the death of ecstasy, the terminus of dreams, the return to reality.

It was during my first year on a new job that I met one Gary Morrison, skydiver extraordinaire (world famous for his death-defying leaps into the blue and a 25-knot wind on a hamburger bet. Results: one wrenched shoulder and a greasy hamburger on the way to the hospital). Gary and I struck up a friendship over a common bond, our preoccupation with the sky. Of course, even now, I glance sideways at those whose lot it is to leap out of perfectly good airplanes, which Gary has now done over a thousand times, but they *are* flyers, of sorts, and walk and talk just like real people.

During a lunch-time conversation one fateful day, Gary asked me if I had ever flown a "jump plane." Reluctantly, for fear of losing out on a new experience, I replied that I hadn't. My fears were unfounded, however, as I was informed that 300 hours and a commercial pilot's ticket, to satisfy the insurance company, and a check-out by the aircraft's owner were all that was required to become the "jump pilot."

The following Saturday, after satisfying all the above requirements, I became the newest, nervousest, grinningest, skydiverhaulingest Cessna 205 pilot in the state of Nevada.

Satisfying the insurance company and the 205's owner was nothing compared to the checkout I got by the members of the Las Vegas Sport Parachute Club. Since my initiation, I have learned that most skydivers don't really like to fly, and use the airplane only for transportation from terra firma to the exit point. One of the aforementioned, the World Famous, confessed that while he didn't mind the takeoff and climb, he broke out in a cold sweat at the prospects of riding the aluminum chariot all the way back to the ground. Strange people these jittery jumpers.

My instructions included keeping the wings level while making five to 15 degree heading corrections on the jump run, slowing the airplane as much as possible over the drop zone, reducing power to the minimum when the jumpmaster called out, "Cut," and above all, not stalling the airplane with jumpers hanging out in the wind. There was one last word of caution: "One of us had to stand on the tire of the right main gear. Be sure to lock the wheel, that is, be sure you have the brakes on."

At my first "lift," I wasn't sure anybody was going along. The entire club stood around in a huddle, fidgeting and casting suspicious glances in my direction. Finally, Gary took the bull by the horns and picked out four volunteers. We were on our way.

Takeoff and climb to altitude were routine. I tried to make myself busy and professional under the close scrutinization of these very attentive pilot critics, but one can tap the oil pressure glass and rub fingerprints off the mixture control

USPA File Photo

knob just so many times without arousing suspicion.

When the jump run came I was about five miles from the drop zone and not at all on the jump run heading. I had to fly from south to north, passing over the drop zone and extending about a mile beyond, make a 180 and line up on what I thought was the proper heading. Above the din of the roaring engine, rushing wind and shouts of the assemblage in the cabin, I could make out, "40 right." Have you ever tried making a 40-degree turn with the rudder, cross-controlling to keep the wings level?

The rest of the jump run, punctuated by shouts of "five right," etc., went fine until I heard the "Cut." I had before, and have since, flown airplanes in, uh, non-standard configurations; nothing had prepared me for the effect of a thousand pounds of five full-grown skydivers poised in the slipstream on one side of the airplane, all shouting such things as . . . well, anything but "Geronimo." The 205 had exchanged its normally passive, forgiving nature for that of a snarling, bucking aerial bronc. I had the controls almost to the stops and was having a lot of fun trying to keep her "straight and level." As if all this wasn't enough, someone out there was still shouting out heading corrections, all to the left, of course.

Meanwhile, we're dropping like the well-known streamlined brick and my left leg is starting into convulsions from the strain of holding in the rudder. The sudden, violent pitch-down of the nose indicated that it was all over. I was left alone with a sweet-faced, smiling aluminum puppy.

Back on the ground I was congratulated for not entirely muffing the lift and assured that experience would permit me to come out at the proper altitude, at the proper distance from the drop zone, on the proper heading. The only major criticism I received was from Gary, who had been on the wheel. "You forgot to lock the wheel," he said and I assured him that I would henceforth remember that particular detail.

At the end of the first day I felt like an elevator operator. I had more ups and downs than a third-grader's yo-yo. Flying the jump plane, however, was becoming fun; I *was* beginning to level out at the right place and have the skydivers over the proper point at the proper time, without wasting precious time flying all over southern Nevada getting lined up on the jump run. At the end of the day, the only criticism I received was, again, "You gotta remember to lock the wheel." The next several weekends provided enough experience to "lock the wheel" at least 63.5 percent of the time.

Came the day of the critical jump run. For mysterious reasons known only to skydivers, this particular jump run had to be precise. It was to be from 12,500 feet AGL and the five jumpers were going to attempt something called a "Star." The answer to my "whazzat?" was that it would all be explained to me, on paper, at a later date and right now I was to concern myself with getting the airplane into the right airspace at the right time.

The drama was unreal. Grim faces were attentively turned to jumpmaster Morrison (who would not trust this delicate operation to one of lesser experience) while he leaned out the door and fixed steely eyes on the drop zone, which had long ago passed from my sight under the 205's nose.

We're on the jump run, close to the drop zone, tension is unbearable and heading corrections are coming fast ... "FIVE RIGHT" ... "FIVE RIGHT" ... "THREE LEFT" ... "CUT!"

Close the cowl flaps, ease the throttle back to just above idle ... don't let the engine cool too fast ... slow her down to 80, keep the wings level, watch the heading ...

Nothing is happening! All the jumpers are still sitting in the cabin, staring at me.

USPA File Photo

That is, all the jumpers except Morrison. Morrison is doing something very funny ... he's out there on the wheel and running like a squirrel in a cage. There is fire in his eyes and his wildly working lips are trying to form words ... something like ... "LOCK THE ..." My reactions were like lightning. As soon as the realization penetrated my brain, I crammed on the toe brakes. The wheel stopped turning, but Morrison didn't stop running; that is, he didn't stop running until his legs were stretched out in front, his posterior had bounced off the tire and he was rapidly descending, backwards. The last impression I got was of a face like a thunderstorm, fading rapidly, and an outstretched arm terminating in an obscene gesture.

I managed to get the four remaining jumpers out over the drop zone with no problems. Morrison had missed the drop zone about a mile and was now wandering around somewhere in southern Nevada. I decided to terminate activity for the day and returned direct to Las Vegas, foregoing a landing at Jean, where we had been conducting operations.

Some weeks later, after all had been forgiven and I had been taken back into the fold, I was treated to another insight into the peculiar behavior of the species, *hard-hatted, goggle-eyed sky-plummeter*. It seems that once upon a time this particular group had been involved in a full stall, and resultant inverted flight, while hanging out on the strut just prior to exit (Ahhhh, the warning about not stalling the airplane). I guess that experience served to prove their conviction that flying just isn't safe. We had just passed through 6,000 feet AGL when a gust caused the stall-warning light and buzzer to activate. Apparently, the sensor on the wing's leading edge became stuck and I managed to unstick it only after several porpoising maneuvers. Turning around to apologize for the inconvenience to my passengers, I found that I was alone! Like a covey of over-sized quail, my denizens of the sky had taken flight, I was later told, at the first beep of the stall indicator. Conversations with other jump pilots has revealed that this phenomenon, known as the "premature exit," manifests itself not infrequently.

One of my most memorable experiences came on one of those beautiful, sunny Nevada days in January. The ground temperature was in the high seventies and I anticipated a pleasant day of 7,500 foot AGL lifts. Donning a well-worn leather jacket and regular street clothes, sans hat and gloves, I picked up the five jumpers at McCarran Field. I should have known then that something was amiss: we normally flew out of Jean or Thunderbird Field.

The fully-suited jumpers looked like an Alaskan expedition as ski-masks, motorcycle-type gloves and several layers of clothing were evident under their jump suits. Reluctantly, I asked for flight instructions and was told that they were doing an exhibition jump over Indian Springs (a small town and Air Force Base about 40 miles northwest of Las Vegas). Altitude: 12,500 feet AGL that meant 16,000 feet MSL!

Things went pretty well through the flight to Indian Springs, the 2,500 feet AGL pass to drop the wind indicator, and the first few thousand feet of the climb. Then I started getting cold. By 11,000 feet MSL I was freezing and our rate of climb had dropped to a little over 300 fpm. By 11,500 feet MSL two of the jumpers decided that they had had enough and asked for a jump run. After they exited, rate of climb went back up to about 500 fpm, decreasing steadily. By 15,000 feet MSL I could no longer feel my hands and feet, rate of climb had dropped to about 150 fpm and I could get only one word out before catching my breath while talking to the AFB's control tower.

Is there a descriptive word beyond cold and freezing? If not let's make up one, say . . . frold. Well, by the time we reached 16,000 feet MSL and turned on the jump run, I was frold! Again, Morrison was the jumpmaster and the final minutes before exit went something like this:

Morrison: "Five right."
Owens: Nothing.
Morrison: "FIVE RIGHT!"
Owens: Nothing.
Morrison: "OWENS?!"
Owens: "nnnnnnnnnnnn."
Morrison: "#!%&! FIVE RIGHT!"
Owens: Makes aircraft nose bobble slightly to left.
Morrison: "#$%&! I SAID #$%&! FIVE RIGHT #$%&!"

Owens: Correct somewhere to the right.

Morrison mumbles something about getting out while we're still in the vicinity and three figures pass my peripheral vision.

Alone at last! I stiffly closed the cowl flaps, kept just enough power to keep the engine from cooling too fast, pointed the 205's nose at Tonopah highway and pegged the airspeed indicator. At red-line speed, I am happy to report, there was no flutter and the wings stayed on.

My childhood was spent in the White Mountains of Arizona where average winter temperatures are somewhere near zero and snow piles up six and eight feet; but nowhere, no time, have I ever been so cold as sitting in the cockpit of that flying icebox. Back at the field, I swilled gallons of coffee and created a dense cloud of cigarette smoke before being able to do more than grunt and point to what I wanted.

As if the very act of operating the airplane isn't enough, the jump pilot is soon treated to the peculiar humor of the jumpers. After gaining some confidence in my abilities, the group's sadistic side started to show. On climbs to altitude in the now-closed cabin, a peculiar thing started to happen. It seemed that each of the body-flingers delighted in saving up his gaseous emissions for an entire week and then releasing them during the climb, despite a placard above the door warning against such a practice.

Another nerve-shattering practice was to reach over and pull the mixture control knob to the "off" position just as they made their exit. Imagine serenely flying along at about 10 grand, feeling the familiar nose-down lurch as the jumpers leave the plane, and the engine quits! The first time it happened to me I was busier'n a one-legged man at a butt-kicking contest until I found the cause. Back on the ground the chuckling group

USPA File Photo

denied any part in such an ill-advised maneuver and explanations ranged from a shrug of the shoulders to, "Yeah, well, uh, musta got caught in my rig when I leaned forward." From that flight on I made it a practice to place my hand over the mixture control on the jump run.

The jokes, however, weren't all one-sided. My own "get-even" was to wait until we were climbing over the rugged Spring Mountains and slump over in the seat "asleep." I would be shaken "awake" but pretend drowsiness all the way to altitude. This kept the jumpers' minds off any mischief and gave me REVENGE!

Another very effective method was the imaginary "near-collision." Throwing the left arm up in front of the face, banking sharply with the right, then peering out the side window with a murderous look kept the jumpers occupied with discussion of the "damned idiot that almost ran us down" all the way through the climb and exit.

Other pilots weren't as subtle. Some were downright evil. Well aware of the phenomenon of the "premature exit," one pilot would wait until the climbing routine was well established and an acceptable drop zone lay below, then subtly reach down and place the fuel selector in "off." When the big silence came he would grab the control wheel with both hands and scream, *"Oh my God, we're out of gas!"* He tried timing the resultant full-to-empty cabin one time, but found that, in this case, the hands (of the watch) were not quicker than the Aiiiiieeeee's (sorry 'bout that).

There are, sometimes, hazardous aspects of flying jumpers. One time a student's reserve parachute deployed while he was sitting by the 180's open door; pulling the student out *through* the door post, the canopy wrapped around the tail, and the student dangled from his shroud lines. The pilot managed to save the airplane after the wind tore the canopy loose, and the student descended under his main canopy. But even considering those hazards, flying the jump plane for skydivers is a most enjoyable, rewarding experience, if common sense and good safety practices are followed. Besides being thoroughly familiar with the airplane, including its peculiarities at full gross, door off, bodies hanging out in the wind, etc., the pilot should make himself thoroughly familiar with FAR Part 105, which deals with skydiving.

The pay is poor (ranging from nil to a daily cup of coffee) but some of the fondest memories of a pilot's life will be made during those exhilarating flights. Keep the wings level, wear the proper clothing, listen to the jumpmaster and, above all . . .

LOCK THE WHEEL!

HOPES AND CHALLENGES FOR A NEW MPA

by David Gustafson
18 Virginia Rd.
Barrington, RI 02806

Edited for Space
Reprinted by Permission
SPORT AVIATION (December 1975)

Try to imagine a 78 foot wing that weighs only 75 pounds. Attach it to an airframe and tail group that weighs another 75 pounds and you're on the way to visualizing Joe Zinno's unique Man Powered Aircraft, the Olympian ZB-1.

Joe hopes to capture the £50,000 prize offered by British industrialist Henry Kremer. The Pounds Sterling go to the first man or group of men who can fly a figure-8 course around two pylons a half-mile apart and cross the start and finish lines at least 10 feet off the ground. Man power only. No rubber bands. No stored energy of any kind. No pushing by ground crew.

Joe's not the first to try. At least 16 MPA's have lifted off since 1960, but none have come close to challenging Kremer's purse. The Jupiter went 1,171 yards in a straight line to the end of the runway and the Puffin finally made a 180 degree turn after 90 attempts but crashed shortly thereafter. At least seven others have been turned into firewood.

Among the firsts incorporated in the unusual and fascinating structure of the Olympian are: (1) a dual airfoil in the wing that produces high lift inboard and minimizes the diving moments outboard, (2) the entire outer 5 foot section of the wing moves by means of a single, spring-loaded cable, producing a full wing-tip aileron, (3) the 8½ foot prop is geared for variable pitch (climb and cruise settings for an MPA?), (4) drive power is developed through a freshly designed reciprocating pedal system that's geared to produce 230 rpm in cruise and a maximum of 295 revs. Joe's also making use of an inverted airfoil on the stabilator and looks forward to accomplishing flat turns with an enormous rudder.

Few designers have been as weight stingy as ole Scrooge Zinno in fabricating a prototype. His ribs are made of balsa strips, fitted to a box spar of 1/32" ply with spruce spar caps, and skinned with clear mylar plastic. The fuselage is given its odd shape with aluminum tubing that's so thin a child could bend it. The boom running back to the tail group was originally a spruce-plywood construction but an untimely photo call along with the effects of some gusty winds on those huge tail surfaces created a wooden pretzel, so Joe opted for the additional weight of a single aluminum tube.

Final assembly of the ZB-1 is taking place at the abandoned Quonset Naval Air Station in Rhode Island. The 8,000 foot runway there is long enough to give Joe a crack at two records on the first day of flight tests: first MPA to fly in the United States and the world's distance record. As soon as the plane has left the ground it will become an immediate candidate for a museum. If it can be pumped through the Kremer course one time that will definitely be the last flight.

JOE ZINNO'S MAN POWERED AIRCRAFT FLIES

Reprinted by Permission
SPORT AVIATION (July 1976)

Edited for Space Photos by the Author

by David Gustafson
18 Virginia Rd.
Barrington, RI 02806

On April 21, [1976] at 8:16 A.M., EST, Joe Zinno lifted his Olympian ZB-1 off the ground at Quonset Naval Air Station (abandoned) in Rhode Island and became the first man in North or South America to fly from level ground by manpower alone.

Joe put over 7,000 man-hours and $5,000 into the four years it took to design, build and fly the Olympian.

On the night of April 15 a high pressure system moved in and promised good weather along with calm winds at sunrise. The FSS backed the prediction, so the first call was set for 6:30 the following morning. It turned out to be gorgeous. There were six happy people: three men on the ground crew, Joe, a guy with a Polaroid, and myself with a couple of cameras.

The enormous hangar doors were pushed open, the Olympian was carefully revolved 90 degrees and Joe climbed into the cockpit for his first attempt at moving the finished plane. He pumped the reciprocating pedals and the nose-wheel and bicycle wheel began turning—in spite of the fact that the bicycle tire was nearly flattened by the addition of Joe's 150 pounds on top of the airframe's 150 pounds.

Anyway, it worked. He was moving forward. The prop, which was geared directly to the bicycle wheel, swung smoothly and quietly. The big bird eased through the hangar's shadow and into the warm morning sunlight.

We were all feeling the tension of anticipation. Then the wind swung around and hit the Olympian headon from where she sat. Joe studied the tetrahedron for a minute and said "let's try a taxi run. Straight out from where she is right now." Strictly "unofficial," of course.

Joe climbed back into the cockpit and fastened his stiff seatbelt. The canopy was lifted into place and pinned shut. With one man on each wing tip for guidance, Joe got 'er rolling across the wide ramp. The speed quickly increased to the point where the crew was running. The tail group was kept level as the plane advanced steadily for almost 400 feet before it was stopped. Joe got out and we all agreed the ZB-1 moved just fine. The crew picked up the plane and walked a 180 (the gearing prevents backing up).

ZB-1 was taxied back to the hangar and everyone seemed pleased with the performance, including the designer, builder, and pilot.

As implied, neither the bicycle wheel nor the small nosewheel are steerable, but that wasn't presenting a problem in the early taxi runs. It was assumed the plane could be launched in a straight line and once airborne the wing-tip ailerons and huge rudder would provide good control. It worked on paper.

A short discussion about minor kinks was followed with an agreement to reconvene at 5:00 A.M. the next morning.

I called Joe that evening to confirm the scheduling, which he did emphatically. His courage was up to full.

Dawn was still only a suggestion when I pulled around the big hangar and found to my total surprise that there were over a hundred parked cars and about 250 people standing around—half of them carrying sophisticated camera equipment. Joe, who woke up at 1:00 after barely an hour's sleep, had been on the phone through the dark of night keeping his promise to this TV station and that newspaper.

It was a carnival.

When Joe pulled up, he was immediately surrounded by microphones, movie cameras and flood lights. Just getting the padlock off the hangar door took seven minutes.

An hour later, 6:30, the Olympian was on the field and facing into the wind. Ten minutes later the huge prop was swinging like a windmill. Cameras were rolling. Sunlight glistened on the mylar. The plane's speed increased. The headwind was steady at 4 mph. The nosewheel eased off and the tail group dropped. It dropped lower, finally touching the ground and slowing the plane. Then the tail went up. Speed went up. One of the wings lifted, bringing down an outrigger. Its tiny plastic wheel made a loud, scraping noise. The speed diminished. One of the wing-tip runners got the outrigger back in the air. The tail bounced again. The plane slowed and stopped 600 feet from where it started. The press converged on the cockpit. "Why didn't it fly?" asked a dirty dozen. "Only a test," snapped Joe.

Everyone backed off and a second run started. The sequence was predictable: rotation; tail thump. Tail and speed both went up. Rotation, outrigger contact. More effort. The wings showed lift and strained upward, increasing the dihedral. But not enough. It clearly needed an additional 5-6 mph to get off. The Olympian stopped again.

Inside the canopy, Joe was sweating and breathing hard. "Need vent holes in this greenhouse," he said between puffs. For a man of 52, he was doing well, but understandably, the single hour of sleep, the pressure of the press corps, and the physical strain were beginning to show.

The plane was returned to the starting point. Joe made a total of six runs that day, and while we could see the rotations as well as obvious lift in the bicycle tire, Joe just wasn't strong enough to get a lift-off in the docile winds. On the other hand, there were trim problems that needed attention. Generally the results were considered cautiously promising. Joe was undaunted.

Unfortunately, confused TV and newspaper accounts that night and the following morning went from "better luck next time" to dismal accounts of abject failure. The press had expected "instant record."

In the week that followed, Joe rested, relaxed and tended to several mods. He vented the cockpit, shortened the outriggers, straightened the nosewheel, adjusted the ailerons, moved the seat back, geared down the elevator control to slow the rotation, added foot straps to the pedals, changed seatbelts, and rode like hell on his bicycle.

Joe called the night of April 20 and said in a calm voice: "We're go at 5:30 tomorrow morning." "OK."

It was light when the hangar doors were reopened. There were a few clouds in the sky and rain was forecast for the end of the day. A cold, 6 mph wind was building strength. 10 mph was Joe's limit. This time around there were only about 50 people.

The first taxi test was impressive. The headwind was up to 7. One of those police radar units was stationed at the end of the runway. The operator announced the plane's ground speed by walkie-talkie to a crewman running with the Olympian. "Seven!" he cried out. The wings were really high and straining. Joe aborted. He was out of runway.

Fifteen minutes later Joe made his second try for the day. He came even closer to flight. The concentration evident on the faces of the bystanders was intense. The run was smooth, but now the vents in the greenhouse were letting in too much air, ballooning the mylar and tearing it off the aft section of the fuselage. The holes were patched over. The ailerons were going out of trim.

Third run: no lift off. By this time Joe was ready to put it away. The winds were occasionally pushing 10 mph. He was so close though, and most people sensed it. "Let's give it one more shot. This time we'll tape the ailerons down." The trailing edges were high, and since they were *full* wing-tip ailerons, they produced a negative force that dragged down the whole wing. Someone produced some gaffer's tape and the ailerons were secured.

The Olympian, driven solely by designer/builder/pilot Joe Zinno's muscle power prepares for its first brief flight.

It was assumed the rudder was large enough to control the plane if flight occurred.

Once again, Joe climbed in. It was 8:15 A.M. The Olympian was aimed slightly toward the left side of the runway, directly into the wind.

Silently and slowly the huge machine developed inertia. At first the long balsa prop swung lazily like one of those ceiling fans in a Bogart movie. The ground crew then changed from a walk to a trot, then a run. The prop became a blur. The aircraft began a very gradual, unscheduled left turn. The headwind was 8 mph. The radar man cried "eight" mph forward motion. And accelerating. The giant wings stretched upward. Then, so gently it was really hard to believe it, that bicycle wheel left the ground. I bent over to make sure. The Olympian was flying! Cries went up: "It's flying; it's off; he's got it in the air." I stood there paralyzed, trying to crush my camera as I whispered to myself in amazement, "He did it, he did it, the s.o.b. did it." (Like any respectable amateur, I forgot to take pictures.)

It was beautiful. It was thrilling. It was fast.

Joe flew right into the history books.

His wing man with the walkie-talkie had enough presence of mind to holler "right rudder!" Joe thinks he might have responded.

It wasn't enough, however.

The left wing drooped and touched the ground. The wind got under the right wing and abruptly lifted it way up. Then it came down. The fuselage bounced. The plane swung left. Lateral motion started. The noise was awful. The outrigger collapsed. The left wing dragged. The bicycle wheel collapsed. The plane stopped.

Incredible silence.

Not only was Joe alright, he was elated. He leaped out and then leaped up and down. "We did it. It worked." There was a lot of back slapping and handshaking. Joe kissed his wife, Fran. She cried. Everyone cheered.

Hours later, Joe recounted: "I was definitely airborne. I could feel it floating and then slipping. I got so excited I forgot to pedal. I couldn't think."

The sounds of the ground loop were worse than the actual damage. An outrigger was bent, one rib was cracked, there was a dent in the trailing edge of the wing, the bicycle wheel—already scheduled for replacement—was shot, and the mylar needed a couple rolls of Scotch tape.

To be continued . . .

JOE HUGHES: UPDATE

Reprinted by Permission
AIR PROGRESS (May 1976)

by Keith Connes

In the January issue of Air Progress, we ran a story about Joe Hughes, whose aerobatic act with wing walker Gordon McCollom ended tragically when Gordon was killed during the climactic inverted ribbon pickup at Reno last August. The resulting mail from our readers indicated a great amount of interest in Joe–an interest shared by this writer–so when I learned that Joe had selected a successor to Gordon, I visited him and his new wing walker, Steve Trevor.

We sat over coffee in the Meadowlark Cafe. The little coffee shop sits on the perimeter of one of those country airports whose continued existence you worry about, because it's awfully close to high-priced urbanized real estate. The runway is a modest 2100 ft. and no wider than it needs to be, with the traditional power lines at one end. You see lots of Cubs, Champs and–for the swanky set–Apaches, decorated with casual paint jobs and graffiti. A sign outside the cafe says "Everybody Welcome," and you believe it.

This air show at Phoenix was put on to attract people for a crowd scene for CBS TV. They were shooting a pilot film for a series to be titled "Spencer's Pilots."

Something happened to foul up our camera plane, so we had to take these closeups on the ground.

Steve is a nice looking, slender, personable guy. He's a biological sciences major at the University of California and he wants to go to medical school and eventually work in cancer research. Doing stunts on the wing of a Stearman is strictly a temporal thing, for maybe a summer or two.

I asked Steve how he got involved in aerobatics.

Trevor: "I was a close friend of Gordon for the past 10 years. He introduced me to gymnastics and we used to work out together. Then about two years ago, I saw Gordon perform with Joe—and I told Gordon that if he ever wanted to quit the act, I'd like to give it a try."

Then I asked Joe how he went about choosing a successor to Gordon.

Hughes: "After the accident, I had about a hundred people applying to me for the job. I needed someone who was an athlete, under 30 years of age, and not big or heavy. A major problem in putting a man on the wing is drag; you just can't have a big man out there. Another consideration is character. Our show begins the moment we land at the airport and doesn't end until we leave. The man I work with must be the master of himself, on the ground as well as in the air. And he must be someone I can identify with. Remember, in the air we have to communicate without words.

"Steve phoned me a week or two after the accident. We'd met once before, but frankly, I didn't remember him. But when he told me he was a gymnast and lived close by, I wanted to meet him. Well, we sat over coffee and talked and I knew right away he was the man.

"We spent the first two days in the hangar, with Steve going through the routine on the rack until he knew it thoroughly. Then we flew, and everything he'd learned had to be modified to take the wind effect into account."

Were there any problems?

Hughes: "For one thing, when Steve started doing the horizontal handstands, his body began to oscillate back and forth, in a sort of whiplash effect. We went down and talked it over, and then my wife remembered that Gordon had prevented this by scissoring his legs instead of keeping them strictly parallel."

Trevor: "We didn't know precisely how Gordon had worked his handholds, so we had to figure them out. You see, the rack didn't come with a set of directions."

A few weeks later, I went to see a preview of the act. Joe and Art Scholl and others were putting on an air show at Phoenix as part of a pilot film they were doing for CBS television.

It was Steve Trevor's first time on the wing in front of a crowd.

He gave a flawless performance.

Joe Hughes has himself another wing walker.

Joe and Steve doubled for the actors in "Spencer's Pilots," but a stunt man went into the lake — and then went into the hospital. (He jumped too soon.)

Steve Trevor and Joe Hughes.

I, AN ACTUAL CUSTOMER, FLY AN ACTUAL BD-5

Reprinted by Permission Edited for Space
AIR PROGRESS (SEPTEMBER 1976)

by John H. Boynton

Newton, Kansas, could be just another of thousands of western-plains farm villages. A mile or two out of town, one drives through friendly fields of wheat and maize populated by an occasional group of dairy cows. What sets Newton apart, of course, from all the other farm villages is the fact that Jim Bede chose to locate his brain-storming aircraft development facility at the nearby and otherwise obscure county airport.

The BD-5, and flight evaluations of same, are not exactly press-stopping news. Indeed, a long series of articles covering virtually every phase of the BD-5 saga, from promises to production, have appeared, yea glutted, the aviation media in recent years. It's to the point now that such a story elicits "Oh yeah, another 'I flew the BD-5' episode" or "Looks like some more promises from the 'Bede guys' again" out of the saturated general aviation readership. So why this article? you ask. First, I am a living, breathing, bonafide customer for the BD-5. Not a paid staff writer for some magazine nor a member of the Bede promotion staff. I don't have to sell either aircraft or magazines. So far as I know, there has not appeared an article on flight evaluation from a production-version priority holder. Second, to my knowledge, *no one* has flown the Xenoah-powered BD-5 and published an impression of same.

Jim Bede is an enigma in aviation circles. He's probably an enigma in any circle. But that, too, is not news. What *is* news is that, almost without exception, those who work for him at Bede Aircraft

Corporation all retain one or more of his "blue-sky" personality traits (or handicaps, whichever the case). Mr. Bede's office, on the south end of the west side of the field, overlooks his "eminent" domain, a row of bright-blue steel prefab buildings stretching northward nearly the length of the field. The largest contains the waiting lobby for those ambitious and enthusiastic buyers and builders who have yet to see a finished product but still continue to visit in droves, only to hear the standard disclaimers and "he's in conference right now" from a well-coached receptionist. This building also contained a warehouse-full of kit parts that have long since been shipped, *sans engine*, to a world-wide network of kit builders. Many have taken up hang gliding and are simply the proud owners of a two-thousand-dollar weather vane. A few have tried motorcycle and skimobile engines with varying success. One or two took the initially offered Hirth.

Further down the airfield is the Bede Flight Test Center, that beehive of activity that contains the engineering-flight-test personnel, the seasoned shop craftsmen, and of course the Jet Demonstration Team. The receptionist there is also well trained to screen out spies, gawkers, and irate customers. It's not a large building, but how large must it be to contain a handful of BD-5s?

I first talked with Rich Cranmer, that lucky individual appointed to handle a series of customer (read here rubberneck) checkouts in preparation for the Oshkosh EAA Convention last August. I had been chosen, as a priority holder, to be checked out with that group of about fifty but was unable to comply with their pre-Oshkosh schedule because of other commitments. Since I was now a "special" case, Rich sent me to Mr. Bede himself, who had just returned the night before in his flat-tired Baron (both fact and pun intended). He did have an airless nosewheel on landing.

Jim Bede, as expected, was open-shirted and open-handed. He's got to be the friendliest harried company president around, and his broad smile and rachety mile-a-minute speech are totally disarming. We talked about company objectives (to fill the void left by the biggees), new ideas (an inflatable hang-glider), and why there wasn't a really fast two-place aircraft (the BD-7 has that option). Then I popped the question. Yes, he wanted me to fly N503BD if it was free from testing. And yes, Les Bervan, Chief Test Pilot, had said he would finish some fuel-flow tests by mid-afternoon that day. But, it looked like it might be too windy. We both looked at the straight-out windsock. Well, I allowed as to how I had nearly a 1,000 hours, most of it in Bonanzas, and that the wind had *never* bothered me (I lied). I then pointed out that the wind was right down the runway and not *too* gusty . . . even ought to help with relative takeoff and landing speeds. Mr. "easy-talking" Bede said he'd let Rich and Les decide that. Rich talked to Les and I was set up for a verbal rundown at 1:30 that afternoon, "but don't expect to fly today. It's too windy."

With almost two hours to kill, I drove back to Newton and had lunch at a pancake house. As the wind was whipping the tall trees outside, I realized just how desperately I wanted to fly this object of my patience for over two years. The wind, and a promise of a frontal passage the next morning, could eliminate any chance of flying on this trip altogether. I finished my sandwich and drove the two miles to Bede Aircraft.

The rundown was simple enough. Les Berven sat behind a cluttered desk and

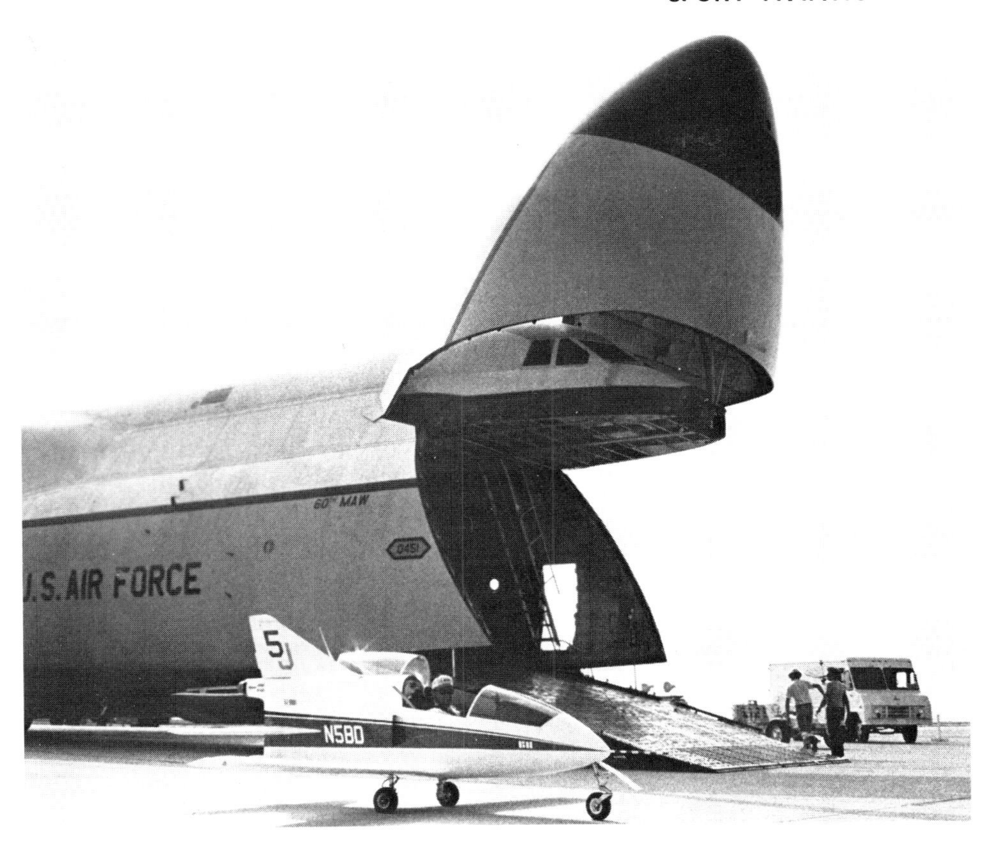

talked from cues on an outline of the BD-5 Operations Manual, a document he had first written over a year ago. He related his frustration in trying to keep current a manual that was constantly barraged with design changes, each one seeming to have a domino effect on everything else in the book. Preflight checkout had no surprises, since there's not much airplane to look at. Start procedures were no more complicated than those for my 350 Honda. Push the button and work the throttle.

Emergency procedures boiled down to a simple engine-restart if it conked out, and that was (are you ready?) another push on the start button. "If that fails, bring the throttle back to a ½-inch and try again. If that doesn't work, glide it back to Newton," he said nonchalantly like someone who had done it many times before. He concluded by saying landing procedure was pretty much the reverse of takeoff, with flaps extended at 110. It all seemed so simple I thought I was being deceived.

After our talk and the almost casual manner with which Les had preempted an obviously impossible dual-check-out in a single-seater, he looked once again at the wind. Les was more concerned about my own confidence in landing under twenty-knots-gusting-to-twenty-five conditions than he was the wind itself. I detected that and quickly allowed as to how it looked like it had died down a bit, that it was still smack down the runway, and that I'd probably be landing a little hotter than normal anyway. Evidently I convinced him of my confidence, because he got one of the hangar people to run me down the runway a few times in the truck-mounted trainer, which I had flown previously in June. These second runs produced the same bad tendencies . . . over-control in pitch and a disbelief that touchdown can be so uncomfortably close to the ground. Afterward, Les emphasized that it was much easier to "fly the real thing" than the trainer, since the latter represented the former in slow-flight configuration only. I took this as gospel and began my walk-around in the hangar of "the real thing."

As I looked down at this refugee from Macy's toy department, Les added some final cautions. "It has a strong tendency to weathervane on the runway, but you only need to tap the opposite brake once or twice to get it back. After takeoff, get the flaps up quickly and climb out at 105." "What's quickly?" I asked. "Oh, three or four seconds after you lift off and the gear is up. That should be about 20 feet and 80 miles an hour."

Now, getting into the BD-5 has also been documented well in the literature, but yet one more data point might be interesting. I am not a small person, weighing about 195 and measuring just under six feet. The prescribed procedure is to place one hand on the canopy hoop and one foot in the seat from the left side (an old cavalry custom, remember?). It's a new kind of aviation yoga. But once you're inside, with both feet vanished into a hole under the panel, the soft contoured seat gives one the impression of settling into a chaise lounge for some extended "Bede hangar flying" instead of operating "the real thing." Indeed, at some point, an instrument-rated BD driver is most assuredly going to fall asleep waiting for a clearance. Wonder how the FAA handles airplanes picked up by runway sweepers at four in the morning?

Once bodily installed, I waited for Les to roll me out of the flight-test hangar onto the ramp for startup. The Xenoah fired up immediately. It sputtered like you'd expect from a Yamaha or a Suzuki, but it was quite smooth and reasonably quiet with the canopy closed. That was surprising as well, since the pilot sits a scant inch-and-a-half from the forward cylinder. Throttle response, as with a bike engine, was instantaneous and comforting.

Les told me, again in his characteristic nonchalant manner, that he'd be doing some desk work and would monitor 122.95, the frequency already facing me on the tiny transceiver.

The stiff wind virtually pushed me the length of the taxiway, and I was soon all alone at the threshold of the enormously wide runway. As far as I was concerned, I was all alone in the world. That same feeling one gets when his first flight instructor says "It's all yours."

Full power into a raging wind. Acceleration was positive with the three-jug Xenoah. I used up only about 400 feet to get to the stated rotation speed of 65, but it didn't rotate. I don't know why, and I forgot to ask later. It rotated instead at about 70 and after a hundred feet or so more of bumping from the tar joints. Once airborne, the little ship climbed briskly and reached in no time the gear-retraction point of 20 feet and 80 indicated. The wheels came up with a healthy yank on the T-handle, ominously placed right between the legs, and pitch-up during and after retraction was barely noticeable. That's easy to understand considering that the three tiny gear barely protrude from the underside. Flap retraction immediately after the gear was up produced a more pronounced aerodynamic pitch-up perturbation, but it was never uncomfortable.

With the aircraft cleaned up, the climb rate immediately rose to over 1,400 fpm . . . not bad for a warm day and 2,000 feet pressure altitude. I decided at that point that enough had been written about aerodynamic characteristics and that I'd concentrate on engine-related performance and just plain handling impressions.

At 7,500 feet with the Newton airport safely below and the Kansas countryside disappearing in all directions over a hazy, cloudless horizon, I leveled off in preparation for accelerated maneuvers. He had cautioned me about doing aerobatics on this, the first, flight. So I stuck to tight level turns at about 75 degrees bank angle, aborted entries into loops, rapid roll rates caught at wings vertical, and wingovers. The effect of using rudder to facilitate a roll maneuver was quite significant. That is, a pure uncoordinated roll rate was not spectacular, but stick-roll combined with initial rudder produces a very fast roll rate because of aerodynamic cross-coupling. This non-aerobatic series was sufficient to convince me of one thing. The BD-5 prop version is the closest anyone will ever get to the feel of a military jet without actually riding in one. I have been fortunate enough to sit front seat in a T-33 on five different occasions, and the view through the windshield, the hand on the stick, and the push through the butt in accelerated maneuvers brought it all back for me in the BD-5. It's a real sensation!

Well, I was supposed to do stalls, wasn't I? I'd almost forgotten amid the fun of cranking around the sky. First, clean. No problem. A definite buffeting before breaking straight ahead. I tried another. Same thing. Then full flaps, throttle back a third time. This one produced a slight dip to the left and a secondary stall once the nose dropped slightly. Then I remembered Les had said to release stick force once it breaks. Not *push* it forward. I tried that method, and he was right. With the gear also down, the same characteristics were evident, and again I over-reacted with the stick to get the same secondary stall. The wing dipped a little more sharply, but right aileron and increasing airspeed brought it right back. It was all so simple.

Looking at my watch and the fiber-optic fuel indicators in the right wing, which contained the tank I'd been using to this point, I decided it best to switch. The fuel control is an uncomplicated torque-tube jutting past the right bicep and disappearing into the bulkhead forward of the engine compartment. Everything on the BD-5 is this simple. The engine never hesitated during the switchover.

With about a half-hour of anxiety-free flying left, I concluded I needed some more tight turns and wingovers. In a wingover, the application of power at the beginning of climb and the withdrawal of same at the top are amazingly reponsive with the smooth little Xenoah. Since the drive system provides for override of the prop, there's no drag when the power is pulled off for the downside of this maneuver, and the aircraft builds up speed rapidly. I pulled up again as if entering a loop, added power again, and aborted the loop at about 40 degrees pitchup. In the zero-g of the abort, I wondered if the engine would sputter, but it didn't. Never even whimpered.

When I pulled away from the ramp, I was certain Les' casual attitude meant he wouldn't worry about me and for sure wouldn't call to check my status. But he must have chickened out, because the speaker squawked out "Five-oh-three-bee-dee, everything okay?" I assured him it was and that I was about to begin descent. With all his time in the -5, it probably never occurred to him that I was just plain having fun and didn't want to end it.

An experienced fighter pilot at this point, I cranked it over into a steep bank and peeled off just like in one of those Gary Cooper movies. Power back, top rudder, find the field again. Rate of descent increased to about 2,000 fpm, just like in a jet, and I got down to pattern altitude in no time. Leveling off, I began to set up for a high downwind, power off all the time. Adjacent to the runway threshold, I pulled in half flaps and descended to about 1,000 AGL before turning base leg. At this point, I dropped the gear so I wouldn't forget like a previous reporter had done.

The ship trimmed up nicely on final after the last 20 degrees of flap were dropped. I still hadn't added power, and I was slowly coming down to the over-the-fence speed of 90, as instructed. With high hopes of touching down without adding power, I glided over the fence at 85 but with too little altitude and too much distance to the runway. So I, too, chickened out. With the wind gusting and giving me a few squirrely gyrations, I wasn't about to preserve my honor and make the dead-stick Les had requested I try for. Half-throttle produced an immediate engine surge and acceleration response. It also pushed the nose down slightly, as Les had cautioned in the briefing. That perturbation, nonetheless, surprised me, since it is opposite to what you get in other aircraft, and I over-controlled like I had in the trainer. Thus, I had a built-in oscillation in pitch when I started my flare and pulled the power off, which caused the aircraft to stall high. As the nose dropped, so did the plane, and the nosewheel touched a split-second before the main wheels. That in turn bounced the nose back into the air and I was flying again. Reacting quickly, I added some power, pulled the stick all the way back, and cut the power, this time less than a foot off the runway. The angle-of-attack was higher than before, and the aircraft stalled neatly onto the main gear and stuck when the nose fell through. The same stiff wind slowed me down immediately, and I executed a 180 on the deserted runway to get back to the hangar.

It would be fruitful to discuss trim settings at this point, since pitch sensitivity at the slower speeds seemed to be directly related to them. For takeoff, the recommended 1/3-travel setting made rotation and climbout almost effortless with regard to stick force. For my weight, and therefore center-of-gravity position, I found I needed almost full forward trim in a cruise configuration at nearly maximum throttle. And on final with only idle power, I found I needed full aft trim, when the gear and flaps were down, to take out the excessive stick force. Thus, the design of the trim range seems to be pretty close without any excess capability. I did, however, get the feeling that both takeoff and landing could be accomplished at a neutral trim setting without unmanageable stick forces. That's more than I can say about landing a Bonanza, since I've never been able to do it without aft (or pitch-up) trim.

As I taxied to the ramp, I had the fleeting thought "I've flown the BD-5 and lived!" When I got to the hangar, someone came out to roll me in, so I swung in an arc and snapped the mag buttons off. The Xenoah sputtered down instantly. With the tail held up, I climbed out the same way I had inserted myself, and Les strolled out the door to greet me. "Now you can say 'I've flown the BD-5 and lived!' " he quipped.

"So, what about the BD-5?" you, the unconvinced reader, might ask. "What's a one-place mock-jet-fighter good for besides fun?" "What happens when the kicks wear off and I still have to make payments?" They're all good questions. But as a priority holder with the expectation of getting my airplane sometime next summer, I feel compelled to defend my original intentions. First of all, I don't have an airplane of *any kind*. When I fly, I rent. So the BD-5 I'm buying won't be a second, fun-machine aircraft to fill the void between hops in a turbo-Bonanza or a Learjet. It will be my only source of air transportation outside an airline terminal gate.

"But you can't take anyone with you" is your immediate reply. No I can't, but for those few times when I want to, I can still rent the same Mooney or Bonanza I've been renting all along. The cost is the same, whether I own a BD-5 or not. I just won't be able to prorate my overhead across the rental time, but Jim Bede claims overhead on a -5 will be negligible.

"Well, you've rationalized the whole thing because you just like to tool around the sky in your new little toy," you say. "You said yourself, it feels just like a jet fighter." And I have to think a minute about that. By God, you're probably right. Life is a pretty brief experience. If we deny ourselves *all* the fun life has to offer, it gets even shorter, and dull to boot. People don't need motorcycles, or RV campers, or expensive sports cars, or sailboats, or ski weekends either. But they buy them. And they love them. That's how I think the BD-5 priority holders will feel about their little ship when they get it.

Oh well, so I *am* taken in by the mystique and genius of Jim Bede and his barnstorming-days type of dreaming. So I *do* have a latent compulsion to be a fighter jockey and clear the sky of MiGs. Maybe I *am* a sucker for something that combines the supersonic look of an XKE with the design integrity of an Aerostar and the performance of the best unblown singles. Yes, I flew the BD-5 and lived . . . lived to want some more of it. I'll stay mesmerized by Jim Bede's smooth talk of endless promises and even better engines and super performance and new gimmicks and option discounts. And someday in the autumn of my life, a letter will arrive in the mail telling me the balance on my contract is due and I can come to Newton in a month to pick up my plane.

And I'll dig out this article, look at the publishing date, and be glad I wrote it.

HOMEBUILT CRASHES RAISE QUESTIONS

Reprinted by Permission
AIR PROGRESS
Edited for Space
(May 1976)

by Valerie Petrie

Were the homebuilders negligent? Does the design have problems? Or was it four more cases of pilot error?

When four aircraft of the same type crash within less than a month of each other, there are bound to be a lot of people wondering if there's something more than tragic coincidence involved. Last February 8th, Paul Saltzman Jr. of Evansville, Indiana crashed in his BD-5 shortly after takeoff. Just three days later, Walter Lansing Jr. crashed in his newly acquired BD-5-J while attempting a takeoff. Both Saltzman and Lansing were killed. On January 18th, William McClenahan crashed during takeoff in a BD-5 and as of this writing is still in serious condition with head and back injuries.

The fourth crash, another jet, occurred March 8; the pilot parachuted to safety.

The NTSB Report is still in the works, but anxious pilots and homebuilders want the facts now. Without drawing conclusions as to the direct causes of these crashes, Air Progress asked the FAA Accident Investigators on the scene for whatever facts were available concerning these accidents. These are the results:

Paul Saltzman Jr. was 34, a BD-5 dealer, and, as of his latest FAA medical, in good physical condition. He was a rated commercial pilot with a total of 5,300 hours of flying time, the bulk of which was in American Yankees and Cessna 150s.

Saltzman enthusiastically sold BD-5 kits and had flown his own BD-5 with the Hirth engine for the first time in October, 1975. During this time he accumulated a total of 15 hours in the BD-5 without incident. Saltzman then decided to install a more powerful Kawasaki motorcycle engine.

When interviewed earlier by the Evansville Courier Press regarding his new BD-5 modification, Saltzman commented "It's so new you just worry because there are so many things that could go wrong."

By February 8th he had completed his Kawasaki engine installation and arrived at Dress Regional Airport to make his first flight in the newly modified BD-5.

According to FAA Investigator Paquette, witnesses reported that the plane made a seemingly normal liftoff from the 8,000 foot long runway, began a left turn to crosswind, then spun once and nosed straight down. The plane crashed approximately 500 feet off the departure end of the runway and burst into flames.

Paquette said that there was no evidence of an engine failure or mechanical malfunction and reiterated that the BD-5's experimental certification places the full responsibility for the aircraft's safe construction and operation upon the owner/pilot of that aircraft.

There was no indication prior to the accident that Saltzman was having trouble with the aircraft, but the lack of a transmitter on board the BD-5 would have made it impossible for him to contact anyone if he had suspected a problem.

Saltzman's interview with the Evansville Courier had indicated that over 5,000 BD-5 kits had been sold, 30 in the Evansville area. The investigating GADO was soon swamped by calls from concerned homebuilders regarding the incident.

The second crash occurred at the Bede Factory in Newton, Kansas just three days after the Saltzman accident. Walter I. Lansing Jr. was a private pilot with a total of 860 hours flying time, all in light single engine piston aircraft.

An allowable percentage of his BD-5J was built by Jerry Kibler, the same man who reportedly built much of writer Richard Bach's BD-5J. Lansing finished building the plane and then turned it over to Bede's chief flight test pilot, Les Berven, for evaluation.

According to FAA investigator Art Richardson, of the Kansas City GADO, test pilot Berven had put a total of 2.7 hours of flight evaluation into Lansing's BD-5J and pronounced it one of the best models he had flown yet. Berven then gave Lansing the appropriate ground training on the systems of the aircraft as well as a cockpit briefing. This portion of the preparation lasted approximately two hours. Lansing had received two or three hours of accumulated time in the BD-5 truck-mounted trainer over the last few weeks as well.

On the morning of the accident, Lansing had reportedly done "lots" of high speed taxiing practice runs. It was to be Lansing's first actual flight in his BD-5J, and both Bobby Bishop and Corky Fornoff of the BD-5 jet team were there to wish him well.

The expert witnesses reported that during the takeoff phase, Lansing's BD-5J lifted off and began to porpoise. It then returned to the runway, impacting initially on the nose gear, which was damaged but remained intact. The mains then

hit the runway, catapulting the aircraft up 20 to 30 feet in the air. At the top of the bounce in a level or slightly nose low attitude the plane snapped half inverted to the right and started down. The right wingtip appeared to impact first, following immediately by the nose. Lansing's crumpled aircraft then slid to a stop at the edge of the 7,000 ft. long runway.

Witnesses and the FAA reported that there was no evidence of engine trouble, and that the engine itself received little or no damage. The cockpit area also remained surprisingly intact. The porpoising effect is reportedly not uncommon in BD-5s when fully loaded with fuel. In some cases, this loading also results in a slightly aft CG. And as former Air Progress editor Richard Weeghman reported in his pilot evalation of the BD-5J, (Feb. 1974 issue) the aircraft is very sensitive to a light touch on the controls, and the pilot's sense of speed is amplified by the aircraft's tiny size.

The third unhappy incident involving a BD-5 occurred in Arlington, Washington, a suburb of Seattle. It had been raining for weeks, and William McClenahan was itching to test fly the Xenoah powered BD-5 he and a group of friends had just completed. The first clear day came on January 18th.

A friend of McClenahan's quoted the commercially rated CFI as saying, "I'm going to fly today come hell or high water" prior to the ill-fated flight. According to FAA Investigator Miller, witnesses saw McClenahan take off and climb to approximately 200 feet. At this point, the sound of the engine stopped and McClenahan apparently elected to attempt to turn and land on a nearby taxiway rather than use the remaining 2,300 feet of runway. The BD-5 then entered a series of uncontrolled falling leafs and hit the ground hard in a flat attitude.

There was quite a bit of downward deceleration damage, and the aircraft broke just aft of the canopy. The engine also came out of its mounts, but there was little apparent exterior damage to the wings. McClenahan suffered severe head and back injuries.

The Xenoah engine McClenahan and his friends installed in their BD-5 was the first to be used in a non-factory flight. At first it was rumored that the apparent engine failure was caused by the same fuel starvation problems McClenahan and his friends had been having during the taxi tests they conducted in early January.

It seems the group of homebuilders had been using a silicon-type bathtub sealer to seal in the fuel gauges instead of the special cement recommended by the Bede Factory. The homebuilder's substitute had been dissolving in the gasoline and blobs of it had clogged up the fuel pump on various occasions.

FAA Investigator Miller said, however, that after the accident, an inspection of the fuel pumps and lines revealed no foreign materials in the fuel. He also stated that the homebuilders had removed one of the BD-5's two fuel pumps and plugged the holes where the fuel lines went prior to McClenahan's flight. They also used a fuel line that was considerably smaller than the one provided by Bede.

When I spoke with John Hall, Director of Marketing at Bede Aircraft, (Jim Bede was not available) he said that the homebuilders had apparently not complied with a Bede notice that warned them not to fly with the original Xenoah ignition coil.

The fourth and most recent crash took place as we went to press on March 8th. Again, it was a jet, and again, it was at the Bede factory at Newton, Kansas.

Factory test pilot Bobby Bishop was carrying out spin tests on a BD-5J which had been fitted with the new GAW-1 airfoil wing. Although details are sketchy, the tests had apparently been proceeding normally, with spins of increasing number of turns being carried out to each side. After making three turns of a left spin, the airplane entered a flat spin mode; Bishop bailed out at 4000 feet, breaking an ankle.

Comments from Bede representative, John Hall regarding the Saltzman and McClenahan crashes emphasized the fact that in both cases, the BD-5 homebuilders had installed engines and other equipment that was not tested nor approved by Bede Aircraft. In cases such as this, the homebuilder is acting totally on his own as an aircraft manufacturer and a test pilot.

As far as the BD-5J crash of Walter Lansing Jr. is concerned, Bede representative Hall stated that it was "most unfortunate." Bede's requirements for applicants with as little as 100 hours of flying time who want to fly the BD-5J jet are two hours of ground school, a couple more hours of truck-trainer and some high speed taxiing.

AVIATION YEARBOOK INDEX

a

b

c

d

e

i

j

l

m

n

o

p

q

r

s

t

U

W

Z